Series on Chemistry, Energy and the Environment

Prospects for Li-ion Batteries and Emerging Energy Electrochemical Systems

Series on Chemistry, Energy and the Environment

ISSN: 2529-7716

The aim of this series on "Chemistry, Energy and the Environment" is to bring together authoritative contributions where the multidisciplinary character of Chemistry and its relationship to energy and the environment can be illustrated. In each volume, the latest advances in bio-, energy- and environmentally-related fields are described in a way that unambiguously shows the major contributions from chemists at the top of their field who describe new and improved technologies for the future.

Published

Vol. 4 *Prospects for Li-ion Batteries and Emerging Energy Electrochemical Systems*
edited by Laure Monconduit and Laurence Croguennec

Vol. 3 *Advanced Green Chemistry*
Part 1: Greener Organic Reactions and Processes
edited by István T. Horváth and Max Malacria

Vol. 2 *Elaboration and Applications of Metal-Organic Frameworks*
edited by Shengqian Ma and Jason A. Perman

Vol. 1 *Perovskite Solar Cells: Principle, Materials and Devices*
edited by Eric Wei-Guang Diau and Peter Chao-Yu Chen

Volume 4

Series on Chemistry, Energy and the Environment

Prospects for Li-ion Batteries and Emerging Energy Electrochemical Systems

Edited by

Laure Monconduit
Institut Charles Gerhardt de Montpellier (ICGM) - CNRS, France
Réseau Français sur le Stockage Électrochimique de l'Énergie, France
ALISTORE-ERI European Research Institute, France

Laurence Croguennec
Institut de Chimie de la Matière Condensée de Bordeaux (ICMCB) - CNRS
Réseau Français sur le Stockage Électrochimique de l'Énergie, France
ALISTORE-ERI European Research Institute, France

Series Editors

Karl M. Kadish
University of Houston, USA

Roger Guilard
Université de Bourgogne, France

World Scientific

NEW JERSEY • LONDON • SINGAPORE • BEIJING • SHANGHAI • HONG KONG • TAIPEI • CHENNAI • TOKYO

Published by

World Scientific Publishing Co. Pte. Ltd.
5 Toh Tuck Link, Singapore 596224
USA office: 27 Warren Street, Suite 401-402, Hackensack, NJ 07601
UK office: 57 Shelton Street, Covent Garden, London WC2H 9HE

Library of Congress Cataloging-in-Publication Data
Names: Monconduit, Laure, editor. | Croguennec, Laurence, editor.
Title: Prospects for li-ion batteries and emerging energy electrochemical systems / edited by Laure Monconduit (Institut Charles Gerhardt de Montpellier (ICGM) - CNRS, France, Réseau Français sur le Stockage Electrochimique de l'Energie, France, ALISTORE-ERI European Research Institute, France), Laurence Croguennec (Institut de Chimie de la Matière Condensée de Bordeaux (ICMCB) - CNRS, Réseau Français sur le Stockage Electrochimique de l'Energie, France, ALISTORE-ERI European Research Institute, France).
Other titles: Prospects for lithium ion batteries and emerging energy electrochemical systems
Description: New Jersey : World Scientific, 2017– | Series: Series on chemistry, energy, and the environment ; volume 4 | Includes bibliographical references and index.
Identifiers: LCCN 2017047295 | ISBN 9789813228139 (hardcover : alk. paper : pt. 1)
Subjects: LCSH: Lithium ion batteries. | Lithium cells. | Electric batteries. | Sodium ions. | Electrochemistry.
Classification: LCC TK2945.L58 P7645 2017 | DDC 621.31/2424--dc23
LC record available at https://lccn.loc.gov/2017047295

British Library Cataloguing-in-Publication Data
A catalogue record for this book is available from the British Library.

For any available supplementary material, please visit
http://www.worldscientific.com/worldscibooks/10.1142/10658#t=suppl

Typeset by Stallion Press
Email: enquiries@stallionpress.com

Printed in Singapore

Contents

Preface

Since the birth of lithium-ion batteries (LiBs) in the 1990s, electronic applications are continuously becoming more energy and power demanding, with weight and volume getting increasingly restricted. Early LiBs consisted of layered oxide $LiCoO_2$ at the positive electrode and graphite at the negative electrode. LiBs are still at the centre of researches to challenge recent concerns relative to the storage of renewable energies and specific features. Chapters 1 and 2 describe: (i) some of the most recent promising alternative materials to graphite proposed to date as negative electrode and (ii) the main orientations to increase the energy and stability of layered oxides at the positive electrode to fit the set of requirements depending on the applications targeted. The electrochemical behavior as well as the failure mechanism of these optimized electrodes for LiBs were investigated through the utilization of advanced characterization tools and theoretical calculations able to suggest directions for improving their electrochemical performance.

The enlarged and generalized implementation of the LIBs from portable electronics to large-scale applications leads to the question of the recycling. Although efficient recycling schemes are in place for lead–acid, Ni–Cd, and primary batteries, the more complex recycling of LIBs is still at its infancy. This situation will dramatically evolve in the coming years due to combined effect of huge production increase and better collecting rates. Chapter 3 reports the challenges and legal framework to recycle Li-ion batteries, as well as the perspectives for the recycling business at the horizons 2025.

If the research continues to focus on the improvement of LIBs, alternative battery technologies are also to be widely explored. In fact the prospects for attaining even higher performance by other battery technologies through innovative research has become appealing. Another motivation for developing alternative battery technologies stems from concerns related to a future-limited lithium supply which is associated to enlarged and generalized implementation of the LiBs from portable electronics to large-scale applications. For five years, helped by the chemical similarity between Li and Na, by the early studies on intercalation of alkali metals and due to the "unlimited" sodium mineral resources, we are assisting to a fast progress for sodium-ion batteries (SIBs). SIB positive and negative electrodes, electrolyte materials and the perspective on future SIB development are given in Chapter 4.

The research and development efforts related to alternative battery systems was recently enlarged to multivalent active metals as anodes for batteries like magnesium, calcium, and aluminum which present the advantages of being light, active, and highly abundant metallic elements in earth crust. Both fundamental and practical/technical aspects related to Mg, Ca, and Al as anodes in batteries are presented in Chapter 5.

Conventional LIBs utilize organic liquid electrolytes which pose potential thermal and chemical threats associated to the electrolytes' flammability. Today the safety requirements remain a major hurdle, slowing down the expansion for the foreseen staggering energy storage demands in general, and that of the electro mobility sector in particular. Accordingly, much research has been focused on polymer electrolytes, leading to the fabrication of all solid-state Li-based batteries that could bypass the above-mentioned intrinsic limitations of organic liquid electrolytes. A detailed survey of the research on polymer electrolytes and their industrial-scale applications, their challenges are reported in the Chapter 6.

We would like to thank all the chapter contributors, who are leading experts in their respective field. We hope that by gathering these new developments, readers will get an accurate and deep understanding of the electrochemical energy systems.

Laure Monconduit
Laurence Croguennec
Editors

1 Negative Electrodes for Li-Ion Batteries: Beyond Carbon

Phoebe K. Allan*,†, Nicolas Louvain‡,§ and Laure Monconduit‡,§,¶,||

*University of Cambridge, University Chemical Laboratory, Lensfield Road, Cambridge, CB2 1EW, UK
†Gonville and Caius College, Trinity Street, Cambridge, CB2 1TA, UK
‡Institut Charles Gerhardt de Montpellier (ICGM) — CNRS, France
§Réseau Français sur le Stockage Electrochimique de l'Energie (RS2E), CNRS FR3459, 33 Rue Saint Leu, 80039 Amiens Cedex, France
¶ALISTORE-ERI European Research Institute, FR CNRS 3104, F-80039 Amiens Cedex 1, France
||laure.monconduit@umontpellier.fr

List of Abbreviations

C/*x*	charge rate such that the full capacity is reached in *x* hours
CMC	carboxymethyl cellulose
CVD	chemical vapor deposition
DEC	dimethyl ethyl carbonate
DFT	density functional theory
DMC	dimethyl carbonate
EC	ethylene carbonate
FEC	fluroethylcarbonate
FIB	focused ion beam
GITT	galvanostatic intermittent titration technique
LCO	$LiCoO_2$
LIB	lithium-ion battery
LMO	$LiMn_2O_4$
LTO	$Li_4Ti_5O_{12}$
NMR	nuclear magnetic resonance
OCV	open-circuit voltage
PAA	polyacrylic acid
PC	propylene carbonate
PDF	pair function distribution
PTSI	*p*-toluenesulfonyl isocyanate
PVDF	polyvinylidene fluoride
SEI	solid-electrolyte interphase
SFG-VS	sum frequency generation vibrational spectroscopy
TEM	transmission electron microscopy
ToF-SIMS	time-of-flight secondary ion mass spectrometer
XPS	X-ray photoelectron spectroscopy
XRD	X-ray diffraction

I. Introduction

Conventional lithium-ion batteries (LIBs) are composed of a layered $LiCoO_x$ material cathode and a carbon/graphite anode. Graphite has a layered structure and can be electrochemically reduced in an aprotic organic electrolyte containing lithium salts. Lithium is intercalated between the layers of graphite to form a Li–C alloy. The carbon/graphite anode possesses a theoretical specific capacity of 372 $mAh \cdot g^{-1}$ according to the following lithium intercalation equation:

$$Li^{+} + e^{-} + C_6 \leftrightarrow LiC_6$$

In the $LiCoO_2$/graphite full-cell system, the overall cell reaction is as follows:

$$LiCoO_2 + Li_{n\text{-}x}C \leftrightarrow Li_{1\text{-}x}CoO_2 + Li_nC$$

This equation indicates that the cell reaction is a simple migration of lithium ions between positive and negative electrodes. Nevertheless, the lithium intercalation primarily occurs at a voltage below 0.1 V *versus* Li^+/Li, close to the lithium electroplating potential.[1] For high discharge rates, and harsh operating conditions, graphite electrode can be polarized to such an extent that reactive lithium metal dendrites may grow on the electrode surface and sporadically cause "soft" short circuiting of the cell.[2]

New electrode materials that can improve the performance and cycle life of LIB have been extensively explored. As part of this, considerable effort has been directed toward developing alternative negative electrodes that are able to deliver higher capacities and higher energies than graphite electrodes, particularly when charged at high rates.[3] The reader is pointed toward several reviews that extensively explore the range of materials proposed by researchers to replace graphite in future batteries.[4–6] This chapter describes some of the most promising alternative materials proposed to date: (1) titanates, a material undergoing typical intercalation reactions, and (2) metals and semimetals, which form alloys electrochemically, such as silicon, tin, or antimony.

This study describe how the utilization of advanced characterization tools and theoretical calculations have informed researchers about how these materials behave electrochemically and how improvements in understanding of failure mechanisms of cells are able to suggest directions for improving their electrochemical performance.

II. Lithium Titanates

A. Introduction

Titanium oxides, both those that contain lithium in their starting composition and those that do not, have been proposed as an alternative to carbon/graphite electrodes in LIBs in order to improve specific capacities, high-rate cyclability, and safety.[3,7–9] These materials store lithium through intercalation of lithium into the structure and reduction and oxidation of the titanium during lithiation and delithiation, respectively. This intercalation mechanism is analogous to that of metal oxide cathodes such as $LiCoO_2$, but takes place at lower voltages. In a full-cell configuration, lithium removed from the $LiCoO_2$ cathode is inserted into the titanate host; while the cobalt ions are oxidized, the titanium ions are simultaneously reduced. Among lithium titanium oxides, spinel $Li_4Ti_5O_{12}$ (LTO) is one of the most studied electrode materials as a potential carbon/graphite replacement in LIBs for power applications such as electric vehicles.[8,9] Lithium-free titanium oxides, TiO_2, are still considerably investigated as electrode materials,[10] while their use in commercial lithium-ion systems might appear unfavorable because of high potential and unlithiated negative.[11]

LTO is a lithium intercalation compound with an average lithium insertion/deinsertion voltage close to 1.55 V *versus* Li^+/Li.[9] This is a relatively high voltage for a negative electrode, and this suggests that lithium dendrites formation and the organic electrolyte reduction can be avoided, and that they circumvent many of the safety concerns of graphite anodes. One of the most important features of LTO is that it exhibits *zero strain* during lithium intercalation; there is almost no volume change during the lithium insertion and extraction processes of lithium during battery cycling. Because of this, lithium can be inserted and removed many thousands of times with very little loss in capacity.[12]

B. The LTO Structure

LTO, also written as $Li_{1.33}Ti_{1.67}O_4$, $Li_{4/3}Ti_{5/3}O_4$, or $[Li]_{8a}[Li_{1/3}Ti_{5/3}]_{16d}[O_4]_{32e}$, is an important member of the solid solution family $Li_{3+x}Ti_{6-x}O_{12}$ ($0 \le x \le 1$) which takes the spinel structure.[13] The spinel structure has the

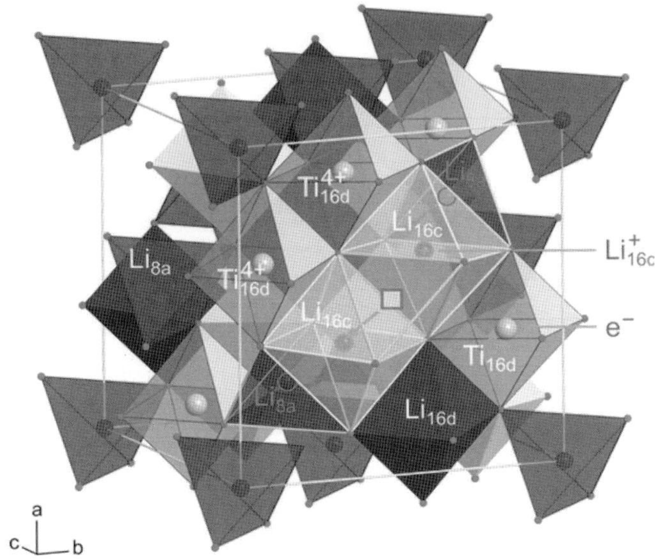

Figure 1. Crystal structure of $Li_{4+x}Ti_5O_{12}$ (space group $Fd3m$); at $x = 0$, the Li ions occupy 8a and 16d sites. With $x > 0$, they start to occupy the 16c sites; simultaneously, Li ions on 8a are shifted toward the 16c site. Taken from Ref. 22 with permission of the American Chemical Society.

general formula AB_2O_4, with oxygen ions arranged in a cubic close-packed array occupying the 32e crystallographic sites (Figure 1). The structure contains both tetrahedral (8a, 8b, 48f) and octahedral (16c, 16d) sites that are partially occupied by the A and B cations. LTO crystallizes in a cubic spinel structure in the $Fd3m$ space group with a unit cell parameter of a = 8.352–8.370 Å.[13–17] At room temperature, the tetrahedral 8a sites are fully occupied by Li^+ ions, whereas the octahedral 16d sites are randomly occupied by 1/6 Li^+ and 5/6 Ti^{4+} cations.[18] The remaining octahedral (16c) and tetrahedral (8b, 48f) sites are empty and available, and suitable, for lithium insertion/deinsertion. The occupancy of lithium ions in specific crystallographic sites within the pristine LTO structure is an area of ongoing research.[9] Lithium ion mobility within the pristine LTO has been studied by means of infrared spectroscopy,[19] X-ray and neutron powder diffraction data,[20] and solid-state ^6Li and ^7Li nuclear magnetic resonance (NMR).[21–23] These studies demonstrate that Li^+ ions are mobile; those in 8a sites could shift to the 16c sites with increasing temperature.[19]

C. Lithium Insertion Properties

The lithium insertion properties of LTO under galvanostatic conditions were studied in the 1980s in order to investigate new compositions within the $Li_{3+x}Ti_{6-x}O_{12}$ solid-solution family, initially because members of this family displayed interesting superconductivity at low temperatures.[24,25] The chemical lithiation of LTO was first demonstrated in 1983 by Murphy *et al.*[13] Colbow *et al.* investigated LTO as a cathode material in practical cells for the 1.5-V rechargeable battery market[14] and first measured the galvanostatic charge–discharge properties of LTO/Li half-cells. The material exhibits a plateau in its discharge voltage profile implying a two-phase lithium insertion mechanism[13,14,26,27]:

$$Li_4Ti_5O_{12} + 3\ Li^+ + 3e^- \leftrightarrow Li_7Ti_5O_{12}.$$

The theoretical specific capacity, as calculated from the abovementioned equation, is 175 mAh g^{-1}. The incoming lithium ions are inserted into the 16c crystallographic sites. From a structural point of view, it could be possible to insert more than three lithium ions per mole of LTO; empty 8a crystallographic sites still remain in the structure of $Li_7Ti_5O_{12}$.[27] Indeed, it should theoretically be feasible to reach a specific capacity of 293 mAh g^{-1} for LTO, based on the number of Ti(IV) cations in the structure. However, in order to achieve this, the cell must be discharged to lithiation voltages as low as 0.01 V *versus* Li^+/Li.[27] However, using such a low voltage will negate the safety advantages of using LTO *versus* a regular carbon anode.

Understanding lithium ion transport (diffusion) in and out of the LTO framework is hugely important to know the electrochemical rate capability. Scharner *et al.* used X-ray diffraction (XRD) measurements to conclude that the constant-voltage plateau was due to the presence of two distinct phases that are mutually interconvertible.[26] However, combined neutron and XRD measurements on Li-intercalated LTO found that, somewhat unexpectedly, the "two-phase" representation of Li^+ diffusion within LTO was, in fact, more subtle.[28] At intermediate compositions, that is, when the lithiation to form $Li_7Ti_5O_{12}$ is not yet complete, the two-phase equilibrium is not stable and the system may relax into a solid solution. ^{6}Li and ^{7}Li solid-state NMR studies confirm that the lithium diffusion mechanism in the anode material LTO is complex.[21,22] The

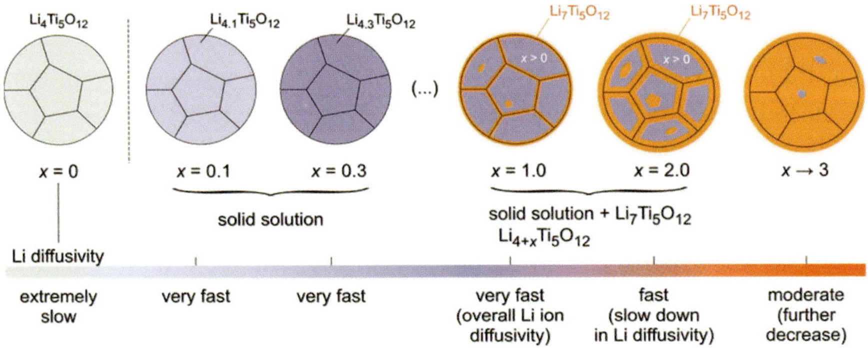

Figure 2. Schematic lithiation model for $Li_{4+x}Ti_5O_{12}$ based on ^{6}Li NMR measurements. Initially a quasi-solid solution is formed up to $x = 0.3$; with increasing Li content, regions rich in Li^+ might be present, with a composition close to the final one ($x = 3$); during further lithiation, these Li^+-rich regions grow, and an almost fully rock-salt-type oxide $Li_7Ti_5O_{12}$ is finally formed. Taken from Ref. 22 with permission of the American Chemical Society.

experiments, outlined in Figure 2, revealed that lithium diffusion is fast at low Li content with the formation of a *quasi*-solid solution. The lithiation mechanism comprises both the formation of a solid solution and the separation of phases into Li-poor ($Li_{4+x}Ti_5O_{12}$) and Li-rich ones ($Li_7Ti_5O_{12}$), while the grain boundary structure, hence the percolation, might influence the lithium diffusion mechanism. Faster charge–discharge rates impose the two-phase behavior and the formation of nanoscale domains of each phase, preventing the solid-solution equilibrium. The nanoscale two-phase mixture results in an increase in electronic and ionic conductivities because the lithiated LTO possesses a much higher lithium diffusion coefficient that its nonlithiated parent, due to Ti^{3+} concentration.[26,29]

D. Electrochemical Properties

Figure 3 displays the typical galvanostatic charge–discharge profiles for LTO anode. The open-circuit voltage (OCV) is generally higher than 2.5 V *versus* Li^+/Li, in an LTO/Li half-cell with a nonaqueous carbonate-based lithium electrolyte.[13,14,31] LTO features a rapid decrease of the voltage from the OCV until approximately 1.5 V *versus* Li^+/Li during the initial discharge (Figure 3a).[12] The voltage remains constant during the lithiation, when the formation of a solid solution and the separation of

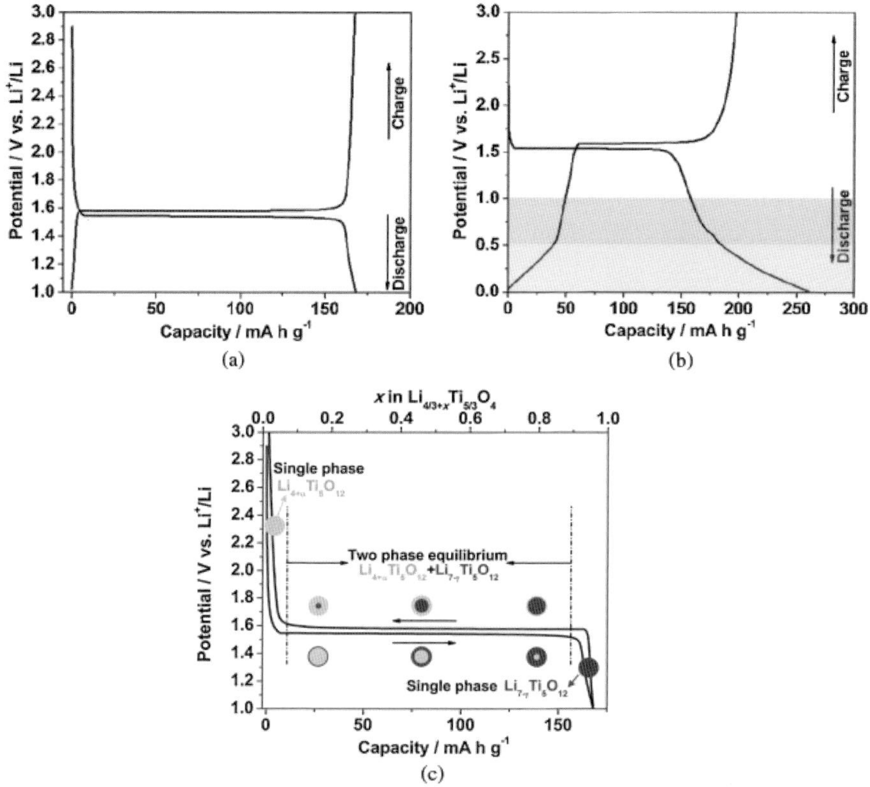

Figure 3. Typical galvanostatic charge/discharge voltage profiles of LTO/Li half-cells using a current density of 1 C between 3.0 and 1.0 V *versus* Li^+/Li (a), between 3.0 and 0.01 V *versus* Li^+/Li (b), and illustration of the two-phase mechanism during the electrochemical process (c). Taken from Ref. 9 with permission of Elsevier, and adapted from Ref. 30.

phases, into Li-poor ($Li_{4+x}Ti_5O_{12}$) and Li-rich ones ($Li_7Ti_5O_{12}$), occur (Figures 2 and 3c).[12–14,21,22,26] When the full lithiation is reached, the voltage rapidly drops up to 1.0 V *versus* Li^+/Li (Figure 3a). The reversibility of the process is usually good to excellent on the first galvanostatic cycle between 3.0 and 1.0 V *versus* Li^+/Li, with Coulombic efficiencies higher than 90%.[32] The efficiency decreases drastically when the cutoff voltage is lowered, especially below 0.5 V *versus* Li^+/Li (Figure 3b).[30] The goal of cycling below 1.0 V *versus* Li^+/Li is to take advantage of the remaining 8a crystallographic sites in the structure of $Li_7Ti_5O_{12}$,[27] but below 1.0 V *versus* Li^+/Li, the lithium insertion leads to the construction of the rock-salt $LiTiO_2$ phase at the surface of LTO grains.[33] This phase does not allow the removal of lithium, hence leading to large irreversible capacities (Figure 3b).[34,35] Some additional irreversible capacity obtained below 1.0 V *versus* Li^+/Li also originates from electrolyte instability at low potential.

In full-cell setups, LTO electrodes are cycled up to 30,000 times with 95% capacity retention (Figure 4).[36] LTO can be cycled *versus* different positive electrode materials, leading to a cell voltage of 1.85 V for a

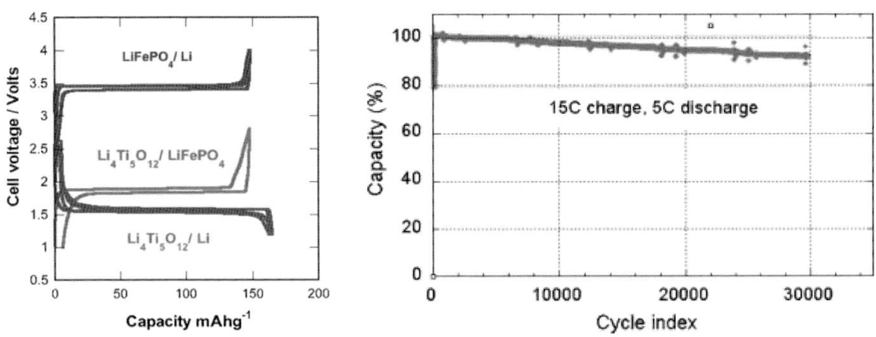

(a) (b)

Figure 4. Voltage–capacity cycle for $LiFePO_4$/Li, LTO/Li, and $LiFePO_4$/LTO coin cells at a C/24 rate; the capacity is given per gram of the positive electrode material (a); cycle life of a $LiFePO_4$/LTO 18650-cell cycled at a charge rate of 15 C (4 min) and a discharge rate of 5 C (12 min) (b). Taken from Ref. 36 with permission of Elsevier.

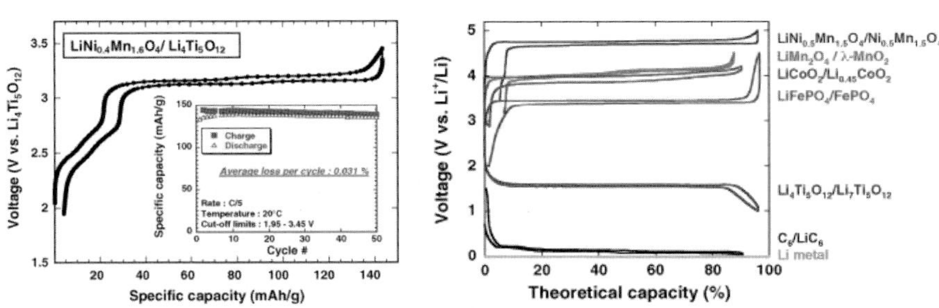

(a) (b)

Figure 5. (a) Charge–discharge profile of $LiNi_{0.4}Mn_{1.6}O_4$, obtained under galvanostatic mode at C/5 rate between 1.95 and 3.45 V *versus* LTO/$Li_7Ti_5O_{12}$ (10th cycle); (b) voltage *versus* capacity for several electrode materials, obtained at 20 °C with a cycling rate of C/5. Taken from Ref. 42 with permission of Elsevier.

$LiFePO_4$/LTO cell[36] and 3.15 V for $LiNi_{0.4}Mn_{1.6}O_4$/LTO (F).[37] The safety and reliability advantages of LTO-based LIBs have been emphasized by several studies that compared these to graphite-based cells,[38–41] and these remain the main advantages of LTO over graphite.[11]

E. Disadvantages of Lithium Titanium Oxide as Electrode

1. *The Gassing Problem*

Because LTO has an average potential of 1.55 V *versus* Li^+/Li, approximately 1.5 V above the average voltage of the graphite anode lithium insertion, it was long thought that none of the problems of electrode polarization and metal deposition occurring with carbons would occur with LTO. The electrolyte reactivity in cycling, though strongly dependent on the electrode material, is usually observed below 1.0 V *versus* Li^+/Li,[43] and it was first admitted that LTO working potential would prevent electrolyte reduction.[11,44] Moreover, to retain optimum cyclability, LTO electrode materials are usually cycled between 2.5 and 1.0 V *versus* Li^+/Li (*vide infra*), a voltage window that corresponds to the stability window of most LIB electrolytes.[45,46]

However, several problems prevent LTO being considered the most preferable choice for large-scale applications by the LIB industries. The most severe of these is the formation of gas during galvanostatic cycles and storage, especially at elevated temperatures.[47–51] Gas formation is a serious safety concern, made additionally grave by the flammable nature of both the gases and electrolyte. Gas formation is also detrimental to the electrochemical performance of the battery. Gassing is also known to occur in lead–acid batteries as a result of overcharging or short circuits inside the battery.[51,52] However, the origin and underlying mechanisms of gas formation within LTO-based cells are unclear, and the understanding of the roles that the $LiPF_6$-based electrolyte plays in gas formation reactions in LTO-based batteries is far from complete; currently, there are only few reports specifically referring to the gassing behavior of LTO electrodes.[48,53–56] For widespread implantation of LTO batteries in electric and hybrid vehicles and energy storage systems, an effective remedy for this problem must be sought.[47]

Figure 6 displays the charge–discharge voltage profiles of LTO/LCO cells at 20 °C and 60 °C,[57] and the gas concentration of various chemical

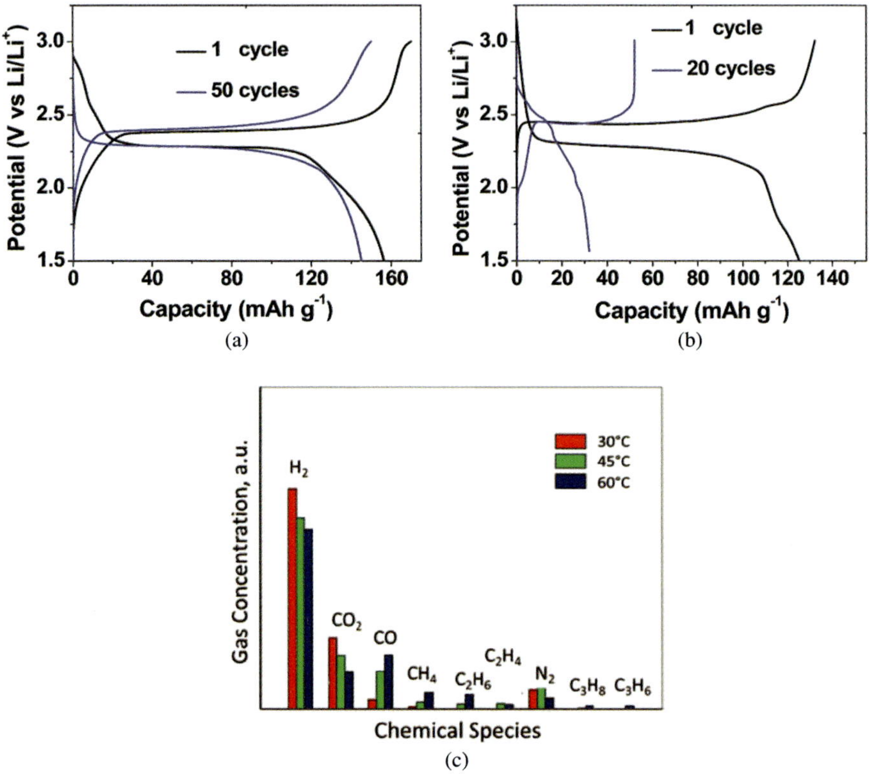

Figure 6. Commercial LTO powder electrode/$LiCoO_2$ cell in $LiPF_6$ 1 M in ethylene carbonate/dimethyl ethyl carbonate (EC:DEC): (a) charge–discharge voltage profiles at 20 °C; (b) charge–discharge voltage profiles at 60 °C (Taken from Ref. 57); (c) concentration of various chemical species in the gases collected from LTO/$LiMn_2O_4$ cells after 5 months at 30 °C (red), 45 °C (green), and 60 °C (blue) as determined by gas chromatography (Taken from Ref. 54 with permission of ECS).

species collected from LTO/LMO cells at 30 °C, 45 °C, and 60 °C after 5 months of electrochemical testing.[54] The specific capacity of the cell over 20 cycles is drastically reduced at 60 °C (Figure 6b), and large amounts of gases are generated (Figure 6c).[54,56] It is important to note that gassing not only occurs during cycling, but also in stored cells.[50] Gas chromatography measurements revealed that hydrogen gas, H_2, is a major constituent, formed from the beginning of the first discharge. CO and CO_2 are subsequently formed during galvanostatic cycling.[54,58] It was initially suggested that H_2 is derived from traces of H_2O that remain in the electrode, while CO and CO_2 result from the decomposition of electrolyte solution initiated at relatively high temperatures by the strong Lewis acid $LiPF_5$ species originating from the decomposition of the electrolyte salt, $LiPF_6$.[53,54,59–62] However, it is unlikely that the trace amounts of H_2O could be responsible for such large quantities of H_2, and such aggressive gassing is not observed in equivalently treated carbon cells. Instead, studies on epitaxial LTO (111) films demonstrated that PF_5 may not play an important role as Ti^{4+}/Ti^{3+} surface species that catalyze the reduction of solvent species on the electrode surface.[50,58,62,63]

To study gas evolution during operation, *in situ* and operando measurements have given a wealth of information, all-the-more important

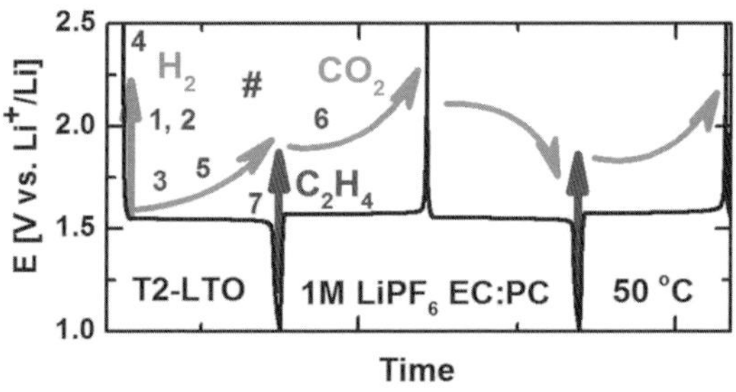

Figure 7. Schematic illustrating the proposed gas evolution processes during the first two cycles of an LTO/Li half-cell with a LiPF6 1 M in EC:PC (propylene carbonate) electrolyte. Numbers (#) referring to the suggested reaction mechanisms in Table 1 are written in dark red. Taken from Ref. 58 with permission of ECS.

Table 1. Summary of gas evolution mechanisms on LTO at 50 °C

Reactions	Number
$2\ H_2O + 2e^- = 2\ OH^- + H_{2\ (gas)}$	1
$2\ H^{\cdot} = H_{2\ (gas)}$	2
$OH^- + EC = \text{radicals} + CO_{2\ (gas)}$	3
$LiPF_6 + H_2O = LiF + 2\ HF + POF_3$	4
$POF_3 + EC/PC/DMC = \text{organofluoride compound} + CO_{2\ (gas)}$	5
$Ti^{4+} + EC/PC/DMC = Ti^{3+} + \text{radicals} + CO_{2\ (gas)}$	6
$2Li^+ + EC = Li_2CO_{3\ (solid)}/(CH_2CH_2OLi)_{2\ (solid)} + C_2H_{4\ (gas)}$	7

Taken from Ref. 58 with permission of ECS.

when complemented by theoretical investigations.[38,47–50,53,54,57,58,61,62,64–74] Figure 7 and Table 1 summarize the proposed gas evolution reactions on LTO electrodes cycled *versus* Li metal at 50 °C as studied using gas chromatography.[58] At 50°C, gas evolution is particularly high on the first galvanostatic cycle (Figure 7). This phenomenon is usually attributed to the formation of the solid electrolyte interphase (SEI) (see Subsection 1 "SEI Management") and electrolyte degradation (Table 1), effects that progressively form a layer that passivates the electrode surface.

F. Improving the Performance of LTO

To prevent the formation of gas, reactions between the electrode surface and the electrolyte must be minimized by controlling the interface between LTO and the surrounding electrolyte solution. Many potential strategies have been employed, all based on the construction of a barrier layer between LTO and the surrounding electrolyte solution.[9]

1. *SEI Management*

Many lithiation reactions take place outside the voltage window in which the electrolyte is stable. Breakdown of electrolyte components on the electrolyte surface forms an electrically insulating solid layer on the anode that allows lithium ion transport through the layer but serves to protect the electrode surface against further electrolyte redox reactions on it. This layer is called the SEI,[75] and results from the redox decomposition reactions of electrolyte components at the interface between particles.

The existence of an SEI layer in LTO-based anode for LIB was confirmed in two studies.[76,77] Initially, hypothesizing that LTO-based anodes induced no SEI between 3.5 and 1.0 V *versus* Li^+/Li where the electrolyte is electrochemically stable, Shu *et al.* cycled LTO/Li half-cells between 5.0 and 0.0 V *versus* Li^+/Li in LP30 electrolyte to determine the products that were formed outside this voltage window.[77] They showed that a passivating layer of approximately 10 nm is formed at the surface of the cycled electrode, containing fluorinated and copolymer species stemming from electrolyte decomposition below 1.0 V *versus* Li^+/Li, and lithium alkylcarbonates from electrolyte oxidation above 3.5 V *versus* Li^+/Li. It must be noted that in such a large voltage window, carbonate-based electrolytes are not stable, and their degradation was expected. Second, within a voltage window where both electrolyte and LTO electrodes are *stable*, El Ouatani *et al.* used X-ray photoelectron spectroscopy (XPS) measurements to probe the SEI layer at the LTO electrode surface.[76] The layer is mostly composed of carbonated chemical species (Li_2CO_3, $ROCO_2Li$) and LiF. Other studies have since reported similar conclusions; there is little doubt that an SEI layer is formed on LTO during cycling.[78–80] The protecting character of the SEI on LTO might, however, be less than what is usually observed for graphite anodes.[72] The SEI composition on LTO electrodes does not seem to differ as between half- and full-cell configurations.[81] In full-cell, for LMO/LTO systems, it was evidenced that manganese from the cathode might migrate and deposit on the anode side, but the effect of this on cycling performance of the anode is unclear.[82,83]

The absence of extrinsic conductive carbon in the LTO electrode formulation appears to improve electrochemical performance at 25 °C and 60 °C.[84] The passivating layer formed at the electrode–electrolyte interface would mostly originate from the presence of conductive carbon in the electrode formulation, and such a layer could prevent lithium diffusion through the electrode. Synchrotron-based XPS measurements confirm that SEI is preferentially formed on the carbon additive rather than on LTO grains.[85]

The specific surface area of LTO particles influences the formation of the SEI, analogous to the well-understood effect in graphite anodes.[86–88] For graphite, an optimal surface area is around 6 $m^2{\cdot}g^{-1}$, and the irreversible capacity on the first cycle reaches 40% for graphite with a surface area of 18 $m^2{\cdot}g^{-1}$. For LTO anodes, the higher the surface, the higher the reactivity of the electrolyte toward this surface. LTO particles with areas lower than 3 $m^2{\cdot}g^{-1}$ provide better cycling performances than particles with higher surface areas.[65,88]

The addition of electrolyte additives is another strategy to control the electrode/electrolyte interface. For instance, *p*-toluenesulfonyl isocyanate (PTSI) was found to suppress interfacial reactions.[89] Vinylene carbonate (VC) is also an additive that promotes SEI formation and prevents further reactivity between electrode and electrolyte.[86]

2. *Surface Coatings*

Coating LTO anode material to prevent direct contact between the electrolyte and the electrode surface has been used as a strategy to prevent gas formation. Carbon was the first coating material proposed,[50,90–92] primarily because gassing is neglectable for graphite-based LIBs.[72] Liu *et al.* synthesized carbon-coated LTO nanowires with high crystallinity, which exhibited 78% capacity retention at the end of 500 cycles at a rate of 1 C and temperature of 55 °C.[90]

Other types of coatings have also been investigated. Surface treatment methods, for example, coating LTO with zinc oxide (ZnO),[68] AlF_3,[93] or with ion-conductive polyimide layer,[94] were found to reduce the interfacial reactions effectively and mitigate gassing to a significant extent. Nakajima *et al.* investigated the influence of surface fluorination by F_2, NF_3, or ClF_3 at high temperature on the cyclability of LTO electrodes.[95–98] The chemical affinity of fluorine for lithium induced the presence of large amounts of fluorine (28–35 atomic%), especially through lithium fluoride, LiF, but did not yield significant improvements of electrochemical performances.

3. *Tailoring Ionic and Electronic Conductivities*

In addition to the problems of gas formation, LTO suffers from poor electronic conductivity and relatively poor lithium diffusion rate. To achieve simultaneous high specific capacity, high rate capability, and good cycling stability, several strategies have been investigated: nanostructuring into zero-dimensional (0D), 1D, 2D, or 3D materials, morphology and texture optimization, carbon coating, doping, and composite formulation.[9]

Chemical doping of LTO was the first strategy to be employed to improve its conductivity, and has been extensively reviewed elsewhere.[99–102] Doping usually involves simple chemical changes to optimize synthesis procedures, by adding other precursors.[103,104] Starting from a mixture of lithium acetate and lithium fluoride, the resulting LTO contains fluoride, according to XPS measurements, though the exact composition is not known.[103] The presence of F^- anions within LTO is proposed due to the presence of Ti^{3+} cations. This improved the electronic

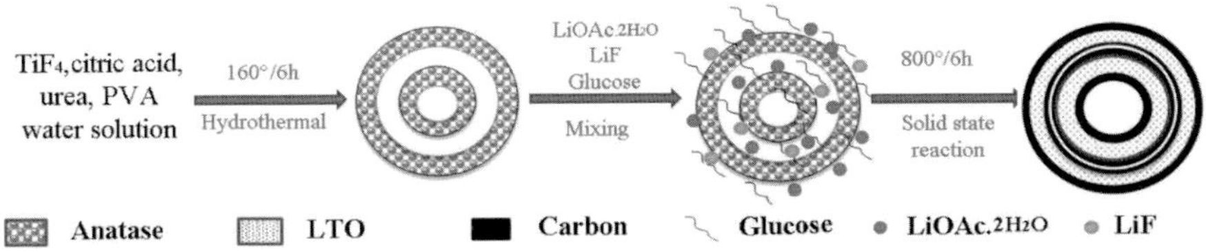

Figure 8. Schematic of the preparation of carbon-encapsulated F-doped LTO composites. Taken from Ref. 103 with permission of the American Chemical Society.

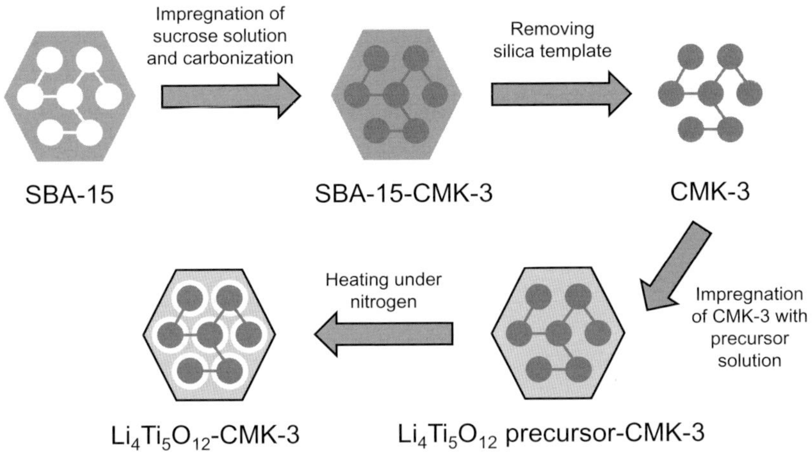

Figure 9. Schematic illustration of the synthesis of mesoporous LTO/C nanocomposite. Adapted from Ref. 113 with Permission of Wiley.

conductivity and, thus, the mixed valence Ti^{4+}/Ti^{3+} LTO shows improved performance. No specific experiment has yet demonstrated this however, and it remains a hypothesis.

The design of hybrid materials made of LTO and carbon is where the major efforts have been placed as a cost-effective way to improve electronic conductivity.[9,105–115] LTO has been grown confined in mesopores, in order to limit the size of crystallites as well as to confine LTO particles deep within a conductive matrix (Figure 9).[105,113] A major disadvantage of this approach may be the high specific surface area of the composite LTO/carbon that is not ideal for SEI management (*vide infra*). A range of other carbon structures have been used to confine LTO particles in a conductive carbon network, including graphene[107,110] and nanotubes.[106,108] Finally, LTO submicron-sized particles coated with carbon layer have been investigated as 0D-based 3D mesostructures.[109,111,113]

G. Critical Perspectives

Graphite is the active material of the majority of the negative electrodes in commercial systems. Replacing graphite with a material that can withstand high charge rates is instrumental in improving LIB lifetime, stability, available power density, and speed of charge. Mesostructured titanium oxides and lithiated titanium oxides such as LTO possess all the features — excellent lifetime, *extra push-over-the-cliff* power, and fast charging rate — required for successful implementation in power applications (vehicles, stationary storage).

Lithiated titanium oxides, however, have drawbacks that have so far prevented their widespread implementation. In LIBs, titanium dioxides, lithiated or not, display an average operating voltage above 1 V *versus* Li^+/Li. Such voltage was originally thought to be high enough to avoid electrolyte decomposition (generally occurring below 1 V *versus* Li^+/Li) and formation of SEI, but studies have now demonstrated that titanium oxides may not be as electrochemically stable as required for large-scale applications in LIBs. *They suffer from the same reactivity and gas formation problems as potential high-voltage cathode materials.* This instability originates from interfacial reactions between the oxide surface and surrounding alkyl carbonate solvents, reactions that are likely to be catalytically activated by titanium oxides. Carbon coatings and electrolyte additives are the only proposed solutions to these problems but *do not solve the gassing issue on the industrial scale*.

Considering that LTO is investigated for large-scale applications, the environmental aspects become an increasing concern. Polyvinylidene fluoride (PVDF) binder is sensitive to moisture and requires the use of organic solvents, which are less environmentally friendly than water. A transition to water-soluble binders such as carboxymethyl cellulose (CMC) or polyacrylic acid (PAA) would be a reasonable solution.[116] Alternative binders have been explored to a very limited degree for LTO. Future work could well involve evaluations of the electrode performance using water-soluble binders. For instance, Leclanché seem to have already chosen the water-based electrode manufacturing for their smart-grid LTO-based battery.[a]

III. p-Block Element–Based Electrodes

Metals and semimetals, such as silicon, tin, and germanium, react with lithium within a specific voltage range to become alloys. When the current is reversed, these materials theoretically de-alloy to reform their original state, allowing continuous cycles of charge and discharge. The reaction of such electrodes takes place as follows:

$$x\mathrm{Li}^{+} + x\mathrm{e}^{-} + X \leftrightarrow \mathrm{Li}_{x}X$$

This reaction can proceed with the addition of multiple electrons per active atom. Therefore, electrodes undergoing alloying theoretically enable capacities up to 10 times higher than the capacity of graphite (375 $\mathrm{mAh \cdot g^{-1}}$); 990 $\mathrm{mAh \cdot g^{-1}}$ for Sn and more than 3500 $\mathrm{mAh \cdot g^{-1}}$ for silicon, for example.[6] While the multi-electron processes are the source of the extremely high theoretical capacities, they are also the source of the main challenges for this type of electrode. Lithium insertion causes dramatic changes to the structure of the electrode and enormous variation in volume during insertion/de-insertion reactions. For example, volume changes of ~300% are expected to take place during lithiation/delithiation for Sn based on the relative volumes of the unit cells.[117] This expansion is generally accepted to be detrimental on all levels. On a particle scale, this volume expansion can lead to stress that causes cracks in the electrode with continuous and excessive breakdown of electrolyte on the new surfaces that are created. On an electrode scale, it can lead to loss of contact between the grains and the current collector, causing electrical isolation of areas of the electrode. These factors combine to give a significant and irreversible loss of capacity in the first cycle, and loss of capacity in each cycle, suggesting that the lifetime of the battery is extremely limited.

[a] Leclanché Energy Storage Solutions. TiBox — High End Energy. http://www.leclanche.com/technology-products/products/tibox/ (accessed Oct 26, 2016).

Since the late 1990s, there has been extensive research into methods of limiting these effects by reducing the size of particles from micron- to nanoscale or by compositing the active compounds with elements that do not react with Li; this will limit the volume expansion and buffer its effects. While these methods have met with some success in improving the electrochemical performance of these electrodes, the mechanism of lithiation/delithiation of many of these materials is only partially understood. To develop a strategy to prevent early cell failure allowing the practical use of intermetallic as anodes for LIBs, understanding their lithiation and delithiation mechanisms as well as their failure mechanism at the atomic level and the electrode level is of critical importance and remains the most significant ongoing challenge to the field. The phases formed within the battery are not necessarily crystalline phases formed in thermodynamic conditions, and understanding their structure presents a challenge to conventional characterization techniques. Hence, there has been an increasing emphasis placed on characterizing the materials using advanced structural methods, using both postmortem analysis of electrodes, where the structure of the material is studied after removal of the cell, or *in operando* methods, where structural characterization takes place during battery operation. In the following sections, the mechanistic understanding gained by this method for several of the most promising alloying anode materials is considered.

A. Silicon

Silicon is as attractive as a negative electrode due to its high abundance in the earth's crust and high theoretical gravimetric capacity — 3579 $mAh{\cdot}g^{-1}$ based on the formation of $Li_{15}Si_4$. The ability to form alloys with lithium electrochemically was first demonstrated by Huggins *et al.*,[118] but despite the promise of this chemistry, silicon exhibits low cycling efficiency and permanent capacity losses that are mainly attributed to the effects of the significant volume expansion during lithiation (~300%), particle pulverization, parasitic side reactions of the electrolyte at low voltages, and loss of electrical connectivity within areas of the electrode.[119]

1. *Mechanisms of (De)Lithiation in Silicon Anodes*

a. *Atomistic-scale lithiation mechanisms*

Huggins *et al.* first explored the lithiation of silicon using Coulombic titration of Li into Si electrodes at 415 °C. A stepped voltage profile is observed with four plateaus corresponding to two-phase reactions forming phases with the nominal compositions $Li_{12}Si_7$, Li_3Si_3, $Li_{13}Si_4$, and $Li_{22}Si_5$ (Figure 10). The room temperature electrochemistry of crystalline silicon however is clearly different;[120] on the first lithiation, it is dominated by a relatively flat, slightly decaying, process at 100 mV and at 375 mV on delithiation. The flat voltage plateau is indicative of approximately two-phase behavior; the silicon reacting to form a second, lithiated phase. Galvanostatic intermittent titration technique (GITT) measurements were performed by Ryu *et al.*, where the electrode was allowed to relax to thermodynamic equilibrium between a series of current pulses giving an idea of the true voltage of the lithiation processes. These confirm that these voltage processes are not true plateaus but gradually slope.[121] TEM measurements have been able to clearly identify a sharp reaction front between

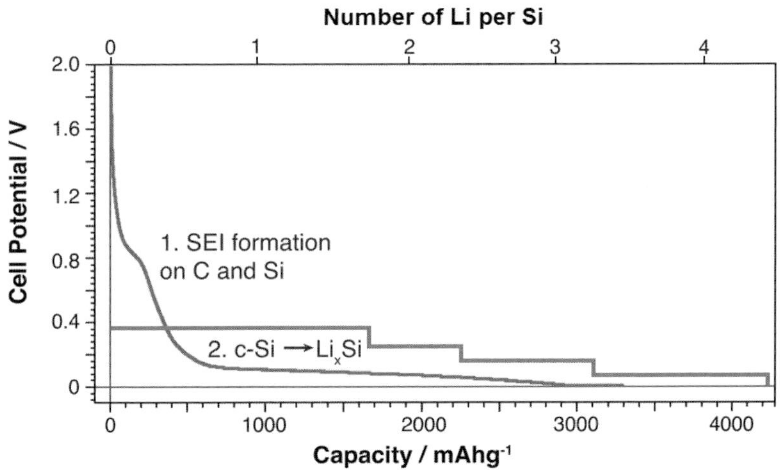

Figure 10. Coulometric titration curve for the Li–Si system at 415 °C (gray line)[118] and galvanostatic charge/discharge of an Si/C powder electrode at room temperature (blue line). Adapted from Ref. 124 under a Creative Commons CC-BY license.

the lithiated silicon and pristine amorphous silicon particles.[6] The presence of this sharp reaction front is likely to be the result of the large activation energy for breaking up the crystalline silicon, meaning that a large local concentration of lithium is required before Si–Si bond breakage is kinetically favored.

At the beginning of the year 2000, the first characterization of silicon electrodes using XRD was performed. The *operando* XRD measurements for crystalline Si, where the sample was cycled inside a battery cell with an X-ray transparent window, allowing characterization during operation, showed the loss of Bragg peaks as lithium was added to the electrode. Although these measurements were not able to probe the nature of the amorphous product formed as a result of the lithiation, they were able to propose a stoichiometry of $x = 3.5$ for the lithiated Li_xSi phase.[122,123] The structure of amorphous lithiated phases has been probed using techniques that are sensitive to the local arrangements of atoms, and are therefore able to give insight into the local silicon arrangements. Pair distribution function (PDF) analysis shows that the silicon is first broken up into larger silicon clusters (Figure 11), and later to silicon dimers and isolated silicon atoms. The kinetic difficulties associated with breaking up the crystalline silicon mean that there is a competition between the reaction of lithium with the crystalline silicon and with the more kinetically facile reaction with the already partially lithiated clusters. Solid-state NMR measurements showed that this results in a range of silicon connectivity being present during the lithiation; this prevents the crystallization of intermediate species as this would require the migration and rearrangement of silicon clusters in order to form ordered arrangements. The most lithiated phase in the Li–Si phase diagram, $Li_{22}Si_5$, is not formed, even at very low voltages. Instead, the metastable crystalline phase — a kinetically trapped phases with positive free energy above thermodynamic equilibrium — $Li_{15+x}Sn_4$ is crystallized from the amorphous intermediate phase at low voltages (~50 mV). This phase tolerates some off-stoichiometry to become over-lithiated at the end of lithiation, where $x > 0$. Again, kinetic considerations are likely to play a large part in explaining this observation; within the $Li_{15}Si_4$ structure, all silicon atoms are equivalent and are surrounded by lithium atoms in an arrangement similar to that of an amorphous compound featuring isolated silicon ions in a "sea" of lithium. Very

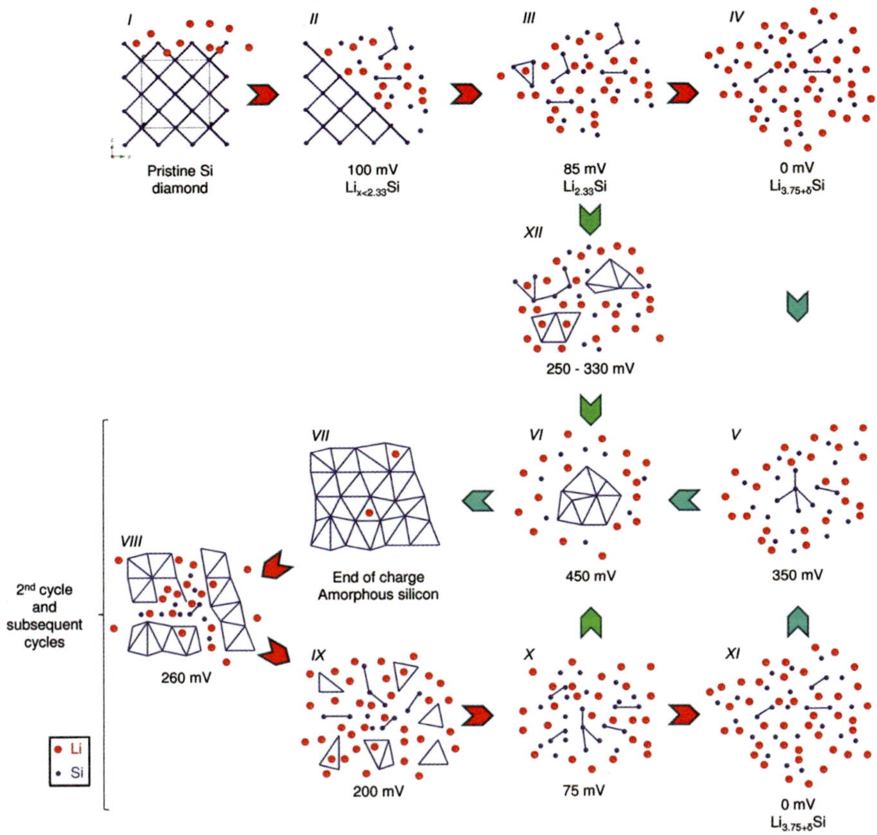

Figure 11. Illustration of mechanisms by which silicon is lithiated and delithiated. I to IV correspond to lithiation of crystalline silicon discharged down to 0 V. Red arrows denote discharge steps, while blue and green arrows denote charge steps. V to VIII correspond to the delithiation of fully lithiated silicon, and VIII to XI correspond to the relithiation of amorphous silicon to form fully lithiated silicon beyond the first cycle. Adapted from Ref. 127.

little rearrangement of the electrode is required in order to form a crystalline structure of this material. The voltage at which $Li_{15+x}Si_4$ is formed, and indeed whether it is formed at all, appears to be a complex affair dependent on stresses induced by the electrode substrate or from the presence of SiO_2 species on the electrode surface[124]; for example, thin films of silicon less than 2 μm in thickness do not appear to crystalize at all.[122,125] At higher temperatures, or when the electrode is held at low voltage for a long time, $Li_{15}Si_4$ can be transformed into thermodynamically stable phases.[126]

Delithiation of the silicon electrode results in the formation of amorphous silicon. This means that in all but the first lithiation, the "active" electrode is in fact amorphous silicon and not crystalline silicon. In order to understand how these electrodes work in practical systems, structural characterization experiments need to probe at least the first two (de)lithiation processes. The lithiation of amorphous silicon shows a sloping voltage profile, where two processes can be resolved in the differential voltage (dQ/dV) plot at approximately 250 and 100 mV, corresponding to the following reactions[127]:

$$\text{a-Si} + x\text{Li}^+ \text{ à a-Li}_x\text{Si } (x \sim 2.0) + x'\text{Li}^+ \text{ à a-Li}_{x+x}'\text{Si } (x' \sim 1.5)$$

In comparison to the first lithiation, the more open structure of the amorphous lithium network allows all silicon to react to form larger clusters during the first process (Figure 11). The second process breaks up these clusters into dimers and isolated silicon ions. After many cycles, the intensity of the 250-mV peak in the dQ/dV plot becomes less pronounced, indicating that the formation and breakup of the larger clusters is impeded, most likely to be a significant contributor to failure in these electrode.[124]

b. *Electrode-scale lithiation mechanisms*

While lithium kinetics on the atomistic scale determines the lithiation of silicon electrode to a large extent, the large volume expansion experienced by silicon electrodes also has effects on an electrode scale. Microscopy has played an increasingly important role in understanding this morphology evolution as well as the mechanism of failure during cycling and has been applied to the various forms of silicon including crystalline nanowires, amorphous nanoparticles, and thin films. From these analyses, detailed elemental mapping, including the distribution of lithium, is possible, both inside a single particle and throughout the full depth of the electrode. During the first lithiation of micron-sized silicon particles of crystalline silicon, after initial lithium consumption to form the SEI layer, a core–shell mechanism was observed and its evolution upon cycling was followed using a focused ion beam (FIB) implemented in the analysis chamber of a state-of-the-art time-of-flight secondary ion mass spectrometer (ToF-SIMS). The lithiation of the silicon particles starts with a two-phase transformation between crystalline silicon and the amorphous Li_xSi phase. Fast lithiation paths through the amorphous Li_xSi phase through the pure silicon core have been observed by scanning electron microscopy (SEM) and transmission electron microscopy (TEM) analyses.[128] These fast diffusion paths result in a two-phase mechanism that proceeds until the silicon core is consumed. In a second stage of lithiation, the lithiated silicon particles become homogenous and the addition of further lithium proceeds through a solid-solution mechanism, where the entire lithiated phase has a single stoichiometry. The mechanism was observed for all particles in the electrode, independent of their position in the electrode relative to the current collector. Cross-analysis with Auger spectroscopy allowed the lithium concentration in the shell to be quantified; the amorphous Li_xSi composition varies between ($2.2 < x < 3.5$).[129]

Cracking of silicon electrodes due to mechanical stresses on the electrode can cause the irreversible loss of connection between the current

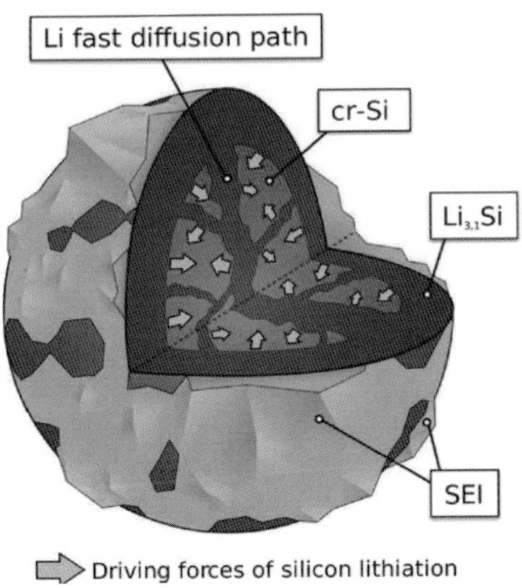

Figure 12. Illustration of a partially lithiated particle during the first lithiation. Thesis Arnaud Bordes, 2016, University of Pierre et Marie Curie, Paris, France.

collector and the particles, leading to electrically isolated, "dead" areas within the electrode, which can no longer contribute to the capacity. Fundamental studies of the mechanical stress on particles caused by the volume changes during lithiation/delithiation focus mainly on individual Si particles, wires, or pillars. There is evidence from TEM studies that the direction of lithiation of silicon also shows crystallographic directionality; the swelling of single nanowires was kinetically favored in the <110> direction.[130] This results in inhomogeneous deformation near the interface between anisotropically lithiated domains.[131]

In real systems, silicon anodes are composed of clusters of active silicon particles with conductive carbon additive and a binder held within a confined space. Interparticle interactions between silicon particles and with other components in the electrode are likely to have a large contribution to the overall electrode stress. These effects have been studied in well-defined crystalline Si nanopillar pairs during lithiation using ex situ SEM and *in situ* TEM techniques.[132] Where the Si pillars are mechanically clamped to a rigid surface, compresses stresses were found to change the reaction kinetics; preferential lithiation occurred at free surfaces and lessened the tensile stress concentrations within the silicon itself, compared to that in unclamped particles. From this, it could be anticipated that the silicon particles within clusters in electrodes may actually be more fracture resistant compared to model systems. However, although compressive stresses appear to enhance the fracture resistance, it must be noted that the volume expansions associated with full lithiation are considerably higher that what was studied and further investigation is necessary to optimize the ratio of electrode components and empty space in order to minimize fracture effects.

The propensity of crystalline silicon (c-Si) to crack is due to differences in how easily the different crystallographic directions can be lithiated. For amorphous-Si (a-Si) electrodes, there is no such lithiation anisotropy. As a result, both the lithiated phases and deformation processes in amorphous-Si particles and wires are more homogenous than those in their c-Si counterparts. Amorphous silicon electrodes are, therefore, more mechanically robust.

While pristine crystalline silicon is relatively brittle, there is evidence that lithiated silicon is far more ductile. *In situ* TEM experiments showed

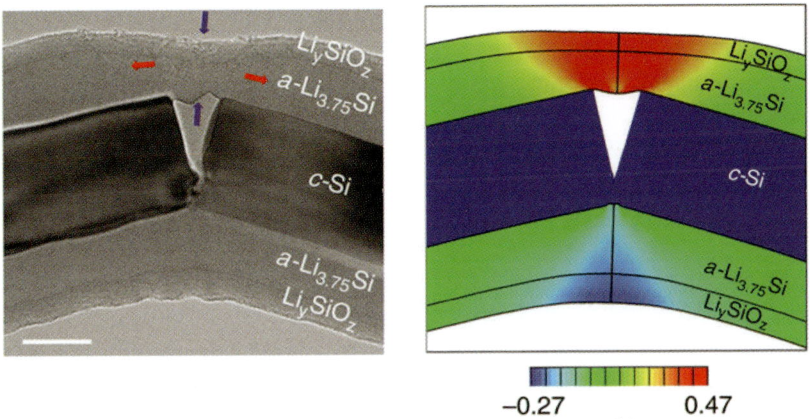

(a) (b)

Figure 13. (a) TEM image showing the brittle fracture of the unlithiated c-Si core and the tensile deformation of the lithiated a-$Li_{3.75}Si$ shell (scale bar, 50 nm). (b) Finite element result showing the simulated elastic-plastic deformation in the nanowire. Adapted from Ref. 133 under a Creative Commons CC-BY license.

that lithium-rich silicon alloys show high damage tolerance — that is, high tensile ductility and high fracture toughness — especially when the Li/Si ratio is above 1.5.[133] *In situ* TEM experiments on a partially lithiated Si nanowire (Figure 13) show a strong contrast between the brittle fracture taking place in the unlithiated Si core *versus* the ductile tensile deformation in the lithiated Si shell. This quantitative characterization and mechanistic understanding of high damage tolerance of Li-rich Si alloys represent a critical step toward the rational design of durable Si-based electrodes for next-generation LIBs.

2. *Failure Mechanisms in Silicon Electrodes*

a *Volume expansion on lithiation*

A large volume of research has been directed toward understanding the fundamental mechanisms of failure in silicon electrodes. Most of these are the result of the volume expansions experienced during lithiation, which are detrimental to performance at many levels. Anisotropic lithiation of crystalline silicon particles leads to inhomogeneous deformation and stress within particles of the silicon and disruption of the SEI layer. On a larger scale, repeated volume expansions and contractions can cause the active materials to lose electrical contact, resulting in cell fade and early cell failure.

b. *Issues relating to the SEI*

Since the lithiation of silicon takes place outside the voltage window where the electrolyte is stable, a key requirement for the long-term cyclability is the formation of a stable SEI. For this to happen, the SEI should be strongly bounded to the electrode surface, but should be flexible enough to accommodate any volume changes without damage. In the case of silicon, it has been widely demonstrated that the SEI formed on the silicon surface is not stable and is insufficiently protective for the electrode from further reaction with the electrolyte.[134] As the electrode expands on lithium insertion, cracks form and new SEI-free silicon surfaces are exposed to the electrolyte. This leads to continuous buildup of SEI, which consumes the lithium and solvent from electrolyte (Figure 14). XPS analyses have been devoted to the characterization of the SEI generated on silicon electrodes with the conventional electrolyte, lithium hexafluorophosphate ($LiPF_6$) in alkyl carbonate solvents. The SEI was found to mainly consist of lithium fluoride, lithium alkoxide, and carbonate and polycarbonate species such as lithium alkyl carbonates and Li_2CO_3.[135–137] *In situ* sum frequency generation vibrational spectroscopy (SFG-VS) was used to probe the molecular composition of the SEI surface species under various applied potentials where electrolyte reduction is expected.[138] In 1.0 M $LiPF_6$ in a mixture of EC and diethyl carbonate DEC, a Si-ethoxy (Si-OC_2H_5) surface intermediate was formed on the Si(100)-hydrogen terminated wafer due to decomposition of DEC. This supports an electrochemical electrolyte reduction mechanism in which the reduction of the DEC molecule to an ethoxy group plays a key role. In addition to providing evidence for the specific chemical composition of the SEI under operating conditions, these results also suggested that the SEI surface composition can vary depending on the chemical termination — and therefore the acidity — of the Si surface. Taken further, this also implies that there are opportunities for the development of "designer" SEI layers,

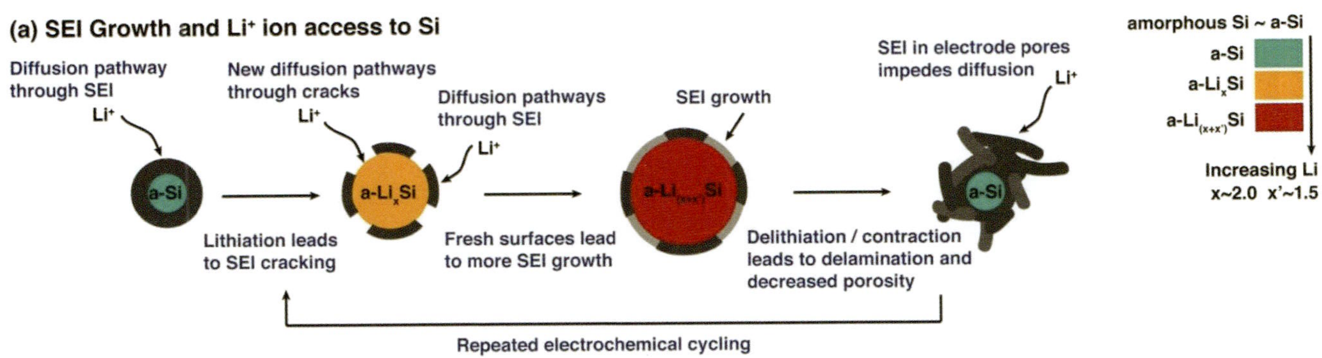

Figure 14. Nanoscale processes of SEI formation involve Li+ ion diffusion through cracks in the SEI, additional SEI growth, and delamination from Si particle surfaces. Adapted from Ref. 124.

where the chemical and mechanical properties are optimized by functionalization of the surface to provide the most stable SEI possible.

The incomplete removal of the lithium when the electrode is charged results in lithium consumption, which is never recovered and reused in charge-storage reactions. The major causes of this in silicon anodes are (1) the loss of contact between particles and current collector forming "dead" areas of the electrode that cannot participate in any further cycles and (2) increased electrode resistance due to the SEI layer resulting in increased voltage polarization between lithiation and delithiation. Capacity loses due to SEI resistance can be significantly improved by lowering the delithiation cutoff voltage or by introducing electrolyte additives, such as fluroethylcarbonate (FEC).[139]

The major breakdown of the $LiPF_6$ salt appears to occur mainly on the first lithiation cycle, and shows very little increase thereafter.[124] However, breakdown of organic solvent species on the electrode surface continues, especially when a binder is present in the electrode. SEM images of FIB cross sections show that after multiple cycles the electrode becomes increasingly dense. The pores that are present in the pristine electrode become clogged with organic species from electrolyte breakdown. This will increase the tortuosity of the electrode, impede lithium-ion diffusion, and increase the resistance of the electrode, increasing voltage polarization.

In addition to losses through the formation of excessive SEI, lithium can also be lost through self-discharge processes. The final lithiation product, $Li_{15}Si_4$, a metastable phase, has been observed to be reactive in the electrolyte and undergo a self-discharge mechanism when the cell is held at the OCV.[127] Lithium from this phase reacts, presumably with the electrolyte, to form species that can no longer contribute to electrode capacity, and therefore is lost from the electrode. This is potentially a main reason of capacity loss in batteries with Si anodes, and further understanding of this process is required in order to avoid these losses.

The majority of studies on silicon-based electrodes are performed in half-cell configuration. A lithium metal counter electrode is used, and provides an essentially unlimited supply of lithium ions to the cell. These studies are ideal for characterizing the processes happening in a single electrode in isolation. In real systems, however, the amount of lithium in the cell is finite, being provided by the ions contained in the electrolyte and the lithium present in the positive electrode. Failure mechanisms in full-cells using silicon negative electrode and $LiNi_{1/3}Mn_{1/3}Co_{1/3}O_2$ positive electrode with finite lithium supplies have been studied using multiple techniques.[140] The electrolyte degradation process exhibited many similarities to what has been observed in the case of half-cells, in particular during the early stages of the cycling, with the formation of inorganic species from the degradation of the organic solvents of the electrolyte. For extended cycling, the major failure mechanism results from all the lithium available for cycling being consumed by these parasitic reactions, and ends up either trapped in an intermediate part of the SEI or in the electrolyte. It is, therefore, no longer available for redox reactions at either electrode. This nevertheless does not actually prevent the further degradation of the organic electrolyte solvents, eventually leading to the formation of lithium-free organic degradation products at the extreme surface of the SEI, which eventually obstructs the porosity of the composite electrode or disconnects the active material particles.

An additional source of lost lithium is through reaction with the intrinsic existence of silicon oxide groups on the silicon surface. Undesirable interfacial interactions can irreversibly consume lithium during the first cycle; the SiO_2 layer is converted to crystalline Li_2O and elemental silicon, a reaction that is irreversible within the voltage range used and therefore consumes lithium from the cell, which is not recovered. Li_2O has both an electrical and mechanical effect on the subsequent cycling. Both the native oxide layer and Li_2O are electronically insulating, which can isolate any particles encapsulated by this layer, and also increase the impedance of the electrode, resulting in low Coulombic efficiency.[141] Li_2O also has higher stiffness than silicon and can therefore contribute to electrode cracking. However, the role of oxygen in the electrode may not be fully detrimental, and in some cases, it has even been suggested to be beneficial. Li_2O is a good conductor of lithium ions, and so electrodes using SiO or SiO_x ($x < 0.5$) nanoparticle as the active material form large amounts of Li_2O and have a short lithium diffusion path and a buffering matrix of the Li_2O that accommodates volume change by the presence of oxygen.[142,143,144]

3. *Improving the Cyclability of Silicon Electrodes*

a. *Electrode formulation*

Improving the cyclability of silicon electrodes requires a method to either reduce the volume expansions taking place within the electrodes, or mitigate the harmful side reactions that these processes induce. The performance of silicon electrodes can be improved by careful electrode formulation, where choice of binder and the ratio of silicon to conductive carbon and binder is optimized. Different binders are preferred for alloying compared to electrode, which undergo intercalation chemistries. In the case of intercalation where negligible volume changes take place, PVDF binders, which form a network of fine threads to hold the electrode together, are sufficient to hold the electrode together.[145] For alloying, binders that cover the electrode surface, protecting it to a certain extent from reaction with electrolyte, and that have strong interactions with the surface that are preserved during electrode expansion, are preferred. Both CMC and PAA that bind to surface Si–O and Si–OH groups have been shown to vastly improve performance in these electrodes.[116] Using a simple geometrical model, Beattie *et al.* proposed that a 1:1:1 ratio of the three components should demonstrate optimum performance by allowing free expansion of particles and, thus, prevent their rearrangement within the electrode.[146] This has been confirmed through electrochemical experiments.[147] However, diluting the active material within the electrode will have a detrimental effect on the gravimetric capacities that can be achieved.

b. *Nanosizing electrode particles*

Stress evolution in Si particles during lithiation/delithiation can be mitigated by nanostructuring the silicon as, for example, nanowires. This approach has been confirmed by TEM to be successful below a critical particle size of 150 nm for crystalline silicon and for larger particles (850 nm) of amorphous silicon.[148,149] The smaller particle size helps in mediating the volume expansion and therefore minimizes particle cracking, particle loss, and the formation of new surfaces for SEI formation.[4,150] However, electrodes fabricated from these materials often themselves have a large electrodes surface areas, which can exacerbate the problems associated with electrolyte degradation. In addition, nanostructuring processes are expensive and difficult to apply on a large scale.

c. *Electrode coating and electrolyte additives*

Another approach to stabilizing the silicon SEI is to chemically direct the interface between the electrode and the electrolyte using electrode coatings of electrolyte additives to get desired properties.

Coating of electrode materials is a strategy which has been widely used to improve the cycling performance in other LIB electrode systems, either for improving ionic or electric conductivity, for limiting volume expansion, or preventing isolation of the particles.[151] In the case of silicon anodes, a further advantage of these modified surfaces is that they can favor the formation of a robust interface between the electrode and the electrolyte. For example, coating silicon particles with aluminum alkoxide (alucone) consumes the native SiO_x layer, thus preventing the formation of Li_2O.[141] The large number of cross-linked Al–O bonds makes this coating mechanically flexible and so able to withstand large volume changes. Moreover, density functional theory (DFT) calculations have demonstrated that during lithiation lithium atoms bind very strongly to the oxygen atoms within the alucone in the most energetically favorable sites. When the film is irreversibly saturated with Li atoms, it becomes electronically conductive in addition to being ionically conductive. This promotes electron transport through the electrode facilitating fast and reversible electrochemical reactions.

Addition of additives into the electrolyte can give a degree of control over the chemical nature of the SEI layer and promote the formation of an SEI with desirable electronic (electronically isolating, ion conducting) and mechanical properties (flexible, able to accompany volumetric changes in the electrode).[152] Fluoroethylene carbonate (FEC) has been shown to be a promising additive for $LiPF_6$-based electrolytes. The decomposition of the FEC at low voltages was investigated by synchrotron-based XPS that gave a chemical composition depth profile over the whole SEI layer. FEC is reduced at higher voltages than EC and DEC, and forms a flexible SEI found to consist primarily of lithium fluoride and $-CHF-OCO_2$-type compounds.[151]. This SEI layer is sufficiently stable to limit the emergence of large cracks and preserve the original surface morphology, as well as suppressing the additional SEI formation from the other solvent. In comparison, the SEI formed from the decomposition of EC and DEC, without the FEC additive present, covered surface voids and led to an increase in voltage polarization.

d. *Association with inactive species*

Formulation of electrodes containing both a species that can alloy with lithium and a species that is inactive in the presence of lithium ions can buffer some of the mechanical stresses generated by large volume changes. Nano-Si has been, for example, embedded in a $Ti_4Ni_4Si_7$ matrix, and this electrode exhibited superior electrochemical performance, with a stable capacity above 900 $mAh \cdot g^{-1}$ after 50 cycles and a high coulombic efficiency of up to 99.7%.[153] The $Ti_4Ni_4Si_7$ matrix is a good mixed electronic and ionic conductor, and isolates nano-Si from the liquid electrolyte, thus preventing parasitic reactions that are responsible for the formation of an SEI.

A Si nitride (SiN_x) electrode was synthesized from Si nanoparticles via a vacuum chemical vapor deposition (CVD) approach within a graphene matrix.[154] This gave an external composition of Si_3N_4 and $SiN_{0.73}$ within the silicon nanoparticles. The electrode phase was initially electrochemically converted to a matrix of Li_3N with embedded nano-Si, in a mechanism similar to that used for SiO_x with lithium. The silicon nanoparticles then undergo further lithiation. A composite of SiN_x and Si (50% Si weight per weight) displayed a capacity around 1400 $mAh \cdot g^{-1}$ that was stable over many cycles. This was attributed to the improved stress management and conductivity of the Li_3N matrix. However, like in the case of silicon suboxide discussed earlier, the formation of the Li_3N matrix will result in first cycle irreversible capacity, and unless the lithium required for this reaction can be sourced from another source other than the cathode, it is unlikely that such an electrode will be commercially viable.

B. Other p-Block Elements: Antimony, Tin, and Germanium

While silicon has attracted the most attention as an alloying negative electrode due to its low cost and impressive theoretical capacity, the other p-block elements such tin, antimony, or germanium have also been explored as alternative electrode materials despite their lower theoretical capacity.

1. *Tin*

Based on full conversion to $Li_{22}Sn_5$, the most highly lithiated phase in the Li–Sn phase diagram, the theoretical gravimetric capacity of elemental tin is 960 $mAh{\cdot}g^{-1}$ and the volumetric capacity is 2200 $Ah{\cdot}L^{-1}$ (cf. 350 $mAh{\cdot}g^{-1}$, 719 $Ah{\cdot}L^{-1}$ for graphite) although in practical systems only around 85% of this is realized.[155] In addition to this, tin is also a major constituent in some of the best-performing binary and ternary alloy systems and, as a result, there has been a large volume of research on both tin and tin-based intermetallics and composite electrodes. The first systematic work on tin electrodes was directed toward application in high-temperature batteries in the 1980s,[156] but interest in tin-based electrodes was piqued by the Fuji Photo Film Celltec Co. Stalion battery, which was announced in 1996. This battery contained a tin-based amorphous oxide anode that contributed to the higher specific capacity compared to the standard graphite-based cell.[157]

At 440 °C, tin reacts with lithium through sequential two-phase reactions to form Li_2Sn_5, LiSn, Li_7Sn_3, Li_5Sn_2, $Li_{13}Sn_5$, Li_7Sn_2, and $Li_{22}Sn_5$.[158] At room temperature, under practical charging conditions, only Li_2Sn_5 and LiSn can be clearly distinguished in the galvanostatic voltage curve and through XRD measurements.[159,160] When further lithium is added to the electrode, instead of discrete voltage plateaus, a long plateau is observed followed by a sloping process to 0 V *versus* Li/Li^+ (Figure 15). XRD, PDF analysis and ^{7}Li NMR measurements show that the tin connectivity is broken down through a solid solution of structurally similar phases that all contain a body-centered cubic arrangement of lithium and tin, finally forming environments locally similar to Li_7Sn_3 and Li_7Sn_2.[161] On the timescale of the electrochemical measurements, the electrode is not able to relax to thermodynamic equilibrium, and therefore the phases that are formed are poorly crystalline. Formation of these kinetically controlled structures in alloying electrodes requires careful consideration; the exact atomic arrangement and ratio of the phases formed are likely to change

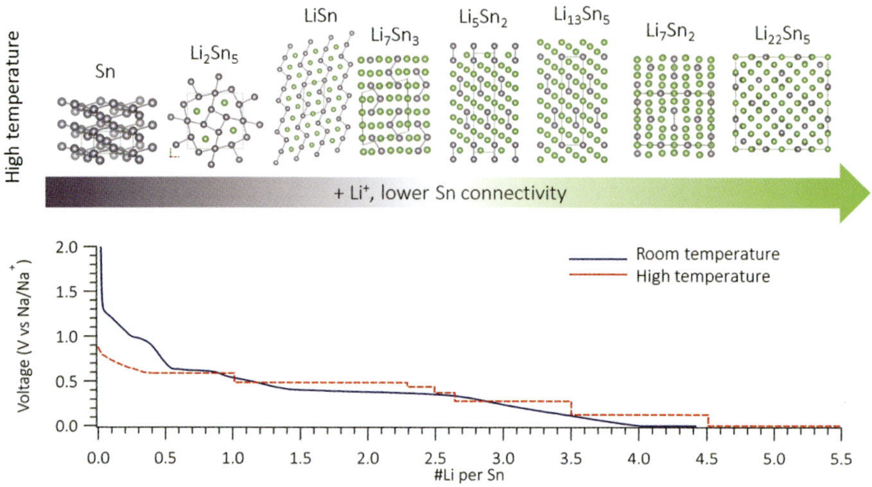

Figure 15. (a) Proposed high-temperature lithiation pathway. (b) Comparison of room temperature and high-temperature electrochemistry. Adapted from Ref. 158.

depending on charging rate and electrode formulation. This is something that must be understood and optimized for use in practical systems.

Electrodes formulated with elemental tin typically show poor cyclability and their lifetime is limited to just a few cycles. While far fewer studies have been dedicated to understanding the failure mechanism in tin electrodes compared to silicon, similar effects of particle pulverization, electrical isolation of areas of the electrode, and continued and excessive SEI growth appear to be the main causes of failure.[162] Multiple-phase transitions between crystalline materials also increase the mechanical stresses on the electrode. The mechanical properties of Li–Sn alloys, including elastic constants such as the orientation-averaged bulk, shear and Young's moduli, and Poisson's ratios, as well as the anisotropy and brittleness ductility, have been explored using DFT calculations.[163] In isotropic Li_xSn alloys, these parameters decrease almost linearly with the increase of Li concentration. This demonstrates that Li–Sn alloys are brittle, in contrast to the trend observed for lithium silicides. The phases that are formed during reaction with lithium are, therefore, ill-equipped to withstand the mechanical stresses associated with multiple-phase transitions.

2. *Antimony*

Antimony has a theoretical capacity of 660 mAh·g^{-1} and a volumetric capacity of 1771 Ah·L^{-1} at an average voltage of 0.947 V, based on full conversion to Li_3Sb. Although this voltage is relatively high for an anode that reduces the overall cell energy density, carefully formulated antimony electrodes exhibit excellent capacity retention.[164]

At first glance, the lithium–antimony system appears to be a relatively simple one. Only two thermodynamically stable compounds, Li_2Sb and Li_3Sb, exist in the phase diagram.[165,166] Considering this, the (de)lithiation mechanism is surprisingly complex. The electrochemical reaction of Li with Sb electrodes is accompanied by a voltage difference of almost 0.6 V between lithiation and delithiation, at a slow lithium (de)insertion rate of C/10 on thick-film electrode.[164] This kind of hysteresis can lead to energy losses during cycling that are unacceptable for commercialization of these technologies. Antimony has, therefore, been used as a model system for studying the fundamental origins of hysteresis in alloying anodes. There are several sources of hysteresis in solid-state electrochemical reactions. First, if the kinetics of atom movement is slow on the timescale of the (dis)charge, the result can be the displacement of voltage curves from thermodynamic equilibrium values. The second major reason is structural hysteresis; if the structural pathway of the reaction is different between the lithiation and delithiation, the voltage profile will be consequently affected.

GITT measurements by Baggetto *et al.* on thin-film antimony electrodes show that at least some of the hysteresis observed for antimony is kinetic in nature.[167] Long rest times during the GITT measurement reduce the voltage difference between lithiation and delithiation to just 60 mV compared to 0.2 V at a rate of C/10, showing that low atomic mobility leads to an increased overpotential during the lithiation step. However, these measurements also highlight a step in the lithium insertion voltage curve which is not present on delithiation, implying that a structural path hysteresis does exist. This additional step is not always resolved in the voltage curve, but structural measurements confirm that this pathway hysteresis takes place even in non-thin-film electrodes. *In situ* XRD measurements by Hewitt *et al.* clearly show the formation of Li_2Sb on lithiation, before transforming on further lithiation to cubic Li_3Sb.[168] On lithium removal, Li_3Sb transforms directly to Sb, bypassing the Li_2Sb intermediate. Solid-state NMR measurements by Chang *et al.* confirm that lithium environments from Li_2Sb are not reformed on delithiation.[166]

First-principles calculations have proposed that the pathway hysteresis has origins in the relative atomic mobilities within the parent phases of the lithiation and delithiation processes, and thus the nucleation driving forces of the phases relative to each other. Lithium mobility in Sb is relatively low, resulting in low lithium overpotentials at reaction interfaces. The nucleation of Li_2Sb is favored over Li_3Sb only at very low overpotential values, but as the overpotentials present in crystalline antimony are indeed quite low, this pathway is favored on lithiation. Li_3Sb, in contrast, has high lithium-ion mobility and the propensity to tolerate a large number of lithium-ion vacancies. This results in an high overpotential on delithiation, indicating that there is a lever effect that favors the nucleation of Sb without first forming the Li_2Sb intermediate phase. The case of antimony shows that overpotentials that arise when the electrode passes through a metastable solid solution can lead to driving forces favoring transformation to phases with a larger change in Li concentration than expected from the phase diagram equilibrium path. Analogous to the silicon system, lithium antimonides are also thought to be able to accommodate a degree of off-stoichiometry; the shift of the Li_3Sb ^7Li resonance in solid-state NMR measurements reported by Chang *et al.* changes as a function of lithiation providing evidence for this.[166] While sizable hysteresis leads to unacceptable energy losses during cycling and is usually considered as a major drawback to hinder commercialization, this hysteresis does not hamper the performance over many cycles in Sb.[164] However, the relative scarcity and expense of antimony may cause more of a hindrance to their application in real devices.

3. *Germanium*

Germanium, despite its relatively high cost, has been the attention of considerable research as a potential high-rate anode. Germanium has a smaller bandgap than silicon ($E_g(Ge) = 0.6$ eV, $E_g(Si) = 1.1$ eV), giving higher electronic conductivity, and the lithium diffusivity in the lithiated phases is 15 times higher than silicon at 360 °C. Calculations suggest that this could increase to approximately 400 times at room temperature.[169,170] These factors could lead to germanium showing enhanced rate performance when compared to silicon. Based on full conversion to $Li_{22}Ge_5$, the highest known lithiated phase ($Li_{22}Ge_5$), the theoretical gravimetric capacity of this system is 1600 mAh·g^{-1}.[155]

At room temperature, the galvanostatic electrochemical voltage curve shows several sloping processes from around 800 mV down to 0 V. This suggests that one-phase, solid-solution behavior is present for at least some of the lithium insertion mechanism. The multiple lithiation processes can be resolved using differential capacity (dQ/dV) plots and show processes around 350–300 mV, and around 200 mV, in addition to the formation of SEI above 500 mV.[171–174]

Upon lithium insertion into germanium, the phase Li_7Ge_3 is formed through a two-phase reaction. This phase has not traditionally been placed in the Li–Ge phase diagram but was predicted to be stable by DFT measurements and confirmed by *ex situ* XRD measurements.[172,175] Li_7Ge_3 contains Ge–Ge dumbbells and, on reaction with further lithium, these dumbbells are gradually broken down during an electrochemical process with a sloping voltage profile. This process forms amorphous species suggesting that XRD measurements in this region provide limited information, but PDF analysis[172] and extended X-ray absorption fine structure (EXAFS)[171] measurements show that the local structure is similar to phases that contain a mixture of dumbbells and isolated atoms. Several of these phases exist on or close to the equilibrium pathway calculated by DFT, and therefore it is likely that a solid solution of these phases exists in this region, as Ge–Ge bonds are gradually broken down (Figure 16). When all Ge–Ge bonds are broken, $Li_{15}Ge_4$ is formed, where the intermediate coordination sphere of germanium is made up entirely of lithium. The diffraction data of Jung *et al.* clearly indicate an expansion of the unit cell of $Li_{15}Ge_4$ at high lithiation, indicating that the system prefers to accommodate additional lithium in an off-stoichiometry $Li_{15+x}Ge_4$ compound rather than form $Li_{22}Ge_5$.[172]

There is some evidence that the delithiation pathway may be different to the lithiation pathway. This is not unexpected in a system following a nonequilibrium reaction pathway. There is no reason why the kinetic

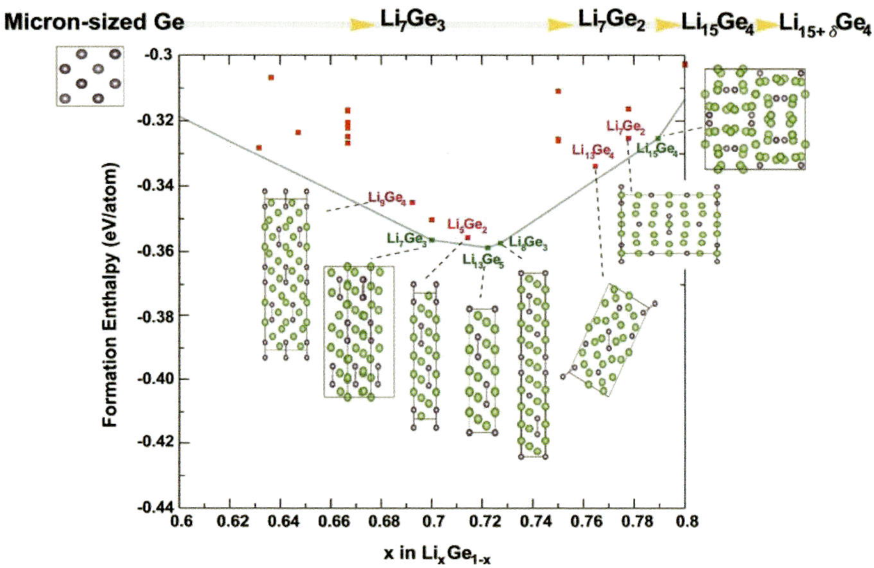

Figure 16. Structural pathway of germanium anodes in LIBs. The main graph shows the lithiation pathway showing the common structural features of possible lithiated Ge phases between Li_9Ge^4 and Li_7Ge_2. The blue line is the tie-line indicating the stable phases at 0 K predicted by DFT. The squares indicate the formation enthalpy of a structure: green squares correspond to the thermodynamic phases; red squares correspond to metastable phases (at 0 K). Adapted from Ref. 172.

considerations will necessarily be the same on lithiation and delithiation — after all, the starting structures would be entirely different and facile transformations are likely to form different structures. A process at around 450–500 mV was characterized by Lim *et al.* for the conversion of $Li_{15}Ge_4$ to amorphous Li_xGe phases using information from EXAFS, which is further converted at around 630 mV to amorphous germanium without the reformation of the Ge-dumbbell phase.[171] However, the ^7Li resonance present in the model compound of Li_7Ge_3 mentioned in the results of Jung *et al.* is present in the *in situ* NMR spectra reported by Tang *et al*, and these authors propose that the mechanistic pathway for delithiation is the same as during lithiation.[176]

A consequence of the presence of this complex kinetically controlled pathway, as for the case of silicon, the lithiation pathway followed by electrodes is a complex one, determined by the interplay of multiple factors including the charging rate, germanium particle size, and the electrode formulation and porosity, that is, the mass ratio of active material, conductive additive, and binder. The voltage at which (de)lithiation processes take place also depends on these factors, which are not consistent between the various studies that have been undertaken so far. Lin *et al.* found that slow lithiation rates (e.g., C/20) led to complete conversion to $Li_{15}Ge_4$ and stable cycling performance over 100 cycles. However, faster rates (C/10) led to incomplete formation of $Li_{15}Ge_4$ with some amorphous Li_xGe remaining. The partial formation of $Li_{15}Ge_4$ resulted in poor capacity retention, but these losses could be mitigated if the lithiation was cut off before the formation of any crystalline $Li_{15}Ge_4$, analogous to the silicon system.[177]

A particular challenge to understanding the Li–Ge system is that differences in data collection strategies and postprocessing of electrodes have the potential to form different structures, particularly in the region under kinetic control. Of particular note, no *in situ* study has reported the formation of any crystalline intermediates between Ge and $Li_{15}Ge_4$, whereas the thermodynamically stable phase crystalline Li_7Ge_3 was unambiguously identified by ex situ XRD. Clearly there is still work to be done on the effects of self-discharge and relaxation of metastable species for germanium and for other alloying anodes. These considerations are of great important for the practical use of these chemistries in real batteries.

4. *Improving the Performance of Tin, Antimony, and Germanium*

As for silicon, methods are directed at addressing the consequences of the large volume expansion of these anodes on their performance. Nanosizing tin particles can improve performance by preventing some of the cracking processes that happen when the electrode expands. Reducing the particle size can also lead to changes in the structural pathways taken on lithiation.[178] For example, cubic α-Sn phase was formed concurrently with the tetragonal β phase below a critical particle size of 17 nm. The origin of this behavior is not understood, but the formation of this phase does not change the subsequent electrochemical pathway.[179]

Over many cycles, particles can aggregate which restricts the benefit of the nanosizing to just a few cycles. Confinement of Sn (or SnO_2) particles in a porous carbon matrix has been shown to stabilize the particles by limiting this particle coalescence. These electrodes have shown improvements in cycling performance; a capacity of 443 $mAh{\cdot}g^{-1}$ after 2000 cycles, with almost 100% coulombic efficiency and good rate capability, has been reported through this method.[180] However, a high first cycle irreversible capacity due to the presence of large surface areas and reaction of lithium with the surface of mesoporous carbon is still an issue.

The preparation of binary compositions, either X'_aX_b composites where both X and X′ form alloys with Li, or M_aX_b composites where M is something that does not form an alloy with Li, is a method that has been shown to significantly improve cycling performance. In the case of M_aX_b, the discharge reaction results in a transformation of the initial phase M_aX_b into a composite formed of the lithiated alloy and from metallic nanoparticles:

$$M_aX_b + nb\mathrm{Li} \rightarrow a\mathrm{M}^0 + b\mathrm{Li}_n\mathrm{X}.$$

The cyclability of Sn_xO composites can be superior to that of the pure metal. Tin oxide, either SnO or SnO_2, undergoes stepwise lithiation, first forming lithium oxide and elemental tin, which then undergoes further, reversible alloying with lithium[159]:

$$\mathrm{SnO} + 2n\mathrm{Li} \rightarrow n\mathrm{Li_2O} + \mathrm{Sn}$$

$$\mathrm{Sn} + n\mathrm{Li} \leftrightarrow \mathrm{Li}_n\mathrm{Sn}$$

The superior cycling performance is attributed to the volume expansion and associated strain being alleviated by distribution of the active nanosized tin phase in the inactive lithium-oxide matrix. The amount of stress associated with lithiation and delithiation experienced by well-separated metallic particles decreased in SnO_x films compared to Sn films, as a porous, easy to lithiate, material forms with cycling.[181] In general, the first reaction can be partially reversible at higher voltages (>1.5 V), leading to higher reversible capacity (~1000 $mAh{\cdot}g^{-1}$), but this always gives rise to capacity fading on cycling beyond 10 cycles.[182] Similar to tin nanoparticles, over many cycles, agglomeration into larger particles in order to minimize the surface energy leads to capacity fade. These effects have been directly observed by TEM measurements (Figure 17). In this study, 3 nm tin oxide nanoparticles were

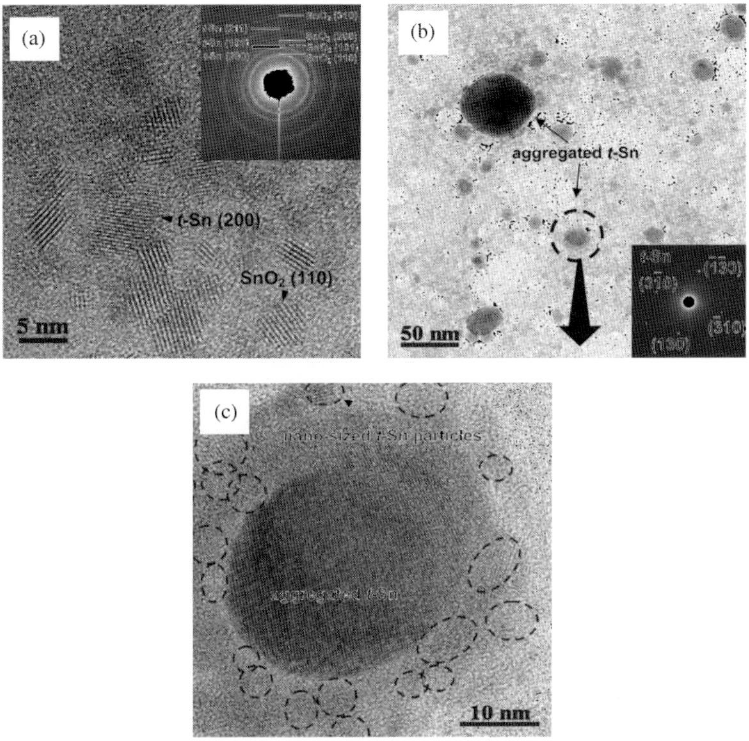

Figure 17. (a) High-resolution TEM image of ~3-nm-sized SnO_2 nanocrystals after 30 cycles; aggregation into the larger Sn particles is observed. (b) High-resolution TEM image of ~8-nm-sized SnO_2 nanocrystals after 30 cycles; aggregation into larger clusters is observed. (c) Magnified image of the circled region in (b). Adapted from Ref. 183.

found to not agglomerate and demonstrated high capacity (740 $mAh{\cdot}g^{-1}$), with negligible capacity fade over 60 cycles. Larger SnO_2 nanoparticles (4 or 8 nm) were observed to agglomerate, causing a rapid drop-off of capacity.[183]

The original studies on tin oxide materials by Idota *et al.* are still some of the best reports of cycling performance for Sn-oxide materials.[157] The total organic carbon (TOC-1) glass material, $Sn_{1.0}B_{0.56}P_{0.40}Al_{0.42}O_{3.6}$, has multiple inactive glass-forming spectator phases, which will both buffer the volume expansion and prevent Sn aggregation. However, as the volume of inactive phases increases, the volumetric capacity decreases, which, in reality, negates much of the capacitive advantage of this type of anode over graphite.

a. *Binary, Ternary Alloys, SnSb, TiSnSb*

Multiphase metals can show superior performance compared to single-phase metal anodes because the material can react with lithium ions over a wider range of voltages. For example, in SnSb, the Sb component can react with lithium at around 0.95 V, whereas the tin can react over a range of voltages between 0.7 and 0 V. At any one time, one phase will be actively lithiated, and the other will act as a buffer. This helps to suppress the strains associated with volume expansion. Sn–Sb alloys have been found to have better performance characteristics than Sn or Sb via a multi-step mechanism as follows:[185]

$$\mathrm{SnSb} + 3\mathrm{Li}^+ + 3\mathrm{e}^- \rightarrow \mathrm{Li_3Sb} + \mathrm{Sn}\ (\sim 0.8\ \mathrm{V})$$

$$\mathrm{Sn} + x\mathrm{Li}^+ + x\mathrm{e}^- \rightarrow \mathrm{Li}_x\mathrm{Sn} \quad x = \frac{2}{5}, 1, \frac{5}{2}, \frac{7}{2} (0.8 - 0\ \mathrm{V})$$

In this case, the nanostructured electrode formed during the initial lithiation of the antimony component prevents the Sn phase from aggregating, suppressing the failure modes discussed earlier, and enhancing cyclability.

Another strategy is compositing the material with inactive metals, which form a finely distributed inactive phase that acts as a buffer to the volume expansion. A highly studied example of this is the ternary alloy, TiSnSb, which can reversibly uptake more than five lithium per formula unit and retain good reversibly cycling over 60 cycles, compared to a rough mixture of the three components that showed significant loss in capacity after only 20 cycles (Figure 18).[184,185]

During the first lithiation, TiSnSb undergoes a conversion reaction, whereas Ti/Sn/Sb composites proceed by a stepwise alloying process, with the Sn and Sb components reacting separately in a sequential manner (Figure 19).[186] In both cases, this results in a mixture of lithium antimonide,

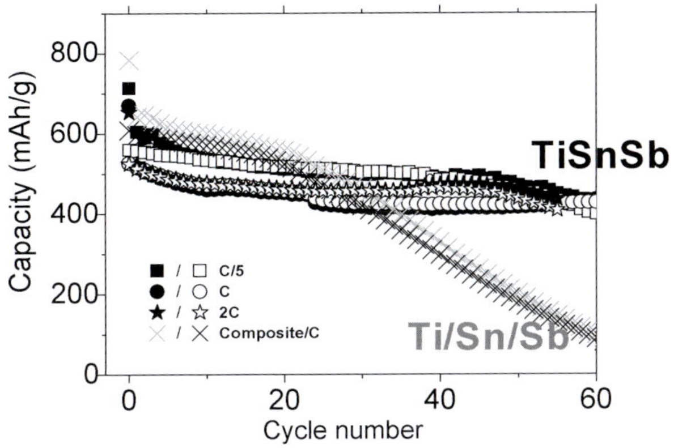

Figure 18. Capacity retention at various cycling rates for TiSnSb cells cycled at C/5 rate between 1.5 and 0 V, and (crosses) for a Ti/Sn/Sb composite cell are shown for comparison. Adapted from Ref. 184 with permission from The Royal Society of Chemistry.

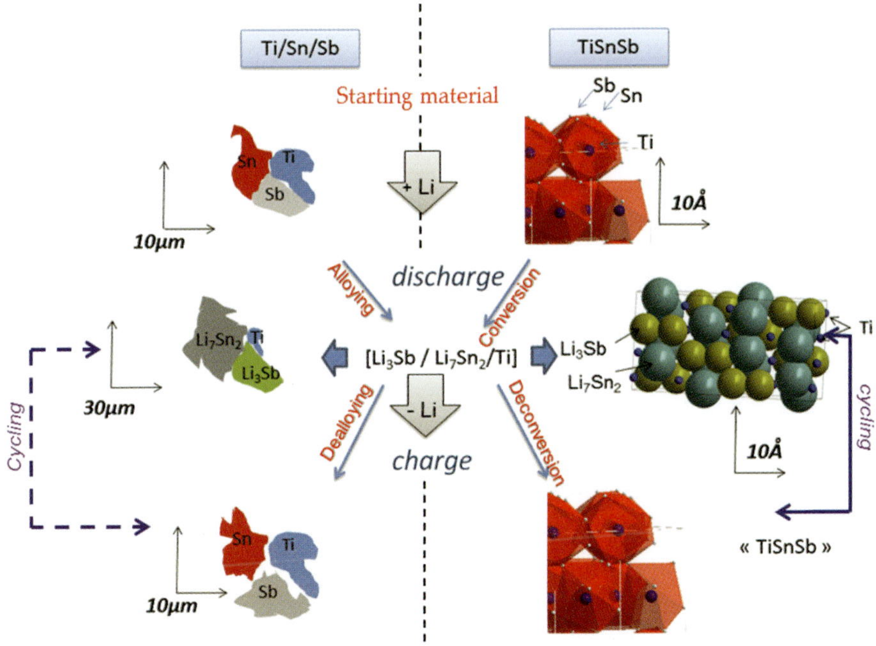

Figure 19. Schematic comparative view of the lithiation mechanism of both TiSnSb and (Ti + Sn + Sb) electrodes *versus* Li. Adapted from Ref. 184.

lithium stannides, and titanium. However, lithium removal highlights differences between the two systems; in the case of Ti/Sn/Sb, Sn and Sb are reformed. In the case of TiSnSb, a phase similar to the original TiSnSb material is reformed; TiSnSb displays a memory effect, which clearly demonstrates that in the lithiated alloy nanosized interfaces between Li_3Sb, Li_xSn, and Ti are present. These will have a key role in controlling the distribution of Li_3Sb, Li_7Sn_2, and Ti particles and ensuring a homogenous distribution in the discharged electrode and avoiding coalescence of the different metals. The formation of metallic Ti nanoparticles embedded in the homogeneous Li_3Sb/Li_7Sn_2 matrix at the end of discharge is also thought to allow rapid mass transport, since the diffusion distances remain short between the redox center, further enhancing performance.[184]

The matrix of reactive metallic nanoparticles and lithiated alloy phases has been shown to be highly reactive and intrinsically unstable. Coupled ^{7}Li NMR and Mössbauer spectrometry experiments show that after a long rest of OCV mode, a phase close to the initial TiSnSb was spontaneously reformed from the discharged electrode. Metallic nanoparticles are known to be extremely reactive, and instability of the large interfaces between nanodomains within the alloy is a likely driving force for reformation of the conversion compounds during charge. Although this does not affect the subsequent cycling behavior, an appropriate formulation of active conversion materials, electrolytes, and the working potential window will be required to enable stabilization of the discharged electrode and thus allow the application of such materials in commercial batteries.[161]

IV. Summary and Conclusions

Graphite is the active material in the majority of current LIBs. While it shows good capacity and stability over many cycles, the capacity is limited and lithium insertion process takes place at low voltages, close to the metal plating voltage, limiting its high-rate charge capabilities. The search for replacement materials has focused on materials that undergo intercalation, analogous to most cathode materials, such as lithium titanates, and materials that undergo alloying reactions.

Titanates offer many exciting properties — high capacity, good cycle life, and the potential for improved high-rate performance. However, studies have shown that the gas formation, possibly catalyzed by surface Ti^{4+} species, is a safety concern that must be mediated before their widespread implementation can be considered. Controlling the electrode–electrolyte interface, either through electrolyte additives or coating of the electrode surface are, to date, the best proposed solutions to the gas formation problem. The development of the solid–electrolyte field, somewhat still in its infancy, has great potential to reduce these problems by removing the liquid electrolyte from cells altogether.

p-Block element–based electrodes show reversible alloying reactions of numerous lithium ions per active material atom, and have the potential to move far beyond graphite in terms of gravimetric capacity; the mostly widely studied material silicon has a theoretical capacity almost 10 times that of graphite. However, the enormous volume expansions that take place when this much lithium is added to the electrode initiate a cascade of negative side effects — electrical isolation of particles, excessive SEI formation, and irreversible losses of lithium — which limit their cycle life. Strategies including nanosizing, surface coatings, and co-alloying with other electrochemically active or inactive materials are being actively researched in order to remove the "roadblock" currently faced by these technologies.

In the search for better negative electrode materials, the use of advanced characterization techniques including XRD, solid-state NMR, and microscopy has provided vital information and the mechanisms of (de)lithiation of these materials, as well as their failure mechanism.

V. References

1. Nishi, Y. *J. Power Sources* **2001**, *100*, 101–106.
2. Buqa, H.; Goers, D.; Holzapfel, M.; Spahr, M. E.; Novák, P. *J. Electrochem. Soc.* **2005**, *152*, A474–A481.
3. Weng, Z. Y.; Guo, H.; Liu, X. M.; Wu, S. L.; Yeung, K. W. K.; Chu, P. K. *RSC Adv.* **2013**, *3*, 24758–24775.
4. Su, X.; Wu, Q.; Li, J.; Xiao, X.; Lott, A.; Lu, W.; Sheldon, B. W.; Wu, J. *Adv. Energy Mater.* **2014**, *4*, 1300882.

5. Zhang, W.-J. *J. Power Sources* **2011**, *196*, 13–24.
6. McDowell, M. T.; Lee, S. W.; Nix, W. D.; Cui, Y. *Adv. Mater.* **2013**, *25*, 4966–4985.
7. Deng, D.; Kim, M. G.; Lee, J. Y.; Cho, J. *Energy Environ. Sci.* **2009**, *2*, 818–837.
8. Chanson, C.; Wiaux, J.-P. *The European Association for Advanced Rechargeable Batteries — Safety of Lithium-Ion Batteries.* http://www.rechargebatteries.org/knowledge-base/batteries/lithium-ion-cell-lilon/
9. Zhao, B.; Ran, R.; Liu, M. L.; Shao, Z. P. *Mater. Sci. Eng. R.* **2015**, *98*, 1–71.
10. Shin, J. Y.; Samuelis, D.; Maier, J. *Adv. Funct. Mater.* **2011**, *21*, 3464–3472.
11. Erickson, E. M.; Ghanty, C.; Aurbach, D. *J. Phys. Chem. Lett.* **2014**, *5*, 3313–3324.
12. Ohzuku, T.; Ueda, A.; Yamamoto, N. *J. Electrochem. Soc.* **1995**, *142*, 1431–1435.
13. Murphy, D. W.; Cava, R. J.; Zahurak, S. M.; Santoro, A. *Solid State Ion.* **1983**, *9–10*, 413–417.
14. Colbow, K. M.; Dahn, J. R.; Haering, R. R. *J. Power Sources* **1989**, *26*, 397–402.
15. Aldon, L.; Kubiak, P.; Womes, M.; Jumas, J. C.; Olivier-Fourcade, J.; Tirado, J. L.; Corredor, J. I.; Vicente, C. P. *Chem. Mater.* **2004**, *16*, 5721–5725.
16. Wolfenstine, J.; Lee, U.; Allen, J. L. *J. Power Sources* **2006**, *154*, 287–289.
17. Allen, J. L.; Jow, T. R.; Wolfenstine, J. *J. Power Sources* **2006**, *159*, 1340–1345.
18. Cava, R. J.; Murphy, D. W.; Zahurak, S.; Santoro, A.; Roth, R. S. *J. Solid State Chem.* **1984**, *53*, 64–75.
19. Pecharroman, C.; Amarilla, J. M. *Phys. Rev. B* **2000**, *62*, 12062–12068.
20. Leonidov, I. A.; Leonidova, O. N.; Perelyaeva, L. A.; Samigullina, R. F.; Kovyazina, S. A.; Patrakeev, M. V. *Phys. Solid State* **2003**, *45*, 2183–2188.
21. Schmidt, W.; Wilkening, M. *Solid State Ion.* **2016**, *287*, 77–82.
22. Schmidt, W.; Wilkening, M. *J. Phys. Chem. C.* **2016**, *120*, 11372–11381.
23. Vijayakumar, M.; Kerisit, S.; Rosso, K. M.; Burton, S. D.; Sears, J. A.; Yang, Z. G.; Graff, G. L.; Liu, J.; Hu, J. Z. *J. Power Sources* **2011**, *196*, 2211–2220.
24. Boyce, J. B.; Mikkelsen, J. C. *Solid State Commun.* **1979**, *31*, 741–745.
25. Johnston, D. C. *J. Low Temp. Phys.* **1976**, *25*, 145–175.
26. Scharner, S.; Weppner, W.; Schmid-Beurmann, P. *J. Electrochem. Soc.* **1999**, *146*, 857–861.
27. Ge, H.; Li, N.; Li, D. Y.; Dai, C. S.; Wang, D. L. *J. Phys. Chem. C.* **2009**, *113*, 6324–6326.
28. Wagemaker, M.; Simon, D. R.; Kelder, E. M.; Schoonman, J.; Ringpfeil, C.; Haake, U.; Lutzenkirchen-Hecht, D.; Frahm, R.; Mulder, F. M. *Adv. Mater.* **2006**, *18*, 3169–3173.
29. Yan, B.; Li, M. S.; Li, X. F.; Bai, Z. M.; Yang, J. W.; Xiong, D. B.; Li, D. J. *J. Mater. Chem. A* **2015**, *3*, 11773–11781.
30. Jiang, S. M.; Zhao, B. T.; Chen, Y. B.; Cai, R.; Shao, Z. P. *J. Power Sources* **2013**, *238*, 356–365.
31. Zaghib, K.; Simoneau, M.; Armand, M.; Gauthier, M. *J. Power Sources* **1999**, *81*, 300–305.

32. Ni, J. F.; Liu, W.; Liu, J. Z.; Gao, L. J.; Chen, J. T. *Electrochem. Commun.* **2013**, *35*, 1–4.
33. Ganapathy, S.; Wagemaker, M. *ACS Nano* **2012**, *6*, 8702–8712.
34. Ge, H.; Li, N.; Li, D. Y.; Dai, C. S.; Wang, D. L. *Electrochem. Commun.* **2008**, *10*, 719–722.
35. Yao, X. L.; Xie, S.; Nian, H. Q.; Chen, C. H. *J. Alloys Compd.* **2008**, *465*, 375–379.
36. Zaghib, K.; Dontigny, M.; Guerfi, A.; Charest, P.; Rodrigues, I.; Mauger, A.; Julien, C. M. *J. Power Sources* **2011**, *196*, 3949–3954.
37. Dedryvere, R.; Foix, D.; Franger, S.; Patoux, S.; Daniel, L.; Gonbeau, D. *J. Phys. Chem. C.* **2010**, *114*, 10999–11008.
38. Chen, K. B.; Yu, Z. Q.; Deng, S.; Wu, Q.; Zou, J. X.; Zeng, X. Q. *J. Power Sources* **2015**, *278*, 411–419.
39. Belharouak, I.; Koenig, G. M.; Amine, K. *J. Power Sources* **2011**, *196*, 10344–10350.
40. Belharouak, I.; Sun, Y. K.; Lu, W.; Amine, K. *J. Electrochem. Soc.* **2007**, *154*, A1083–A1087.
41. Joachin, H. H.; Amiruddin, S.; Li, B.; Kaun, T. D.; Amine, K.; Prakash, J. *ECS Trans.* **2008**, *11*, 13–18.
42. Patoux, S.; Sannier, L.; Lignier, H.; Reynier, Y.; Bourbon, C.; Jouanneau, S.; Le Cras, F.; Martinet, S. *Electrochim. Acta* **2008**, *53*, 4137–4145.
43. Shi, F.; Ross, P. N.; Zhao, H.; Liu, G.; Somorjai, G. A.; Komvopoulos, K. *J. Am. Chem. Soc.* **2015**, *137*, 3181–3184.
44. Abraham, D. P.; Reynolds, E. M.; Sammann, E.; Jansen, A. N.; Dees, D. W. *Electrochim. Acta* **2005**, *51*, 502–510.
45. Sasaki, Y. *Electrochemistry* **2008**, *76*, 2–15.
46. Blomgren, G. E. *J. Power Sources* **2003**, *119*, 326–329.
47. Wu, K.; Yang, J.; Zhang, Y.; Wang, C. Y.; Wang, D. Y. *J. Appl. Electrochem.* **2012**, *42*, 989–995.
48. Lu, X.; Zhao, L.; He, X. Q.; Xiao, R. J.; Gu, L.; Hu, Y. S.; Li, H.; Wang, Z. X.; Duan, X. F.; Chen, L. Q.; Maier, J.; Ikuhara, Y. *Adv. Mater.* **2012**, *24*, 3233–3238.
49. Lu, M.; Tian, Y. Y.; Zheng, X. D.; Gao, J.; Huang, B. *J. Power Sources* **2012**, *219*, 188–192.
50. He, Y. B.; Li, B.; Liu, M.; Zhang, C.; Lv, W.; Yang, C.; Li, J.; Du, H.; Zhang, B.; Yang, Q. H.; Kim, J. K.; Kang, F. *Sci. Rep.* **2012**, *2*, 913(1–9).
51. Yang, Z. G.; Zhang, J. L.; Kintner-Meyer, M. C. W.; Lu, X. C.; Choi, D. W.; Lemmon, J. P.; Liu, J. *Chem. Rev.* **2011**, *111*, 3577–3613.
52. Hammouche, A.; Karden, E.; Walter, J.; De Doncker, R. W. *J. Power Sources* **2001**, *96*, 106–112.
53. Wu, K.; Yang, J.; Liu, Y.; Zhang, Y.; Wang, C. Y.; Xu, J. M.; Ning, F.; Wang, D. Y. *J. Power Sources* **2013**, *237*, 285–290.
54. Belharouak, I.; Koenig, G. M.; Tan, T.; Yumoto, H.; Ota, N.; Amine, K. *J. Electrochem. Soc.* **2012**, *159*, A1165–A1170.

55. Ding, Z. J.; Zhao, L.; Suo, L. M.; Jiao, Y.; Meng, S.; Hu, Y. S.; Wang, Z. X.; Chen, L. Q. *Phys. Chem. Chem. Phys.* **2011**, *13*, 15127–15133.
56. Du Pasquier, A.; Plitz, I.; Menocal, S.; Amatucci, G. *J. Power Sources* **2003**, *115*, 171–178.
57. Guo, J. L.; Zuo, W. H.; Cai, Y. J.; Chen, S. M.; Zhang, S. J.; Liu, J. P. *J. Mater. Chem. A* **2015**, *3*, 4938–4944.
58. He, M. L.; Castel, E.; Laumann, A.; Nuspl, G.; Novak, P.; Berg, E. J. *J. Electrochem. Soc.* **2015**, *162*, A870–A876.
59. Xu, K. *Chem. Rev.* **2004**, *104*, 4303–4417.
60. Sloop, S. E.; Pugh, J. K.; Wang, S.; Kerr, J. B.; Kinoshita, K. *Electrochem. Solid-State Lett.* **2001**, *4*, A42–A44.
61. Burns, J. C.; Sinha, N. N.; Jain, G.; Ye, H.; VanElzen, C. M.; Scott, E.; Xiao, A.; Lamanna, W. M.; Dahn, J. R. *J. Electrochem. Soc.* **2014**, *161*, A247–A255.
62. Kitta, M.; Matsuda, T.; Maeda, Y.; Akita, T.; Tanaka, S.; Kido, Y.; Kohyama, M. *Surf. Sci.* **2014**, *619*, 5–9.
63. Liu, M.; He, Y. B.; Lv, W.; Zhang, C.; Du, H. D.; Li, B. H.; Yang, Q. H.; Kang, F. Y. *J. Power Sources* **2014**, *268*, 882–886.
64. Li, X.; Xu, J.; Huang, P. X.; Yang, W.; Wang, Z. Q.; Wang, M. S.; Huang, Y.; Zhou, Y.; Qu, M. Z.; Yu, Z. L.; Lin, Y. H. *Electrochim. Acta* **2016**, *190*, 69–75.
65. Alaboina, P. K.; Ge, Y. Q.; Uddin, M. J.; Liu, Y.; Lee, D.; Park, S.; Zhang, X. W.; Cho, S. J. *ACS Appl. Mater. Interfaces* **2016**, *8*, 12127–12133.
66. Lu, X.; Gu, L.; Hu, Y. S.; Chiu, H. C.; Li, H.; Demopoulos, G. P.; Chen, L. Q. *J. Am. Chem. Soc.* **2015**, *137*, 1581–1586.
67. Liu, J. L.; Bian, P. W.; Li, J.; Ji, W. J.; Hao, H.; Yu, A. S. *J. Power Sources* **2015**, *286*, 380–387.
68. Han, C. P.; He, Y. B.; Li, H. F.; Li, B. H.; Du, H. D.; Qin, X. Y.; Kang, F. Y. *Electrochim. Acta* **2015**, *157*, 266–273.
69. Gao, J. J.; Gong, B. L.; Zhang, Q. T.; Wang, G. D.; Dai, Y. J.; Fan, W. F. *Ionics (Kiel)* **2015**, *21*, 2409–2416.
70. Fell, C. R.; Sun, L. Y.; Hallac, P. B.; Metz, B.; Sisk, B. *J. Electrochem. Soc.* **2015**, *162*, A1916–A1920.
71. Chancelier, L.; Benayad, A.; Gutel, T.; Mailley, S.; Santini, C. C. *J. Electrochem. Soc.* **2015**, *162*, A1008–A1013.
72. Bernhard, R.; Metzger, M.; Gasteiger, H. A. *J. Electrochem. Soc.* **2015**, *162*, A1984–A1989.
73. Bernhard, R.; Meini, S.; Gasteiger, H. A. *J. Electrochem. Soc.* **2014**, *161*, A497–A505.
74. Wu, K.; Yang, J.; Qiu, X. Y.; Xu, J. M.; Zhang, Q. Q.; Jin, J.; Zhuang, Q. C. *Electrochim. Acta* **2013**, *108*, 841–851.
75. Peled, E. *J. Electrochem. Soc.* **1979**, *126*, 2047–2051.
76. El Ouatani, L.; Dedryvere, R.; Siret, C.; Biensan, P.; Gonbeau, D. *J. Electrochem. Soc.* **2009**, *156*, A468–A477.

77. Hu, Y.-Y.; Liu, Z.; Nam, K.-W.; Borkiewicz, O. J.; Cheng, J.; Hua, X.; Dunstan, M. T.; Yu, X.; Wiaderek, K. M.; Du, L.-S.; Chapman, K. W.; Chupas, P. J.; Yang, X.-Q.; Grey, C. P. *Nat. Mater.* **2013**, *12*, 1130–1136.
78. Borgel, V.; Gershinsky, G.; Hu, T.; Theivanayagam, M. G.; Aurbach, D. *J. Electrochem. Soc.* **2013**, *160*, A650–A657.
79. Lai, C. Y.; Xu, J. J.; Wei, Y. F. *Adv. Mater. Res.* **2012**, *347–353*, 3522–3526.
80. Jiang, J. W.; Chen, J.; Dahn, J. R. *J. Electrochem. Soc.* **2004**, *151*, A2082–A2087.
81. Borner, M.; Klamor, S.; Hoffmann, B.; Schroeder, M.; Nowak, S.; Wursing, A.; Winter, M.; Schappacher, F. M. *J. Electrochem. Soc.* **2016**, *163*, A831–A837.
82. Zhan, C.; Lu, J.; Kropf, A. J.; Wu, T. P.; Jansen, A. N.; Sun, Y. K.; Qiu, X. P.; Amine, K. *Nat. Commun.* **2013**, *4*.
83. Nordh, T.; Younesi, R.; Hahlin, M.; Duarte, R. F.; Tengstedt, C.; Brandell, D.; Edstrom, K. *J. Phys. Chem. C.* **2016**, *120*, 3206–3213.
84. Song, M. S.; Kim, R. H.; Baek, S. W.; Lee, K. S.; Park, K.; Benayad, A. *J. Mater. Chem. A* **2014**, *2*, 631–636.
85. Nordh, T.; Younesi, R.; Brandell, D.; Edstrom, K. *J. Power Sources* **2015**, *294*, 173–179.
86. He, Y. B.; Liu, M.; Huang, Z. D.; Zhang, B.; Yu, Y.; Li, B. H.; Kang, F. Y.; Kim, J. K. *J. Power Sources* **2013**, *239*, 269–276.
87. Placke, T.; Siozios, V.; Schmitz, R.; Lux, S. F.; Bieker, P.; Colle, C.; Meyer, H. W.; Passerini, S.; Winter, M. *J. Power Sources* **2012**, *200*, 83–91.
88. Olivier, J. P.; Winter, M. *J. Power Sources* **2001**, *97–8*, 151–155.
89. Wang, R. H.; Li, X. H.; Wang, Z. X.; Guo, H. J.; He, Z. J. *ACS Appl. Mater. Interfaces* **2015**, *7*, 23605–23614.
90. Liu, J. Y.; Shen, Y.; Chen, L.; Wang, Y. G.; Xia, Y. Y. *Electrochim. Acta* **2015**, *156*, 38–44.
91. He, Y. B.; Ning, F.; Li, B. H.; Song, Q. S.; Lv, W.; Du, H. D.; Zhai, D. Y.; Su, F. Y.; Yang, Q. H.; Kang, F. Y. *J. Power Sources* **2012**, *202*, 253–261.
92. Wen, L.; Wu, Z. Y.; Luo, H. Z.; Song, R. S.; Li, F. *J. Electrochem. Soc.* **2015**, *162*, A3038–A3044.
93. Li, W.; Li, X.; Chen, M. Z.; Xie, Z. W.; Zhang, J. X.; Dong, S. Q.; Qu, M. Z. *Electrochim. Acta* **2014**, *139*, 104–110.
94. Lu, Q. W.; Fang, J. H.; Yang, J.; Feng, X. J.; Wang, J. L.; Nuli, Y. N. *RSC Adv.* **2014**, *4*, 10280–10283.
95. Utsunomiya, H.; Nakajima, T.; Ohzawa, Y.; Mazej, Z.; Zemva, B.; Endo, M. *J. Power Sources* **2010**, *195*, 6805–6810.
96. Kang, X. H.; Utsunomiya, H.; Achiha, T.; Ohzawa, Y.; Nakajima, T.; Mazej, Z.; Zemva, B.; Endo, M. *J. Electrochem. Soc.* **2010**, *157*, A437–A442.
97. Nakajima, T.; Ueno, A.; Achiha, T.; Ohzawa, Y.; Endo, M. *J. Fluor. Chem.* **2009**, *130*, 810–815.
98. Nakajima, T. *J. Fluor. Chem.* **2013**, *149*, 104–111.

99. Yi, T. F.; Yang, S. Y.; Xie, Y. *J. Mater. Chem. A* **2015**, *3*, 5750–5777.
100. Lin, X.; Pan, F.; Wang, H. *Mater. Technol.* **2014**, *29*, A82–A87.
101. Huang, Z.; Wang, D.; Zhang, C. M.; He, D. N. *Prog. Chem.* **2014**, *26*, 1914–1923.
102. Zhang, Q. Y.; Li, X. *Int. J. Electrochem. Sci.* **2013**, *8*, 6449–6456.
103. Ma, Y.; Ding, B.; Ji, G.; Lee, J. *ACS Nano* **2013**, *7*, 10870–10878.
104. Kubiak, P.; Garcia, A.; Womes, M.; Aldon, L.; Olivier-Fourcade, J.; Lippens, P. E.; Jumas, J. C. *J. Power Sources* **2003**, *119*, 626–630.
105. Zhao, E. B.; Qin, C. L.; Jung, H. R.; Berdichevsky, G.; Nese, A.; Marder, S.; Yushin, G. *ACS Nano* **2016**, *10*, 3977–3984.
106. Zhang, P. F.; Chen, M.; Shen, X.; Wu, Q. H.; Zhang, X. E.; Huan, L.; Diao, G. W. *Electrochim. Acta* **2016**, *204*, 92–99.
107. Yu, Z. H.; Wang, L.; Jiang, L. H. *Ceram. Int.* **2016**, *42*, 16031–16039.
108. Tang, Y. K.; Liu, L.; Zhao, H. Y.; Jia, D. Z.; Liu, W. *J. Mater. Chem. A* **2016**, *4*, 2089–2095.
109. Peng, L.; Zhang, H. J.; Fang, L.; Zhang, Y.; Wang, Y. *Nanoscale* **2016**, *8*, 2030–2040.
110. Jia, X. L.; Lu, Y. F.; Wei, F. *Nano Res.* **2016**, *9*, 230–239.
111. Huo, K. F.; Li, X. X.; Gao, B.; Wang, L.; Li, Q. W.; Peng, X.; Zhang, X. M.; Fu, J. J.; Chu, P. K. *Chemelectrochem* **2016**, *3*, 1301–1305.
112. Lin, C. F.; Lai, M. O.; Zhou, H. H.; Lu, L. *Phys. Chem. Chem. Phys.* **2014**, *16*, 24874–24883.
113. Shen, L. F.; Zhang, X. G.; Uchaker, E.; Yuan, C. Z.; Cao, G. Z. *Adv. Energy Mater.* **2012**, *2*, 691–698.
114. Bruce, P. G.; Scrosati, B.; Tarascon, J. M. *Angew. Chemie Int. Ed.* **2008**, *47*, 2930–2946.
115. Yang, Z. G.; Choi, D.; Kerisit, S.; Rosso, K. M.; Wang, D. H.; Zhang, J.; Graff, G.; Liu, J. *J. Power Sources* **2009**, *192*, 588–598.
116. Chou, S.-L.; Pan, Y.; Wang, J.-Z.; Liu, H.-K.; Dou, S.-X. *Phys. Chem. Chem. Phys.* **2014**, *16*, 20347–20359.
117. Beaulieu, L. Y.; Eberman, K. W.; Turner, R. L.; Krause, L. J.; Dahn, J. R. *Electrochem. Solid-State Lett.* **2001**, *4*, A137–A140.
118. Boukamp, B. A.; Lesh, G. C.; Huggins, R. A. *J. Electrochem. Soc.* **1981**, *128*, 725–729.
119. Obrovac, M. N.; Christensen, L. *Electrochem. Solid-State Lett.* **2004**, *7*, A93.
120. Wilson, A. M.; Dahn, J. R. *J. Electrochem. Soc.* **1995**, *142*, 326–332.
121. Ryu, J. H.; Kim, J. W.; Sung, Y.-E.; Oh, S. M. *Electrochem. Solid-State Lett.* **2004**, *7*, A306–A309.
122. Hatchard, T. D.; Dahn, J. R. *J. Electrochem. Soc.* **2004**, *151*, A838–A842.
123. Li, J.; Dahn, J. R. *J. Electrochem. Soc.* **2007**, *154*, A156–A161.
124. Michan, A. L.; Divitini, G.; Pell, A. J.; Leskes, M.; Ducati, C.; Grey, C. P. *J. Am. Chem. Soc.* **2016**, *138*, jacs.6b02882.
125. Li, H.; Huang, X.; Chen, L.; Zhou, G.; Zhang, Z.; Yu, D.; Jun Mo, Y.; Pei, N. *Solid State Ionics* **2000**, *135*, 181–191.

126. Kwon, J. Y.; Ryu, J. H.; Oh, S. M. *Electrochim. Acta* **2010**, *55*, 8051–8055.
127. Key, B.; Bhattacharyya, R.; Morcrette, M.; Seznéc, V.; Tarascon, J.-M.; Grey, C. P. *J. Am. Chem. Soc.* **2009**, *131*, 9239–9249.
128. Wang, L.; Liu, D.; Yang, S.; Tian, X.; Zhang, G.; Wang, W.; Wang, E.; Xu, Z.; Bai, X. *ACS Nano* **2014**, *8*, 8249–8254.
129. Radvanyi, E.; De Vito, E.; Porcher, W.; Danet, J.; Desbois, P.; Colin, J.-F.; Si Larbi, S. J. *J. Mater. Chem. A* **2013**, *1*, 4956–4965.
130. Lee, S. W.; McDowell, M. T.; Choi, J. W.; Cui, Y. *Nano Lett.* **2011**, *11*, 3034–3039.
131. Liang, W.; Yang, H.; Fan, F.; Liu, Y.; Liu, X. H.; Huang, J. Y.; Zhu, T.; Zhang, S. *ACS Nano* **2013**, *7*, 3427–3433.
132. Lee, S. W.; Lee, H.-W.; Ryu, I.; Nix, W. D.; Gao, H.; Cui, Y. *Nat. Commun.* **2015**, *6*, 7533.
133. Wang, X.; Fan, F.; Wang, J.; Wang, H.; Tao, S.; Yang, A.; Liu, Y.; Beng Chew, H.; Mao, S. X.; Zhu, T.; Xia, S. *Nat. Commun.* **2015**, *6*, 8417.
134. Oumellal, Y.; Delpuech, N.; Mazouzi, D.; Dupré, N.; Gaubicher, J.; Moreau, P.; Soudan, P.; Lestriez, B.; Guyomard, D. *J. Mater. Chem.* **2011**, *21*, 6201.
135. Nadimpalli, S. P. V.; Sethuraman, V. A.; Dalavi, S.; Lucht, B.; Chon, M. J.; Shenoy, V. B.; Guduru, P. R. *J. Power Sources* **2012**, *215*, 145–151.
136. Nie, M.; Abraham, D. P.; Chen, Y.; Bose, A.; Lucht, B. L. *J. Phys. Chem. C* **2013**, *117*, 13403–13412.
137. Philippe, B.; Dedryvère, R.; Allouche, J.; Lindgren, F.; Gorgoi, M.; Rensmo, H.; Gonbeau, D.; Edström, K. *Chem. Mater.* **2012**, 24, 6, 1107–1115.
138. Horowitz, Y.; Han, H.-L.; Ross, P. N.; Somorjai, G. A., *J. Am. Chem. Soc.* **2016**, *138*, 3, 726.
139. Yoon, T.; Nguyen, C. C.; Seo, D. M.; Lucht, B. L. *J. Electrochem. Soc.* **2015**, *162*, A2325–A2330.
140. Dupré, N.; Moreau, P.; De Vito, E.; Quazuguel, L.; Boniface, M.; Bordes, A.; Rudisch, C.; Bayle-Guillemaud, P.; Guyomard, D. *Chem. Mater.* **2016**, *28*, 2557–2572.
141. He, Y.; Piper, D. M.; Gu, M.; Travis, J. J.; George, S. M.; Lee, S.-H.; Genc, A.; Pullan, L.; Liu, J.; Mao, S. X.; Zhang, J.-G.; Ban, C.; Wang, C. *ACS Nano* **2014**, *8*, 11816–11823.
142. Rahaman, O.; Mortazavi, B.; Rabczuk, T. *J. Power Sources* **2016**, *307*, 657–664.
143. Wu, S. Q.; Wang, C. Z.; Zhu, Z. Z.; Ho, K. M. *Mater. Chem. Phys.* **2014**, *148*, 1145–1148.
144. Nguyen, D. T.; Nguyen, C. C.; Kim, J.-S.; Kim, J. Y.; Song, S.-W. *ACS Appl. Mater. Interfaces* **2013**, *5*, 11234–11239.
145. Wachtler, M.; Wagner, M. R.; Schmied, M.; Winter, M.; Besenhard, J. O. *J. Electroanal. Chem.* **2001**, *510*, 12–19.
146. Beattie, S. D.; Larcher, D.; Morcrette, M.; Simon, B.; Tarascon, J.-M. *J. Electrochem. Soc.* **2008**, *155*, A158–A163.
147. Bridel, J.-S.; Azaïs, T.; Morcrette, M.; Tarascon, J.-M.; Larcher, D. *Chem. Mater.* **2010**, *22*, 1229–1241.

148. Liu, X. H.; Zhong, L.; Huang, S.; Mao, S. X.; Zhu, T.; Huang, J. Y. *ACS Nano* **2012**, *6*, 1522–1531.
149. McDowell, M. T.; Lee, S. W.; Harris, J. T.; Korgel, B. A.; Wang, C.; Nix, W. D.; Cui, Y. *Nano Lett.* **2013**, *13*, 758–764.
150. Luo, F.; Liu, B.; Zheng, J.; Chu, G.; Zhong, K.; Li, H.; Huang, X.; Chen, L. *J. Electrochem. Soc.* **2015**, *162*, A2509–A2528.
151. Xu, C.; Lindgren, F.; Philippe, B.; Gorgoi, M.; Björefors, F.; Edström, K.; Gustafsson, T. *Chem. Mater.* **2015**, *27*, 2591–2599.
152. Xu, K. *Chem. Rev.* **2014**, *114*, 11503–11618.
153. Son, S.-B.; Kim, S. C.; Kang, C. S.; Yersak, T. A.; Kim, Y.-C.; Lee, C.-G.; Moon, S.-H.; Cho, J. S.; Moon, J.-T.; Oh, K. H.; Lee, S.-H. *Adv. Energy Mater.* **2012**, *2*, 1226–1231.
154. de Guzman, R. C.; Yang, J.; Ming-Cheng Cheng, M.; Salley, S. O.; Ng, K. Y. S. *J. Mater. Chem. A* **2014**, *2*, 14577.
155. Obrovac, M. N.; Chevrier, V. L. *Chem. Rev.* **2014**, *114*, 11444–11502.
156. Wen, C. J.; Huggins, R. A. *J. Electrochem. Soc.* **1981**, *128*, 1181–1187.
157. Idota, Y.; Kubota, T.; Matsufuji, A.; Maekawa, Y.; Miyasaka, T. *Science (80-.).* **1997**, *276*, 1395LP–1397LP.
158. Sangster, J.; Bale, C. W., *J Phase Equilib.* **1998**, *19*, 70–75.
159. Courtney, I. A. *J. Electrochem. Soc.* **1997**, *144*, 2045.
160. Rhodes, K. J.; Meisner, R.; Kirkham, M.; Dudney, N.; Daniel, C. *J. Electrochem. Soc.* **2012**, *159*, A294–A299.
161. Johnston, K. E.; Sougrati, M. T.; Stievano, L.; Darwiche, A.; Dupre, N.; Grey, C. P.; Monconduit, L., *Chem. Mater.* **2016**, *28* (11), 4032–4041.
162. Winter, M.; Besenhard, J. O. *Electrochim. Acta* **1999**, *45*, 31–50.
163. Zhang, P.; Ma, Z.; Wang, Y.; Zou, Y.; Lei, W.; Pan, Y.; Lu, C. *RSC Adv.* **2015**, *5*, 36022–36029.
164. He, M.; Kravchyk, K.; Walter, M.; Kovalenko, M. V. *Nano Lett.* **2014**, *14*, 1255–1262.
165. Sangster, J.; Pelton, A. D. *J. Phase Equilib.* **1993**, *14*, 514–517.
166. Chang, D.; Huo, H.; Johnston, K. E.; Ménétrier, M.; Monconduit, L.; Grey, C. P.; Van der Ven, A. *J. Mater. Chem. A* **2015**, *3*, 18928–18943.
167. Baggetto, L.; Ganesh, P.; Sun, C.-N.; Meisner, R. A.; Zawodzinski, T. A.; Veith, G. M. *J. Mater. Chem. A* **2013**, *1*, 7985.
168. Hewitt, K. C.; Beaulieu, L. Y.; Dahn, J. R. *J. Electrochem. Soc.* **2001**, *148*, A402.
169. Graetz, J.; Ahn, C. C.; Yazami, R.; Fultz, B. *J. Electrochem. Soc.* **2004**, *151*, A698.
170. Fuller, C. S.; Severiens, J. C. *Phys. Rev.* **1954**, *96*, 21–24.
171. Lim, L. Y.; Liu, N.; Cui, Y.; Toney, M. F. *Chem. Mater.* **2014**, *26*, 3739–3746.
172. Jung, H.; Allan, P. K.; Hu, Y.-Y.; Borkiewicz, O. J.; Wang, X.-L.; Han, W.-Q.; Du, L.-S.; Pickard, C. J.; Chupas, P. J.; Chapman, K. W.; Morris, A. J.; Grey, C. P. *Chem. Mater.* **2015**, 150126094946004.

173. Yoon, S.; Park, C.-M.; Sohn, H.-J. *Electrochem. Solid-State Lett.* **2008**, *11*, A42.
174. Baggetto, L.; Keum, J. K.; Browning, J. F.; Veith, G. M. *Electrochem. Commun.* **2013**, *34*, 41–44.
175. Morris, A. J.; Grey, C. P.; Pickard, C. J. *Phys. Rev. B* **2014**, *90*, 54111.
176. Tang, W.; Liu, Y.; Peng, C.; Hu, M. Y.; Deng, X.; Lin, M.; Hu, J. Z.; Loh, K. P. *J. Am. Chem. Soc.* **2015**, *137*, 2600–2607.
177. Lim, L. Y.; Fan, S.; Hng, H. H.; Toney, M. F. *J. Phys. Chem. C* **2015**, *119*, 22772–22777.
178. Im, H. S.; Cho, Y. J.; Lim, Y. R.; Jung, C. S.; Jang, D. M.; Park, J.; Shojaei, F.; Kang, H. S. *ACS Nano* **2013**, *7*, 11103–11111.
179. Oehl, N.; Hardenberg, L.; Knipper, M.; Kolny-Olesiak, J.; Parisi, J.; Plaggenborg, T. *CrystEngComm* **2015**, *17*, 3695–3700.
180. Jahel, A.; Ghimbeu, C. M.; Monconduit, L.; Vix-Guterl, C. *Adv. Energy Mater.* **2014**, *4*, 1–7.
181. Tavassol, H.; Cason, M. W.; Nuzzo, R. G.; Gewirth, A. A. *Adv. Energy Mater.* **2015**, *5*, 1400317.
182. Reddy, M. V.; Subba Rao, G. V.; Chowdari, B. V. R. *Chem. Rev.* **2013**, *113*, 5364–5457.
183. Kim, C.; Noh, M.; Choi, M.; Cho, J.; Park, B. *Chem. Mater.* **2005**, *17*, 3297–3301.
184. Marino, C.; Sougrati, M. T.; Gerke, B.; Pöttgen, R.; Huo, H.; Ménétrier, M.; Grey, C. P.; Monconduit, L. *Chem. Mater.* **2012**, *24*, 4735–4743.
185. Antitomaso, P.; Fraisse, B.; Stievano, L.; Biscaglia, S.; Ayme-Perrot, D.; Girard, P.; Sougrati, M. T.; Monconduit, L. *J. Mater. Chem. A* **2017**, *5*, 6546–6555.
186. Sougrati, M. T.; Fullenwarth, J.; Debenedetti, A.; Fraisse, B.; Jumas, J. C.; Monconduit, L. *J. Mater. Chem.* **2011**, *21*, 10069.

2 Li-Rich Layered Oxides: Still a Challenge, but a Very Promising Positive Electrode Material for Li-Ion Batteries

Ségolène Pajot,*,†,‡,§ Loïc Simonin,‡,§ and Laurence Croguennec*,¶,||,**

*Institut de Chimie de la Matière Condensée de Bordeaux (ICMCB) — CNRS; University of Bordeaux, Bordeaux INP, 87 Avenue Schweitzer, F-33608 Pessac, France
†CEATech Aquitaine, 16 Avenue Pey Berland, F-33607 Pessac, France
‡Université Grenoble Alpes, F-38000 Grenoble, France
§CEA, LITEN, 17 rue des Martyrs, F-38054 Grenoble Cedex 9, France
¶Réseau Français sur le Stockage Electrochimique de l'Energie (RS2E), CNRS FR 3459, F-80039 Amiens Cedex 1, France
||ALISTORE-ERI European Research Institute, FR CNRS 3104, F-80039 Amiens Cedex 1, France
**Laurence.Croguennec@icmcb.cnrs.fr

Table of Contents

List of Abbreviations

2D	bidimensional
ALD	atomic layer deposition
CVD	chemical vapor deposition
DEMS	differential electrochemical mass spectrometry
DFT	density functional theory
DSC	differential scanning calorimetry
EDX	energy dispersive X-ray spectroscopy
EELS	electron energy loss spectroscopy
EPMA	electron probe X-ray microanalysis
EPR	electron paramagnetic resonance
FCG	full concentration gradient
HAADF	high-angular annular dark field
LCO	$LiCoO_2$
MAS	magic angle spinning
NCA	$LiNi_{1-x-y}Co_xAl_yO_2$

NMC	$LiNi_{1-x-y}Mn_xCo_yO_2$
NMR	nuclear magnetic resonance spectroscopy
PLD	pulsed laser deposition
RIXS	resonant inelastic X-ray scattering spectroscopy
SEM	scanning electron microscopy
STEM	scanning transmission electron microscopy
SXAS	soft X-ray absorption spectroscopy
XAS	X-ray absorption spectroscopy
XANES	X-ray absorption near edge structure
XPS	X-ray photoelectron spectroscopy

I. Introduction

Lithium-ion batteries entered the market in the 1990s, and made possible the development of the electronic market with the emergence of a large panel of applications such as portable electronics and power tools, increasingly more and more demanding not only in energy and power, but also in weight and volume restrictions. Lithium-ion batteries are still at the center of researches to challenge recent concerns: transportation with different degrees of electrification, large-scale stationary storage of renewable energies produced by wind and solar farms, and improvement of the grid reliability and stability in helping to compensate for imperfections due to energy mix. All these new applications and demands are pushing further the development of the lithium-ion technology and its market. Indeed, even if considered as mature, improvements are still possible. In this chapter devoted to layered oxides, currently at the positive electrodes in most of the commercial lithium-ion batteries, the possible tracks to follow to further increase their energy, stability, and so on are discussed. Early lithium-ion batteries consisted of the layered oxide $LiCoO_2$ at the positive electrode and graphite at the negative electrode. After more than 25 years of development, the chemistries used remain similar but, for instance, on the positive electrode, the spectrum of the layered oxides used has widened to fit the set of requirements depending on the applications targeted, with the development of complex compositions, blends, composites with surface, or even bulk engineering.

II. The State of the Art

The layered oxides $LiMO_2$ are electrochemically active with reactions mainly being topotactic, that is, occurring without any major volume change and irreversibility in the host structure. The oxidation/reduction processes involve the exchange of electrons coupled to reversible lithium deinsertion/insertion reactions. For most of the layered oxides, their theoretical electrochemical capacity is limited to one electron per transition metal, as only transition metal cations are involved in the redox processes. These layered oxides have bidimensional (2D) structural frameworks with alternated stacking of MO_2 slabs and LiO_2 interslab spaces made up of edge-sharing MO_6 and LiO_6 octahedra, respectively. These interslab spaces are the 2D diffusion pathways for lithium ions in the structure, by jumping from one octahedron to another through intermediate face-sharing tetrahedral sites.[1] By altering the composition of $LiMO_2$ (M = Co, Ni, Mn, Al, etc.), its structural, physicochemical and electrochemical properties can be tailored. Simple or more complex compositions are already on the market, for years for some of them: $LiCoO_2$ (LCO), $LiNi_{0.80}Co_{0.15}Al_{0.05}O_2$ (NCA), $LiNi_{1/3}Co_{1/3}Mn_{1/3}O_2$ (NMC) and so forth. Each of these compositions is chosen for a given application taking always into account the requirements such as cost, safety, energy, power, and cyclability.

As often, the defect chemistry plays a major role in determining not only the structure but also the properties of these layered oxides. Most $LiCoO_2$ materials used in commercial lithium-ion batteries were shown to be in fact overstoichiometric $Li_{1+x}Co_{1-x}O_{2-x}$ (with Li/Co ratio >1 and x > 0) as large-scale industrial syntheses are performed using an excess of the lithium carbonate precursor[2]: (i) to promote the particles' growth for a better chemical stability and a minimized solid–liquid interface with the electrolyte, and (ii) to prevent the formation of the impurity Co_3O_4, which is detrimental to the performances. The introduction of these defects in the structure, trivalent cobalt ions in square-based pyramids with an original electronic configuration $t_{2g}^5e_g^1$, allows suppressing all the phase transitions associated to Li^+/vacancy ordering in $LiCoO_2$, and thus increasing its ionic and electronic transport properties for better performances at high rates.

Researches are still very active on $LiCoO_2$ with the goal to stabilize reversibly the structure of the highly deintercalated Li_xCoO_2 compositions

($x < 0.5$) at high voltage and thus to develop high-energy $LiCoO_2$-based materials, maintaining — or even increasing — their thermal stability in the charged state of the batteries as well as their rate capabilities. Indeed, $LiCoO_2$ is the material of choice at the positive electrode of batteries developed for mobile applications, but even in that field energy demands are increasing with the development of new applications such as electronic tablets. Combining complex and complementary strategies, significant optimization has already been achieved.[3] These material engineering approaches can be (i) the bulk modification of the active material through a partial substitution of cobalt (Al[4], Mg[5], etc.); (ii) the surface modification through the formation of inorganic[6] or polymeric "coatings," being often for the former nanoparticles decorating the surface and the grain boundaries; and (iii) a surface treatment of the electrode using deposition techniques as shown in Figure 1.[7] Partial substitution of the bulk is often combined with surface modifications, either through a further doping of the extreme surface (being thus electrochemically active) or through the formation of a stabilizing interface (allowing lithium diffusion but being not electrochemically active itself).

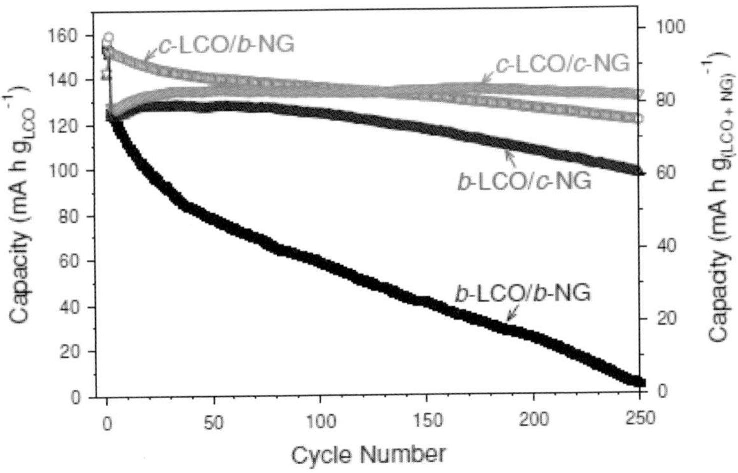

Figure 1. Comparison of the electrochemical performances of $LiCoO_2$/graphite full cells with bare (i.e., uncoated) electrodes (named as *b*-) or with coated electrodes (named as *c*-). The Al_2O_3 coating was performed using atomic layer deposition (ALD). (Taken from Ref. 7 with permission).

The optimized compositions $LiNi_{0.80}Co_{0.15}Al_{0.05}O_2$ (NCA) and $LiNi_{1/3}Co_{1/3}Mn_{1/3}O_2$ (NMC or (333) hereafter) remain expensive due to the cost of both nickel and cobalt; NCA being the most attractive for high rate applications.[8] Indeed, electron hopping among the nickel ions ($Ni^{3+,4+}$) within the MO_2 slabs promotes faster lithium diffusion within the structure, whereas ordering of the Ni, Mn, and Co ions in NMC is detrimental to electron mobility and thus to ionic conductivity. These compositions were developed using the solid-state chemist's background to track the better substitutions for optimized performances.[9] Partial substitution of nickel for cobalt in $LiCoO_2$ was first considered in order to decrease its cost, revealing that $LiNi^{III}{}_{0.7}Co^{III}{}_{0.3}O_2$ was in fact the optimized composition *versus* the end members $LiCo^{III}O_2$ and $LiNi^{III}O_2$, due to the difficulty to control the stoichiometry of the pure nickelate, most often stabilized as $Li_{1-z}Ni_{1+z}O_2$ with an excess of Ni^{2+} ions in the lithium sites that drastically deteriorate the performances.[10] In fact, the combined partial substitution of cobalt and aluminum for nickel was revealed to be the key to control both the 2D character of the structure and the thermal stability of the active material in the charged state of the battery.[11] First, the small size of Co^{3+} and Al^{3+} tailors the crystal field so as to destabilize the formation of Ni^{2+} in the slabs and, as a consequence, in the interslab space. Second, the partial substitution by the electrochemically inactive cation Al^{3+} decreases the amount of unstable Ni^{4+}(and/or Co^{4+}) cations formed at the end of the charge (promoting a better thermal stability) and prevents overcharge as the reversible capacity delivered is intrinsically limited by the restricted number of electrons that can be exchanged on the metals. Third, the stabilization of Al^{3+} in the intermediate tetrahedral sites on increasing the temperature increases the stability of the highly delithiated compositions, as they delay all the phase transitions associated with oxygen loss and to the irreversible formation of spinels. The emergence of the ternary layered compounds, $Li(Ni_xMn_{1-x-y}Co_y)O_2$ found its origin considering first the system $Li[Li_{0.2}Mn^{IV}{}_{0.4}Cr^{III}{}_{0.4}]O_2$ by bringing the redox couple Cr^{6+}/Cr^{3+} into play with the manganese remaining tetravalent.[12] This system displays a multielectron redox center (chromium) responsible for the capacity beside the Mn^{4+} ions, which are inactive during the redox processes and act as stabilizers for the host network. These results have paved the way for the development of a new family of layered transition metal oxides as alternatives to $LiCoO_2$ and $Li(Ni_{1-y-z}Co_yAl_z)O_2$

at the positive electrode of lithium-ion batteries and led to reconsider the early work on the $LiNi_{1-y}Mn_yO_2$ system.[13] Rapidly, $LiNi^{II}_{1/2}Mn^{IV}_{1/2}O_2$ and $LiNi^{II}_{1/3}Mn^{IV}_{1/3}Co^{III}_{1/3}O_2$ were proposed.[14,15] The goal was to combine high reversible capacity with the redox couples $Ni^{2+}/Ni^{3+}/Ni^{4+}$ (exchange of 2 electrons per Ni) and Co^{3+}/Co^{4+}, Coulombic efficiency due to a controlled 2D layered dimensionality for the structure, and a high thermal stability in the charged state due to a large fraction of stable transition metal ions as Mn^{4+}. As mentioned earlier, the main drawback of these manganese-rich compositions is their limited transport properties. Even more complex compositions, $Li(Ni^{III}_{1-2x-y}Ni^{II}_xMn^{IV}_xCo_y)O_2$, were considered in order to combine a mixed valence ($Ni^{2+,3+}$) for a partial electronic delocalization of the nickel network and a significant content of manganese for an improved structural and thermal stability. Changes in the performances with the content such as Ni, Co, and Mn are significant as summarized in Figure 2, and the composition $LiNi_{0.6}Mn_{0.2}Co_{0.2}O_2$ (called 622) appears today as the best choice.[16]

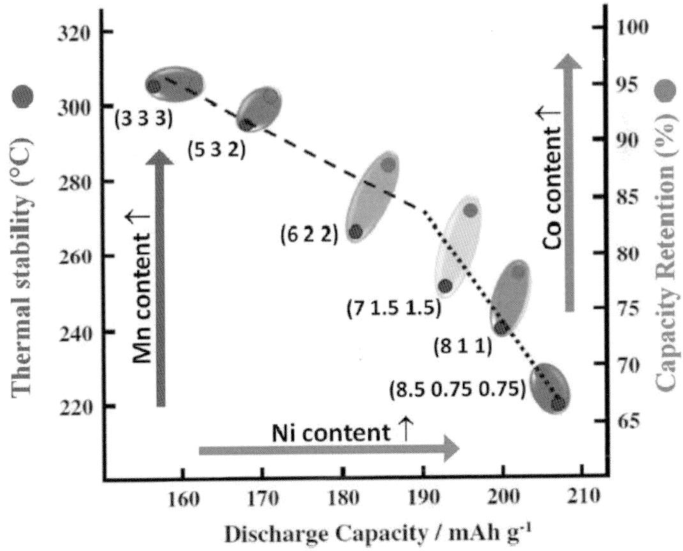

Figure 2. Changes in the discharge reversible capacity (in black), thermal stability in the charged state of the battery (in blue) and capacity retention on long range cycling (in red) for layered oxides of compositions $Li(Ni_xMn_{1-x-y}Co_y)O_2$ with the contents Ni, Mn, and Co given in brackets. (Adapted from Ref. 17).

III. The Layered Oxides $Li_{1+x}M_{1-x}O_2$ Rich in Lithium

In parallel, other layered oxides rich in Li and Mn of general chemical formula $Li_{1+x}M_{1-x}O_2$ were also at the center of challenging researches in the field, as they could deliver, if stabilized, significantly increased energy density with the involvement of the anion in the redox processes.

A. The Intriguing Performance of the Lithium and Manganese-Rich Layered Oxides at the Center of 10 Years of Challenging Researches

1. *From the Average Structure to the Local Structure*

The structure of the compositions, $Li_{1+x}M_{1-x}O_2$, rich in lithium, also known as the overlithiated layered oxides or the Li-rich, derives from those of α-$NaFeO_2$ and $LiMO_2$ with a substitution of Li for M in the slabs MO_2 (Figure 3). The charge compensation is achieved through an increase in the average oxidation state of the transition metal M such as $Li(Li^{+}_{x}M^{3+}_{1-3x}M^{4+}_{2x})O_2$ ($x \leq 1/3$). For $x = 1/3$, Li_2MnO_3 is the most studied and shows an ordering between Mn^{4+} and Li^{+} in the slabs $(Li_{1/3}Mn_{2/3})O_2$ at the origin of a monoclinic distortion described in the $C2/m$ space group. This unit cell corresponds in fact to a $3a_{hex} \times 3a_{hex}$ superstructure in the hexagonal cell of α-$NaFeO_2$. Two octahedral sites are present in the slabs: the $2b$ site is occupied by lithium whereas the $4g$ site is occupied by manganese of different sizes (0.74 and 0.54 Å, respectively). That intriguing compound was considered as electrochemically inactive as lithium deintercalation could not be compensated by the oxidation of manganese being already in its tetravalent state. The decrease in the particle size, inducing also an increasing amount of stacking faults in Li_2MnO_3,[18] was shown to lead to an activation of its electrochemical activity questioning the possible effects of oxygen nonstoichiometry[19] and the formation of spinel-type domains at the particle surface.[20] Indeed, the presence of oxygen defects in $Li_2MnO_{3-\delta}$ induces the partial reduction of manganese (Mn^{4+} into Mn^{3+}) for charge compensation, improving the electronic transport properties in the pristine state and thus the electrochemical performances. More recently it was shown that the formation of highly homogeneous composites with conductive carbon allows activating even highly crystalline Li_2MnO_3.[21] Lithium deintercalation from Li_2MnO_3 was

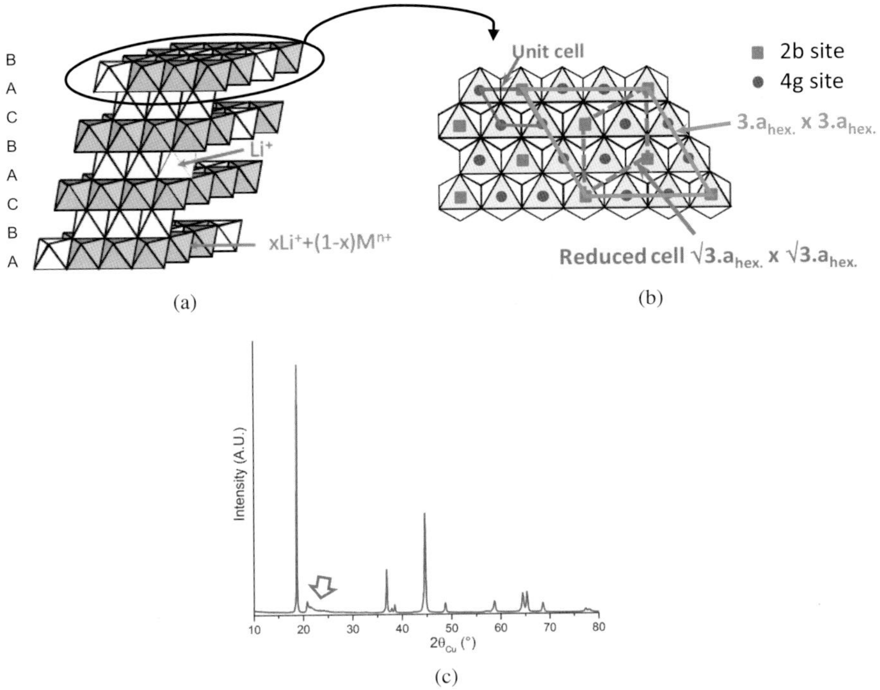

Figure 3. Structural representation of the Li-rich layered oxides $Li_{1+x}M_{1-x}O_2$: (a) the MO_2 slabs are occupied by the transition metal ions M and Li in excess, whereas the interslab space is only occupied by Li; (b) description of the possible cation distribution within the 2D slabs, with the two sites for cations (2*b* and 4*g* in the *C*2/*m* space group description). The small hexagonal cell (in green) describes the typical layered structure $LiMO_2$ (space group *R*-3*m*) with a statistical distribution of the cations in a single crystallographic site. The larger unit cells (in blue, and in red for the reduced one) allow describing an order in the slabs as observed for Li_2MnO_3. It can be an order between the large and the small cations, here 1:2, to stabilize the structure and to minimize the strains. (c) A typical X-ray diffraction pattern of these Li-rich layered oxides with extra lines in the angular range between 20° and 30° 2θ: the broad lines, as observed here, indicate a tendency to the ordering within the slabs, but without any correlation between the ordered slabs along the stacking axis. (Adapted from Ref. 17).

first associated with the removal of Li_2O,[22] and finally revealed to combine the Li and Mn sites with O_2 gas evolution.[23,24] Beyond Li_2MnO_3, the composition of the slabs of the Li-rich layered oxides $Li_{1+x}M_{1-x}O_2$ can be tailored considering a large panel of cations next to Mn and Li (Ni, Co, Al, Mg, etc.). Trivalent cobalt and divalent nickel ions are in fact mostly considered as they are electrochemically active and similar in size to Mn^{4+}

and Li^+, respectively. Indeed, Ni^{2+} is a large cation (0.69 Å) and occupies thus preferentially the 2*b* site, whereas Co^{3+} is smaller (0.545 Å) and occupies thus the 4*g* site next to Mn. The closer to a ratio of 1:2 between large and small cations in the slabs, the more extended is the ordering within the transition metal layers. Nevertheless, due to the complexity of compositions studied, the ordering along the stacking of the layers is barely observed, except for Li_2MnO_3, Li_2RuO_3, and so on.[25–27] Note that as soon as the order is limited (small domain sizes), the structure cannot be described satisfyingly anymore either in the *C2/m* space group or in the *R*-3*m* space group: the first being associated with the fully cationic ordered structure and the second to the fully statistically cationic distributed structure (or to the average structure). The key signature for these structural differences is brought by the additional peaks observed between 20° and 30° ($2\theta_{Cu}$) in the X-ray diffraction patterns (Figure 3): the higher their intensities, the sharper their width, the more ordered are the cations within the structure as these peaks are associated with the superstructure.

As extensively studied by McCalla *et al.* for the Li-Mn-Ni-O and Li-Mn-Co-O systems using the combinatorial approach,[28–31] the material's homogeneity is largely dependent on the synthesis conditions, explaining also all the apparent discrepancies reported on these lithium-rich compounds.[32] They can be described as follows:

— Aggregates with primary particles containing domains of compositions and structures $Li_2Mn^{IV}O_3$ and $LiM^{III}O_2$, explaining the formula often used $(1 - y)Li_2MnO_3.yLiMO_2$ to describe them.[33]
— Aggregates of primary particles being solid solutions (i.e., homogenous cationic distribution within the material) with a tendency to form an order more or less extended between the Li, M^{n+}, and Mn^{4+} ions within the slabs, the size of the cations controlling this order (as just discussed previously).[34–36]
— A mixture of different phases (layered, spinel, and/or rock salt) observed at the scale of the primary particles.[28,29,37] The closer to the surface, the more lithium and oxygen deficient are the phases, in good agreement with the presence of spinel and rock-salt-type domains in the last few atomic layers described in the *Fd*-3*m* and *Fm*-3*m* space groups, respectively.

All the debates concerning the structure of these Li-rich layered oxides come from the fact that the formula $(1 - y)Li_2MnO_3.yLiMO_2$, which describes a simple way how to tailor the formation of Li and Mn^{4+}-rich layered oxides from a chemical point of view, was largely interpreted among the community as describing the structure of the material with local domains being either Li_2MnO_3 or $LiMO_2$. In fact, Li_2MnO_3 domains most often do not exist as their extension is limited to the first coordination sphere as observed in $Li[Mn_6]$ and $Mn[Li_3Mn_3]$ environments, considering also that often they are $Li[(Mn,Co)_6]$ and $Mn[(Li,Ni)_3(Mn,Co)_3]$ for Li-Mn-Ni-O and Li-Mn-Co-O systems.

2. *The Overcapacity: Its Origin and Impacts*

The layered oxides rich in lithium and manganese are very attractive as they deliver significantly improved high specific capacity and, due to Mn, at lower price and for more sustainable compositions. Their theoretical capacity is up to 350 mAh/g (i.e., a significant overcapacity *versus* classical layered oxides $LiMO_2$).[34] To reach this very attractive reversible capacity, it is crucial to "activate" these materials by charging the battery to high potential (up to 4.8 V *versus* Li^+/Li), near the limit of the electrolyte stability *versus* oxidation. Then, they deliver a reversible capacity up to 230–250 mAh/g in the potential window 2.5–4.6 V *versus* Li^+/Li. This performance is exceptional, as shown through comparison in Figure 2, which is significantly higher than that observed for stoichiometric layered oxides $LiMO_2$ ($\leq$200 mAh/g due to a severe loss of capacity in a larger cycling potential window), and as involving the participation of the oxygen anion to the redox processes. Indeed, it is associated to lithium deintercalation and reintercalation reactions partly compensated by the participation of anion to the redox processes. As shown in Figure 4, a long plateau is observed during the first charge, which is not observed for the classical layered oxides $Li_{1-x}M_{1+x}O_2$ ($x \leq 1$). At the beginning of the plateau, all the transition metal ions have reached their upper valence state, here the tetravalent state for Co, Ni, and Mn. From the discovery of these intriguing materials, Dahn and coworkers[34] had proposed a mechanism for charge compensation based on lithium deintercalation balanced by oxygen loss if Li_2O was formed.[22]

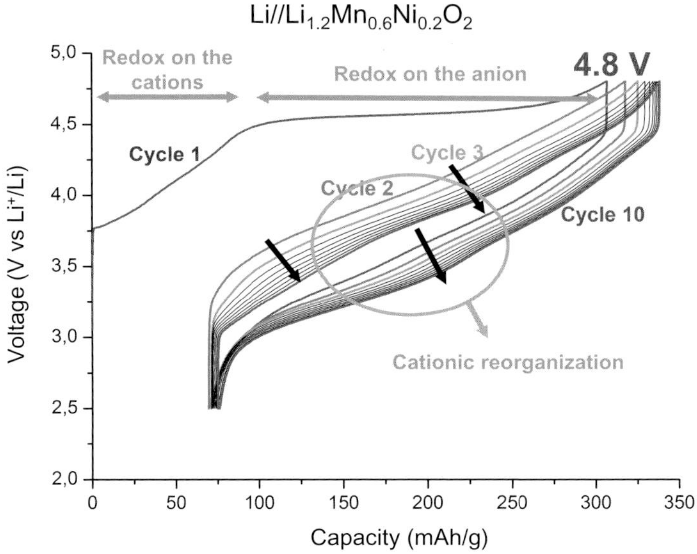

Figure 4. Cycling curves (between the 1st and the 10th cycles) obtained for the lithium batteries $Li//Li_{1.2}Ni_{0.2}Mn_{0.6}O_2$ and showing a long irreversible plateau during the first charge at 4.5 V *versus* Li^+/Li.

Very rapidly the mechanism was considered as being more complex as the experimental oxygen loss was revealed to be an order of magnitude smaller than that expected to balance the lithium deintercalation reaction.[38] In fact for most of the compositions, the only compensation of lithium deintercalation by oxygen loss would have meant most of the transition metal ions in MO_5 coordination environments and an overall diffusion of oxygen through the solid. This exceptional capacity can in fact be explained by the reversible participation of oxygen to the redox processes within the bulk of the material, without major structural modifications. It was demonstrated by Koga *et al.* for the composition $Li_{1.2}Ni_{0.13}Co_{0.13}Mn_{0.54}O_2$,[27,39–41] and directly observed by Oishi *et al.* for $Li_{1.2}Ni_{0.13}Co_{0.13}Mn_{0.54}O_2$ using O K-edge X-ray absorption spectroscopy (XAS) both in the total electron yield mode and in the partial fluorescence X-ray yield mode thus proving the anionic activity even in the bulk.[42] Nevertheless, the stability of the oxidized oxygen species at the extreme surface is obviously small, as it was shown that a partial oxygen loss occurs inducing the migration of transition metal ions from the surface to the bulk, a cationic reorganization and a densification of the metal–oxygen framework.[43–48] The latter induced the formation of a "splayered-type" phase ("midway" between a layered structure and a spinel structure). Indeed, these migrations and structural modifications occur at room temperature, in metastable conditions, and the long range spinel structure cannot be obtained in these conditions. All these structural modifications required to be observed and understood using the combination of advanced microscopy and spectroscopy techniques such as high-angular annular dark field–scanning electron microscopy (HAADF-STEM) and scanning electron microscopy–electron energy loss spectroscopy (STEM-EELS).

The reconstruction at the surface of the primary particles has a huge impact on the potential as it is measured at the surface of the particles and thus driven by the nature of the material present at the surface. The decrease in the potential induced by the splayered structure makes the energy density drop, reducing drastically the performances of that material as energy is the product of the voltage and the reversible capacity. Furthermore, during the next discharge and cycles, to compensate for the oxygen loss, manganese, which was inactive during the first charge, enters in the redox processes through its reduction during the first discharge ($Mn^{4+} \rightarrow Mn^{3+} + 1\ e^-$). Its activity appears at low voltage, below 3.5 V *versus* Li^+/Li. As shown in Figure 5, Li^+ ions that deintercalated at the very end of the charge are only reintercalated at a very low potential on reduction, with an hysteresis of 1 V, showing that the lithium diffusion pathways are very different between first charge and first discharge: the hysteresis is thus not related to a kinetic limitation but to an intrinsic thermodynamic behavior.[49–52] As suggested by neutron powder diffraction and ^6Li magic angle spinning–nuclear magnetic resonance spectroscopy (MAS NMR) analyses of deintercalated and reintercalated compositions recovered after cycling, manganese ions would migrate in tetrahedral sites in the Li layer at high voltage (>4.1 V *versus* Li^+/Li) and would be destabilized during the next discharge only at low voltage (<3.2 V *versus* Li^+/Li) for highly lithiated compositions.[53,54] These manganese ions are mobile as in the vicinity of oxygen vacancies or in very distorted environments formed by the shortening of O-O distances due to the oxidation of oxygen. The mobility of these manganese ions would be at the origin of the hysteresis.

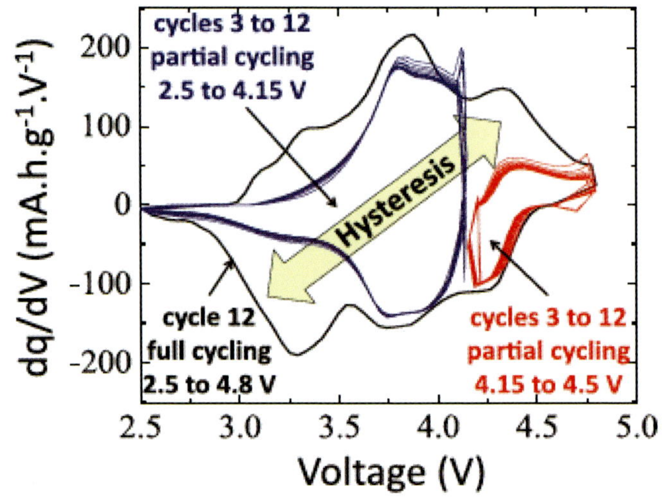

Figure 5. Comparison of the electrochemical signatures obtained in different potential windows for the lithium batteries $Li//Li_{1.20}Ni_{0.20}Mn_{0.60}O_2$: the hysteresis appears to be due to lithium deintercalation at high voltage (~4.4 V *versus* Li^+/Li), the very end of the charge, and to its reintercalation only at very low voltage (~3.2 V *versus* Li^+/Li). (Reproduced from Ref. 52).

Peralta *et al.* compared the electrochemical behavior of different compositions showing a decreasing Li/M ratio from 2 for Li_2MnO_3 to 1.27 for $Li_{1.12}Ni^{2+}{}_{0.32}Mn^{4+}{}_{0.56}O_2$, with in each case the Li/Mn ratio maintained to 2 (Figure 6).[52] They have especially highlighted that the higher the manganese content, the more difficult it is to get the "full" capacity directly from the first cycle. A continuous increase in the reversible discharge capacity occurs during few cycles, concomitantly with the drastic and continuous voltage drop, before being stabilized. The voltage decay proceeds further even if following a smoother regime of decrease. Note that the higher the manganese content, the smaller the electronic conductivity, and thus the more difficult it is to activate the participation of the oxygen to the redox processes. The increasing size of the particles makes even more difficult the spread of this reaction. The formation of Mn^{3+} defects, either within the pristine material or on cycling to compensate for oxygen loss, helps promoting the initiation and the spread of the lithium deintercalation and reintercalation reactions within the particles, but with the drawback — that

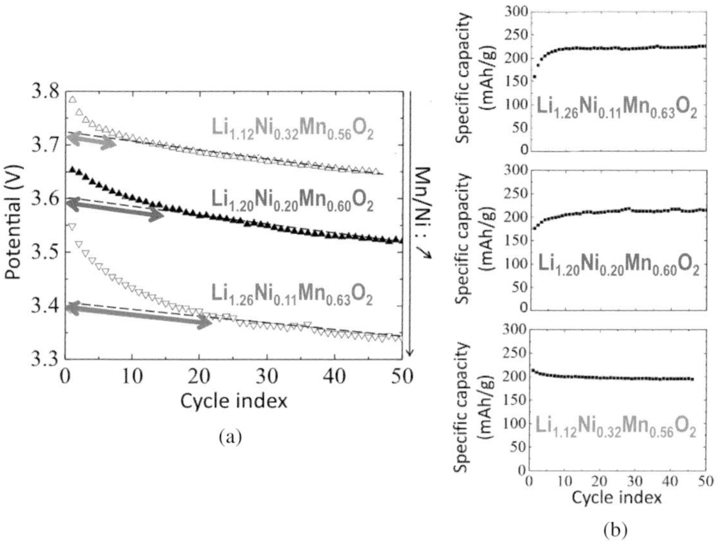

Figure 6. Comparison of three Li and Mn-rich layered oxides $Li_{1.26}Ni_{0.11}Mn_{0.63}O_2$, $Li_{1.20}Ni_{0.20}Mn_{0.60}O_2$, and $Li_{1.12}Ni_{0.32}Mn_{0.56}O_2$: (a) changes in the average discharge voltage upon cycling and (b) discharge capacity *versus* the number of cycles. (Adapted from Ref. 52).

it goes with a more largely cationic migration within the structure and thus voltage decay. This explains the tight relationship between the manganese content, the transport properties, the efficiency of the reaction, and the voltage decay. Partial substitution of divalent nickel for tetravalent manganese helps as it participates, on the one side to improve the transport properties due to the formation of the mixed valence state $Ni^{2+/3+/4+}$, and on the other side to decrease the mobility of the transition metal ions within the framework (as Ni^{n+} is not stable in tetrahedral site).

As a huge part of the electrochemical activity is linked to the participation of the oxygen to the redox processes, and as the irreversible part of this reaction is a surface reaction, efforts have been made to generate specific surface area[39,55] and to tune the particles' morphology.[56] The smaller the surface area and the denser the agglomerates, the less extended is the irreversible oxygen loss due to limited oxygen diffusion within the bulk. As expected, a major impact is observed on the extent of the mechanism involving a structural reorganization at the surface of the material, and thus on the heterogeneity at the local scale of the material obtained after a few cycles, and finally of the magnitude of the voltage drop. Decreasing specific surface area induces better cyclability and smaller irreversible capacity. Even if complex, tailoring the morphology of the Li-rich layered oxides is also a route recently, and interestingly, explored. Oh *et al.*[56] have shown as highlighted in Figure 7 that the controlled synthesis of flake-shaped primary particles not only allows preparing dense aggregates with a small specific surface area to minimize the oxygen loss and structural reorganization, but also promotes fast rate capability due to the preferential growth of the particles within the plane of the easier electron migration.

The extent of the cationic reorganization can also be quite different depending on the stoichiometry and thus oxidation state stabilized for each transition metal ion in the pristine material.[39,46,57–59] High resolution HAADF-STEM and STEM-EELS analyses are crucial to get an in-depth

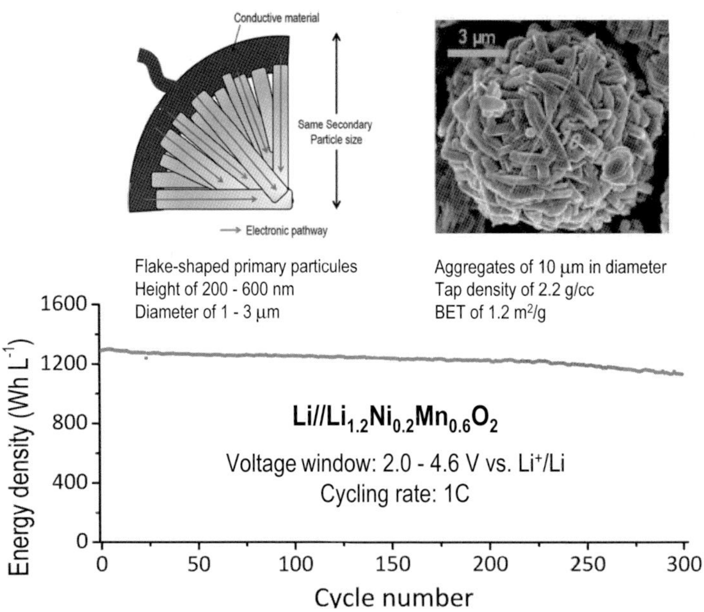

Figure 7. Flake-shaped primary particles in dense aggregates combining small specific surface area with rate capability and thus showing excellent energy retention. (Adapted from Ref. 55).

description of the pristine material and of its changes on cycling, from an atomic and electronic point of view. Indeed, any heterogeneity in the distribution of the transition metal ions or any discrepancy *versus* the Li/M ratio so as to control only the tetravalent state for Mn makes that Mn is active and promotes more extended and faster cationic reorganization. On the basis of the knowledge of Mn-rich layered oxides, the migration of Mn^{3+} on tetrahedral sites to form spinel-type structures has already been largely observed previously for monoclinic $LiMnO_2$[60,61] and orthorhombic $LiMnO_2$.[62] In the Li and Mn-rich materials, two phenomena can be considered, that related to the surface and oxygen loss, and that linked to the presence of Mn^{3+} defects or heterogeneities locally in the pristine material that can initiate quite fast large cationic reorganization within the bulk toward spinel type structure, on the contrary to what is observed when the stoichiometry is fully tailored.[39,45] As reported in references [46,47,63] the composition $Li_{1.2}Ni^{II}{}_{0.18}Mn^{IV}{}_{0.59}Mn^{III}{}_{0.02}Mg_{0.01}O_2$ was shown to contain not only manganese at the trivalent state in the pristine material but also heterogeneities at the local scale, with domains being clearly of α-$NaFeO_2$ type without any ordering in the slabs between the lithium and transition metal ions, and other domains being of Li_2MnO_3 type with an order in the slabs. The anionic redox activity was shown in that case to be rapidly limited with not only a fast voltage decay, but also — as clearly revealed by STEM-HAADF images combined with EELS analyses — a huge and extended cationic migration (preferentially of Mn) resulting in the formation of a concentration gradient in Mn and Ni within the densified domain present at the surface of the particles, the surface being enriched in Ni (Figure 8). The structure stabilized for the first atomic layers observed at the surface of the reconstructed material is "somewhere between" the layered (*R*-3*m* or *C*2/*m*), spinel (*Fd*-3*m*) and rock-salt (*Fm*-3*m*) structures, mostly depending on the composition and structure of the pristine material and on the extent of the cycling. This cationic reorganization occurs at room temperature, thus as expected not perfectly, as it was discussed in reference 45.

Beyond the surface reconstruction of the active layered oxide materials, the oxygen released reacts irreversibly with the carbonate-based solvents of the electrolyte leading to the formation of a surface resistive layer made of decomposition products responsible for increasing polarization, deteriorated performance at high rates, and so on.[64,65]

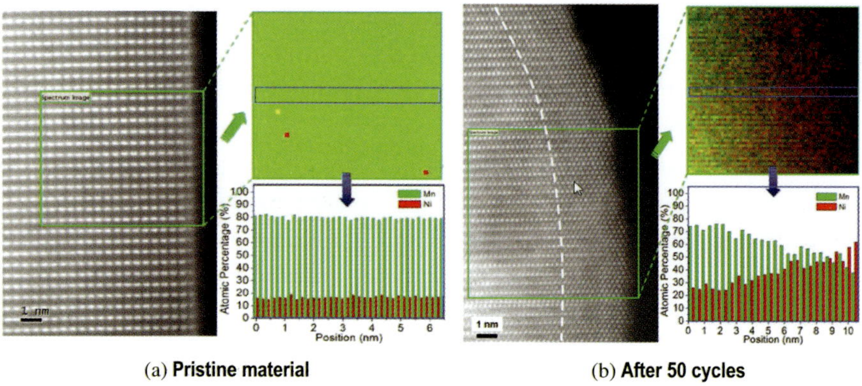

Figure 8. High resolution STEM-HAADF images, the chemical maps, and the corresponding graphs highlight changes in the content Mn and Ni obtained for the material $Li_{1.2}Ni_{0.18}Mn_{0.61}Mg_{0.01}O_2$: a) the pristine material and b) the material after 50 charge and discharge cycles at a *C*/10 rate. The dotted white line materialized the transition between the layered oxide phase and the spinel-type phase localized within the bulk and at the surface of the material, respectively. (Adapted from Ref. 46).

Of course, maintaining the voltage below the voltage plateau prevents any activation and redox on the oxygen, ensuring that in a limited potential window the Li and Mn-rich compositions are stable and deliver performances similar to those of NMC, but at lower cost due to their high Mn content. Nevertheless, even if it remains challenging, it is clear that the mechanism involving anionic redox has to be stabilized in order to take benefit over a long range cycling of the exceptional properties it opens the road for. Substitutions and exotic compositions were explored as possible tracks to stabilize these Li-rich compounds and allow, at least, to get the direct signature of the oxygen redox and further understanding into the anionic redox.

3. *Direct Characterization of the Oxidized Oxygen Species*

In the last 5 years, the composition of the Li-rich layered oxides has been extended to exotic compositions, considering especially the 4*d* transition metal ions such as Ru, Mo, and Sn. These results revealed that by altering the layered oxides' composition, it is possible to stabilize the anionic redox reaction, even at the surface, and as a consequence to maintain a large reversible capacity upon long range cycling without voltage and energy decay. Sathiya *et al.* studied a series of compositions $Li_2Ru_{1-y}Sn_yO_3$ ($0 \leq y \leq 1$) that delivered discharge capacities ranging between 50 and 200 mAh/g for $y = 1$ and 0, respectively.[59,66,67] The voltage decay was shown to decrease with the increasing content in Sn, and to be negligible after 100 cycles for the composition $y = 0.75$ as Sn^{4+} ions are not stable in tetrahedral sites, the required pathway between the slabs and the interslab spaces (Figure 9). On the contrary, by replacing Ti for Sn in $Li_2Ru_{1-y}Ti_yO_3$, the voltage decay was even more pronounced. As soon as the shortening of the O-O bonds occurs due to the oxidation of oxygen to compensate for Li deintercalation along the plateau (Figure 10), Ti^{4+} ions migrate from their highly distorted environments to the neighboring tetrahedral sites where

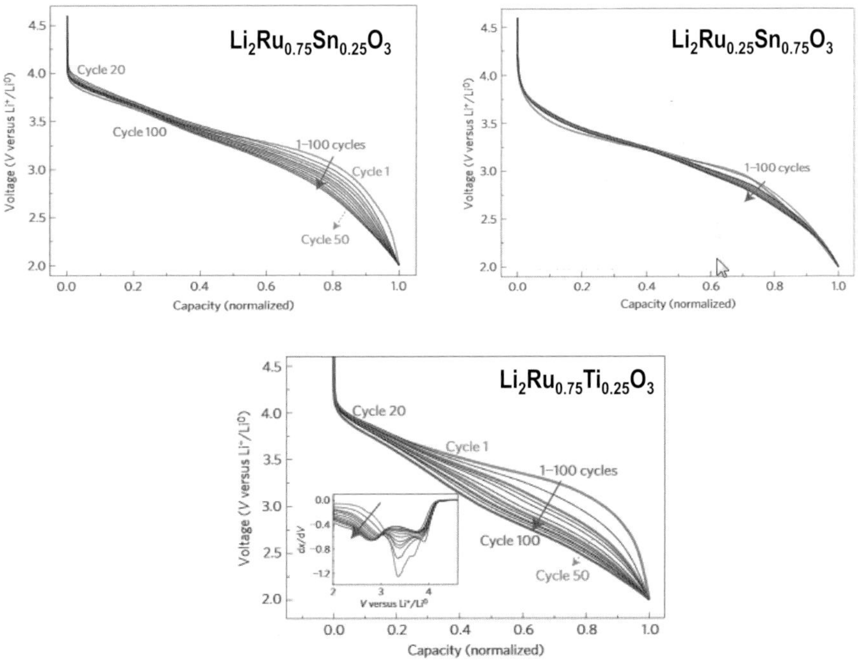

Figure 9. Normalized capacity discharge profiles obtained for $Li_2Ru_{0.75}Sn_{0.25}O_3$, $Li_2Ru_{0.25}Sn_{0.75}O_3$, and $Li_2Ru_{0.75}Ti_{0.25}O_3$ every 10 cycles during the first 100 cycles performed at *C*/5 between 2 and 4.6 V *versus* Li^+/Li. (Adapted from Ref. 67).

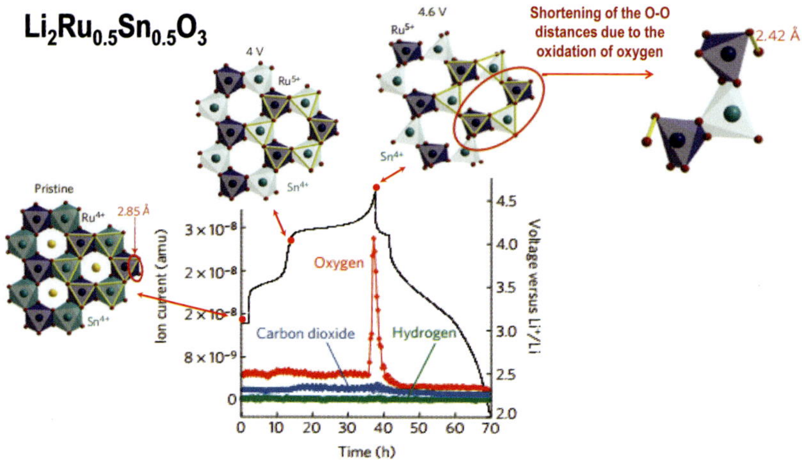

Figure 10. First charge and discharge obtained for a $Li//Li_2Ru_{0.5}Sn_{0.5}O_3$ cell with the gas profiles of CO_2, H_2, and O_2 "detected" at room temperature. Note that the oxygen release was only obvious at the very end of the charge. The description of the $(Li_{1/3}Ru_{1/3}Sn_{1/3})O_2$ slab is given at each step of the charge (Li in yellow, Ru in violet, and Sn in green): before the plateau, Ru^{4+} is oxidized to Ru^{5+} to compensate for the deintercalation of 0.5 Li^+, during the plateau the further Li deintercalation is compensated by the oxidation of the oxygen anions which leads to a shortening of the O-O distances (from 2.85 to 2.42 Å) and to a strong distortion of the Ru and Sn first coordination spheres. (Adapted from Ref. 66).

they can be stabilized. The signature of the anion participation to the redox processes and thus of its oxidation at the end of the charge was obtained combining different spectroscopies.[66] The detection of a component at ~530.5 eV, in good agreement with the presence of oxygen with lower electronic density than in O^{2-} at the surface of the material, was achieved by X-ray photoelectron spectroscopy (XPS). The formation of these oxidized oxygen species within the bulk of the material was confirmed by Electron paramagnetic resonance (EPR) that revealed at high voltage a signal typical of peroxide/superoxide type species (similar to that of CaO_2). This signal was estimated to correspond to 23 ± 6% of the oxygen for $Li_2Ru_{0.5}Sn_{0.5}O_3$ (i.e., a significant amount). The factors governing the stability of the Li-rich layered oxides *versus* oxygen release in their highly deintercalated compositions, and thus the reversibility of the anionic redox, has been studied by theoretical calculations tuning the compositions of Li_2MO_3, considering not only 3*d*, 4*d*, and 5*d* transition metal ions but also a Li/M ratio ranging from 2 to 0.[68] Interestingly, it appears that oxygen release is expected for all the 3*d* transition metal ions, and that the reversibility of the anionic redox would be reached only for Li-rich layered oxides containing the heavier 4*d* and 5*d* transition metals as in that case the antibonding (MO*) states of the d-band and the pure O(2p) localized states of the p-band are expected to be both at the Fermi level. The presence of the O(2p) states only at the Fermi level leads to the formation of pure oxidized oxygen species, either as radicals $(O^-)^\bullet$ or as peroxides $(O_2)^{2-}$. They are highly unstable and reactive, and thus released leading to the irreversibility of the anionic redox.

Recently, complementary insights were presented by Bruce and coworkers[69] and Ceder and coworkers[70] to understand the role of oxygen in the redox activity of typical lithium-ion battery electrode materials such as $Li_{1.2}Mn_{0.54}Ni_{0.13}Co_{0.13}O_2$.[71] Combining different spectroscopies [soft X-ray absorption spectroscopy (SXAS), resonant inelastic X-ray scattering spectroscopy (RIXS), Raman spectroscopy, and X-ray absorption near edge structure (XANES)] to probe directly the oxygen states, lithium deintercalation on the plateau was shown to be compensated through the formation of localized electron holes on the oxygen anions coordinating mainly the Mn^{4+} and Li^+ in excess in the slabs.[69] Indeed, only 9% of the capacity delivered on the plateau is compensated by oxygen loss. Furthermore, the elastic peak observed from the beginning of the plateau in the RIXS spectra, with an excitation energy of 531.8 eV and an increasing intensity, indicates the formation of localized holes on the oxygen atoms (Figure 11), whereas the transition metal K-edge XANES spectra reveal that the oxygen anions reversibly involved in the redox processes are in the vicinity of Mn^{4+} and Li^+ that form more ionic bonds with oxygen than Co^{4+} and Ni^{4+}. Raman spectroscopy, which is highly sensitive to the formation of peroxo-like species, reveals that true peroxo-like species are not formed during lithium deintercalation from $Li_{1.2}Mn_{0.54}Ni_{0.13}Co_{0.13}O_2$. Seo *et al.* studied using DFT how important is the nature of the cations localized around the oxygen anions in order to promote their participation to the redox processes[70] by varying the amount of Li^+ ions around an oxygen from two (in the interslab space) to four (three in the interslab space and one in excess in the slab). In fact, the oxygen anions involved

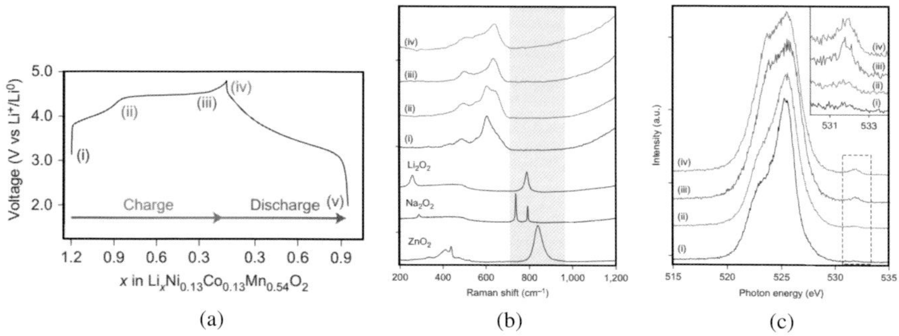

Figure 11. (a) First charge and discharge obtained for a lithium cell Li// $Li_{1.2}Mn_{0.54}Ni_{0.13}Co_{0.13}O_2$ and compositions recovered at different states of charge or discharge [labelled (I) to (V)]; (b) the Raman spectra of compositions (I) to (IV) compared to those of three reference peroxide samples Li_2O_2, Na_2O_2, and ZnO_2; and (c) the RIXS spectra of the same compositions with given in inset the excitation band observed at 531.8 eV. (Adapted from Ref. 69).

in a 180° Li_{Slab}-O-$Li_{Interslab}$ configuration are those prompted to be oxidized, since, as also discussed by Xie *et al.*, their energy is increased and shifted close to the antibonding transition metal d states because along that direction the oxygen 2p orbitals are not hybridized with transition metal d orbitals. Depending on the nature of the transition metal ion M in the immediate environment of oxygen $O(Li_4M_2)$, there is a redox activity of the cations and/or of the anions.

4. *Perspectives for Reversible Redox on Oxygen Ions*

It appears that in these Li-rich 2D structures, it is possible to form a three-dimensional percolating network of vacancies bringing into play the slabs and interslab spaces to ensure the lithium diffusion and to optimize capacity and performances at high rate: $x = 1.09$ would be the percolation threshold for $Li_xM_{2-x}O_2$ (whatever is its structure: layered, spinel, disordered, or rock-salt), while $x > 1.22$ would allow reaching a reversible capacity of 1 Li^+ per transition metal.[72] Combining anionic redox with this infinity of materials known as "Li-rich" opens new horizons for the development of high-energy-density positive electrode materials for lithium-ion batteries. A large panel of compositions enriched in lithium has been already explored: for instance, cation disordered rock-salt structures such as $Li_{1.211}Mo_{0.467}Cr_{0.3}O_2$,[72] $Li_{1.33}V_{0.67}O_{1.33}F_{0.67}$,[73] $(1 - y)Li_3NbO_4 \cdot yLiMO_2$ ($M = Fe^{3+}$, Mn^{3+}, V^{3+}),[74–76] $Li_{1.2}Ti_{0.4}Mn_{0.4}O_2$[76] and Li_4MoO_5[77]; the three-dimensional ordered β-Li_2IrO_3[78]; and the nanostructured rock-salt type $Li_4Mn_2\{O_5\square_1\}$.[79] As for the Li-rich layered oxides, the charge compensation for lithium deintercalation depends on the nature of the transition metal ions in $Li_xM_{2-x}O_2$: for instance, for $Li_{1.3}Nb_{0.3}\mathbf{Fe}_{0.4}O_2$ only oxygen participates to the redox processes, for $Li_{1.3}Nb_{0.3}\mathbf{Mn}_{0.4}O_2$ manganese and oxygen are involved, whereas for $Li_{1.3}Nb_{0.3}\mathbf{V}_{0.4}O_2$ a two-electron cation redox occurs.[76] Interestingly, a Nb-free composition $Li_{1.2}Ti_{0.4}Mn_{0.4}O_2$ was identified to deliver (at 50°C) a high reversible capacity of 300 mAh/g and energy density of 1000 Wh/kg, the latter exceeding that expected for Li_2MnO_3. The results recently reported for β-Li_2IrO_3 have also confirmed not only that reversible anionic redox is possible in three-dimensional structures but also that it can be activated at potentials as low as 3.4 V *versus* Li^+/Li.[78] Furthermore, the three-dimensional structure would be beneficial as it allows, at least for β-Li_2IrO_3, a complete lithium deintercalation without any structural reorganization on cycling. $Li_{1.2}Ti_{0.4}Mn_{0.4}O_2$ (at 50°C) and β-Li_2IrO_3 (with a reversible capacity of 200 mAh/g) appear thus as proofs of concept that open the road for a wider prospection of three-dimensional oxides enriched in lithium, involving anionic redox in stable oxide structures of positive electrode materials for high-energy lithium-ion batteries.

B. The Lithium and Manganese-Rich Layered Oxides as Composite Materials

One of the most challenging, but also promising, positive electrode materials remains the Li and Mn-rich, as its introduction in the next generation of lithium-ion batteries would directly induce a 25% increase in energy density for the full cell. Nevertheless, as discussed just before, its stability is still too limited to consider it ready for commercialization. Two major degradation phenomena can be observed: (i) an irreversible modification of the material on cycling, depending on the composition it is spread into the bulk or strictly limited to the interface with the electrolyte. A partial transition metal dissolution (mainly manganese) can be also observed. (ii) The oxidation of the electrolyte at high voltage that can induce the formation of a resistive layer at the electrode–electrolyte interface and a drastic loss of capacity. The use of the Li-rich layered oxides in lithium-ion batteries is thus facing challenging problems to be solved. Different tracks can be followed, such as (i) driving the potential slightly down through partial substitution in order to mitigate all the electrolyte degradation reactions; (ii) tailoring the particle morphology in order to promote the preferential growth of less reactive crystallographic oriented surfaces[56]; (iii) modifying the surface of the active material through the formation of an inorganic or organic coating, electrochemically active or not, formed before or on cycling through the addition of additives in the electrolyte, in order to stabilize the structure at the surface of the active material and to limit the reactivity at the solid–liquid interface; (iv) covering the surface of the electrode in order to minimize any degradation reaction with any component of the electrode (active material, carbon additive, and polymeric binder); and (v) prospecting material engineering through the formation of core-shell or concentration gradients. The investigation of these five tracks is interesting as fundamental science and understanding are required in order to step further in the direction of the material engineering in the field of materials for lithium-ion batteries. Furthermore, the benefit of the protection of the active surfaces or of material engineering can be a partial answer not only to surface reactivity, but also to limitations in thermal stability and rate capability.

1. *The Benefit of Blends and Coatings*

The blends used as positive electrode materials for lithium-ion batteries are macroscopic mixtures of active materials, that is, made at the scale of the aggregates. Nevertheless, they have already been shown to answer many of the concerns faced by the complex chemistry involved in a lithium-ion battery, with electrode materials and electrolytes being often at the limit of their chemical stability. They lead in some cases to the reduction of the irreversible capacity and to the increase of the cyclability, energy and power densities, and thermal stability of the batteries.[80,81] For instance, mixtures made of the spinel phase $Li_{1+x}Mn_{2-x}O_4$ (the main phase) and of a layered oxide such as $LiCoO_2$,[82] $LiNi_{1/3}Mn_{1/3}Co_{1/3}O_2$,[83] or $LiNi_{0.8}Co_{0.2}O_2$[84] (the minor phase) have been shown to mitigate the manganese dissolution within the electrolyte and thus to improve the electrochemical performances. These layered oxides would act as scavengers of H^+ formed on cycling, before the detrimental formation of HF at the origin of the transition metal dissolution. The combination of the carbon-coated olivine $LiFePO_4$ with layered oxides has also been reported to be efficient in high rate conditions, with $0.5Li_2MnO_3.0.5LiNi_{0.44}Mn_{0.31}Co_{0.25}O_2$,[85] $LiCoO_2$,[86] and $Li_{1.17}Ni_{0.25}Mn_{0.58}O_2$[87] as layered oxides. The benefit of blends is that the total weight of the material is electrochemically active, nevertheless the nature of the mixture or engineering of that blend (a simple mixture, a mixture layer by layer, etc.) can by itself have a huge impact on the actual properties.

To reach the attractive reversible capacity of lithium-rich layered oxides the cycling has to be performed beyond 4.6 V *versus* Li^+/Li, in a potential window in which the stability of the carbonate-based liquid electrolytes is limited and thus detrimental to the surface stability of the active material, for instance, *versus* transition metal dissolution. The development of active materials with a coating is thus an interesting strategy. Depending on their chemical nature, they can act as scavengers for HF and prevent any surface dissolution,[88] or they can mitigate the catalytic effects of oxidized transition metal ions *versus* electrolyte decomposition. They can also improve the electronic conductivity of the composite and thus the performance at higher cycling rates, or the thermal stability in the charged state of the battery. Note that whatever its nature and morphology, the coating must not be detrimental to the lithium diffusion through the active material and electrode.[89,90] These inorganic coatings can be oxides such as Al_2O_3,[88–90] CuO,[91] MgO,[92] Li_3VO_4,[93] ZrO_2,[94,95] and so on or fluorides such as CaF_2[96] and AlF_3. These coatings can be differently prepared and distributed over the active material.[92] The coverage can be total as a thin layer at the surface of the single particles present, either at the surface of the aggregates only or even within the aggregates (with thus a penetration in all the grain boundaries). Deposition methods revealed to be the technique of choice to form uniform films: the chemical vapor deposition (CVD) allows the formation of a very homogenous coating impregnating the surface and the porosity of the powders/electrodes but with a low deposition rate; the pulsed laser deposition method (PLD) and the atomic layer deposition (ALD[7]) are much more adapted to prepare these coatings as the former is already used to prepare solid electrolytes in all-solid-state microbatteries, and the latter is widely used in the OLED's field to prepare at low temperature thin layers as barriers against water permeation, with controlled thickness and composition without defects. The coverage can also be partial, with small particles at the surface of the aggregates and within the grain boundaries. Depending on the synthesis method used to prepare this coating, either it is a second phase next to the active material, or it reacts partially during its formation at the surface of the active material to form a composite coating-interphase-active material, the interphase resulting from the partial interdiffusion of the cations between the coating and the active material. Depending on the reactivity of the phase chosen for the coating and despite an often limited covering of the surface, the benefit can be significant and "total" as for instance HF reacts preferentially with these particles of large surface area that acts as scavengers and stabilize the material.

Most of the coatings are dead mass, and their weight has thus to be limited in order to maintain a good reversible capacity delivered by the electrode and, as discussed already, not to be at the origin of an increasing resistive layer within the electrode, blocking the lithium diffusion and electron percolation. Their weight is often limited to a few percents. In order to overpass this drawback, new systems are under development, such as core-shells or concentration gradients, in which all the material is active, from the surface to the bulk. The goal here is thus to combine the good properties from the phase(s) present within the bulk with those from the phase(s) present more at the surface, in the external part of the aggregates or of the single particles, in order to develop an optimized material combining energy and power densities as well as thermal stability.[97–100] The main problems faced by these composites are (i) the possible strong volume variations that can be observed on cycling between the different components of the composite, which could be at the origin of decohesion within the aggregates and electrodes,[101,102] and (ii) the complex syntheses that required to be developed to control their composition, morphology, and tap density. In the following section, the focus of the discussion is on the synthesis of these new materials formed from material engineering.

2. *The Concentration Gradients*

The full concentration gradients (FCGs) were initiated for the first time for the layered oxides in the field of positive electrode materials for lithium-ion batteries by Sun *et al.* in 2009.[103] These compounds are prepared as spherical aggregates with interlocked primary particles. The composition of the particles present at the surface is different from that of the particles present in the bulk, with a continuous change in the transition metal concentrations from the surface to the bulk as schematized in Figure 12. The target of such an engineering is to combine different properties within the same materials, fundamental properties for the development of optimized positive electrode materials: high energy and/or power density, chemical and thermal stability, and so on.

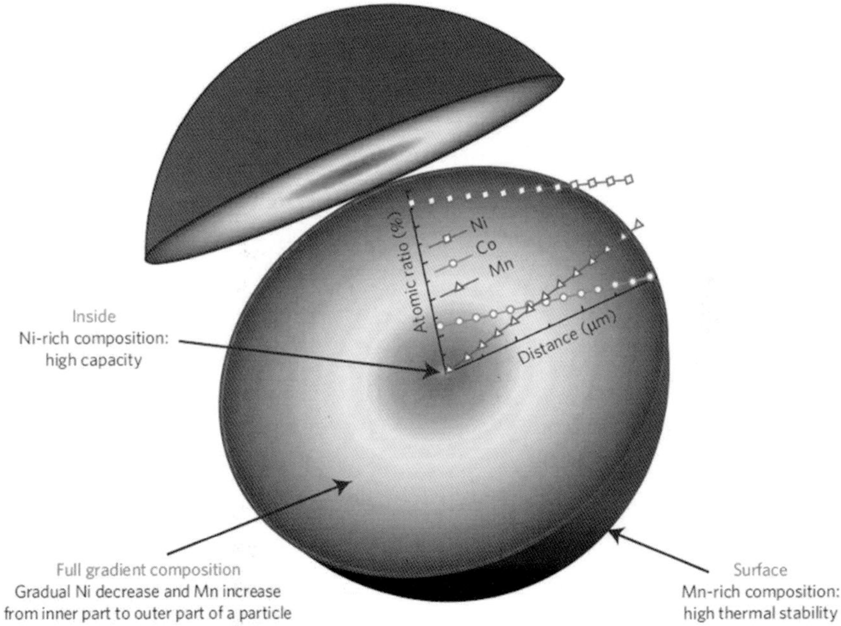

Figure 12. Schematic representation of a spherical aggregate for the layered oxide of average composition $LiNi_{0.75}Mn_{0.15}Co_{0.10}O_2$ and showing a concentration gradient: the latter implies a decrease in the concentration of Ni and Co from the bulk to the surface, and on the contrary an increasing concentration in Mn. (Adapted from Ref. 104).

a. *Their synthesis by coprecipitation*

Most of the concentration gradients were obtained for the Ni-rich layered oxides. Only few syntheses were reported for Li and Mn-rich layered oxides. All the concentration gradients were obtained by coprecipitation. Depending on the target composition and structure, the synthesis conditions can vary with the pH, nature and concentration of the precursors in the pristine preparations, feeding rates of the transition metal preparations within the precipitation solution, temperature of the reactor, stirring rate and temperature of the thermal treatment performed to obtain in a second step the oxide. The tables given hereafter summarize a large panel of compositions, with the synthesis conditions used to prepare them as concentration gradients: the nickel-rich layered oxides are given in Tables 1 and 2 and the manganese-rich layered oxides in Table 3. The intermediate compound formed from the precipitation can be a mixed carbonate or hydroxide, prepared with the targeted concentration gradient in transition metals. The pH required for the precipitation is of the order of 11 for the nickel-rich layered oxides and of 7.5–8 for the manganese-rich layered oxides. The temperature of the thermal treatment applied to obtain the layered oxides is close to 750°C for the Ni-rich and above 900°C for the Co- and Mn-rich. Furthermore, the diameter of the aggregates is typically of the order of 8–15 μm, a size big enough to control the formation of the concentration gradient; the size of the primary particles being often well below 1 μm. That synthesis process in solution allows an intimate mixing of all the precursors and favors a better statistical and homogeneous distribution of the transition metals at the atomic scale. Two pristine sulfate solutions are prepared, the first one corresponding to the composition of the surface. During the precipitation, the first sulfate solution is introduced continuously in the second one, which is then introduced into the reactor. All the solutions are maintained under stirring. The concentration of the transition metal ions in the second solution is thus continuously modified in order to prepare the gradient. Separately, a basic solution made of ammonium hydroxide (NH_4OH), used as chelating agent, and of sodium carbonate ($NaCO_3$) [or sodium hydroxide (NaOH)], used as precipitating agent, is added in the reactor such as to maintain the pH to the target value that allows the precipitation of Ni, Mn, and Co species all

Table 1. Summary of the conditions used to prepare concentration gradients of layered oxides rich in nickel as reported in different papers

Reference	Target average composition (reported as a number in Figure 13)	Bulk composition	Surface composition	Tank volume (L)	Concentrations of the transition metal solution and the basic solutions	pH	Stirring conditions (temperature, speed, atmosphere)	Calcination: temperature, time, (lithium precursor)	Size of the aggregates (*primary particles*) in the intermediate precursor/oxide
103	$LiNi_{0.64}Co_{0.18}Mn_{0.18}O_2$ **(1)**	$LiNi_{0.80}Co_{0.10}Mn_{0.10}O_2$	$LiNi_{0.46}Co_{0.23}Mn_{0.31}O_2$					780°C, 20 h ($LiNO_3$)	14 μm/14 μm
105	$LiNi_{0.72}Co_{0.18}Mn_{0.10}O_2$ **(2)**	$LiNi_{0.80}Co_{0.20}O_2$	$LiNi_{0.55}Co_{0.15}Mn_{0.30}O_2$					750°C, 20 h (LiOH)	14 μm/14 μm
106	$LiNi_{0.67}Co_{0.15}Mn_{0.18}O_2$ **(3)**	$LiNi_{0.80}Co_{0.15}Mn_{0.05}O_2$	$LiNi_{0.57}Co_{0.15}Mn_{0.28}O_2$			11	50°C, 1000 rpm	750°C, 20 h (LiOH)	15 μm (*few nm*)/15 μm (*~1 μm*)
107	$LiNi_{0.83}Co_{0.07}Mn_{0.10}O_2$ **(4)**	$LiNi_{0.86}Co_{0.05}Mn_{0.09}O_2$	$LiNi_{0.68}Co_{0.12}Mn_{0.20}O_2$	4	2 mol/L 2 mol/L NaOH		N_2	730°C, 15 h (LiOH)	12 μm (*few nm*)/12 μm (*< 0.5 μm*)
104	$LiNi_{0.75}Co_{0.10}Mn_{0.15}O_2$ **(5)**	$LiNi_{0.70}Co_{0.10}Mn_{0.20}O_2$	$LiNi_{0.86}Co_{0.10}Mn_{0.04}O_2$					750°C, 20 h (LiOH)	12 μm/10 μm
108	$LiNi_{0.54}Co_{0.16}Mn_{0.30}O_2$ **(6)**	$LiNi_{0.66}Co_{0.04}Mn_{0.30}O_2$	$LiNi_{0.51}Co_{0.19}Mn_{0.30}O_2$	4	4 mol/L NaOH		N_2	875°C ,10 h (LiOH)	12 μm (*1–3 μm*)
109	$LiNi_{0.64}Co_{0.16}Mn_{0.20}O_2$ **(7)** $LiNi_{0.59}Co_{0.16}Mn_{0.25}O_2$ **(8)** $LiNi_{0.51}Co_{0.16}Mn_{0.33}O_2$ **(9)**	$LiNi_{0.75}Co_{0.05}Mn_{0.20}O_2$ $LiNi_{0.70}Co_{0.05}Mn_{0.25}O_2$ $LiNi_{0.62}Co_{0.05}Mn_{0.33}O_2$	$LiNi_{0.62}Co_{0.18}Mn_{0.20}O_2$ $LiNi_{0.57}Co_{0.18}Mn_{0.25}O_2$ $LiNi_{0.48}Co_{0.19}Mn_{0.33}O_2$	4	3 mol/L NaOH		N_2	780°C, 10 h (LiOH) 845°C, 10 h (LiOH) 920°C, 10 h (LiOH)	12 μm (*1–3 μm*)

(*Continued*)

Table 1. (*Continued*)

Reference	Average composition (reported as a number in Figure 13)	Bulk composition	Surface composition	Tank volume (L)	Concentrations of the transition metal solution and the basic solutions	pH	Stirring conditions (temperature, speed, atmosphere)	Calcination: temperature, time, (lithium precursor)	Size of the aggregates (*primary particles*) in the intermediate precursor/oxide
110	$LiNi_{0.60}Co_{0.15}Mn_{0.25}O_2$ **(10)**	$LiNi_{0.70}Co_{0.10}Mn_{0.20}O_2$	$LiNi_{0.50}Co_{0.20}Mn_{0.30}O_2$	4	2 mol/L 4 mol/L NaOH	11.7	1000 rpm, N_2	850°C, 10 h (LiOH)	12 μm (*100–300 nm*)/12 μm
111	$LiNi_{0.65}Co_{0.08}Mn_{0.27}O_2$ **(11)**	$LiNi_{0.89}Co_{0.01}Mn_{0.10}O_2$	$LiNi_{0.61}Co_{0.09}Mn_{0.30}O_2$		4 mol/L NaOH		N_2	840°C, 10 h (LiOH)	10 μm/10 μm
112	$LiNi_{0.60}Co_{0.15}Mn_{0.25}O_2$ **(10)**	$LiNi_{0.70}Co_{0.05}Mn_{0.25}O_2$	$LiNi_{0.58}Co_{0.17}Mn_{0.25}O_2$	4	1.2 mol/L 4 mol/L NaOH 2 mol/L NH$_4$OH			825°C, 10 h (LiOH)	12 μm (*1–3 μm*)
	$LiNi_{0.59}Co_{0.16}Mn_{0.25}O_2$ **(8)**	$LiNi_{0.70}Co_{0.05}Mn_{0.25}O_2$	$LiNi_{0.57}Co_{0.18}Mn_{0.25}O_2$					845°C, 10 h (LiOH)	
	$LiNi_{0.56}Co_{0.19}Mn_{0.25}O_2$ **(12)**	$LiNi_{0.70}Co_{0.05}Mn_{0.25}O_2$	$LiNi_{0.55}Co_{0.20}Mn_{0.25}O_2$					875°C, 10 h (LiOH)	
113	$LiNi_{0.60}Co_{0.20}Mn_{0.20}O_2$ **(13)**	$LiNi_{0.72}Co_{0.17}Mn_{0.11}O_2$	$LiNi_{0.61}Co_{0.15}Mn_{0.24}O_2$	4	2 mol/L	11.5	55°C, 1000 rpm, N_2	500°C, 10 h to 700°C, 800°C, 875°C, 15 h (LiOH)	10 μm (*<0.5 μm*)/10 μm (*~1 μm*)
114	$LiNi_{0.65}Co_{0.13}Mn_{0.22}O_2$ **(14)**	$LiNi_{0.72}Co_{0.11}Mn_{0.17}O_2$	$LiNi_{0.60}Co_{0.12}Mn_{0.28}O_2$	40	4 mol/L NaOH 2 mol/L NH$_4$OH		N_2	830°C, 10 h (LiOH)	
115	$LiNi_{0.80}Co_{0.06}Mn_{0.14}O_2$ **(15)**	$LiNi_{0.96}Co_{0.01}Mn_{0.03}O_2$	$LiNi_{0.64}Co_{0.06}Mn_{0.30}O_2$	40	4 mol/L NaOH 2 mol/L NH$_4$OH		N_2	760°C, 15 h under O_2 (LiOH)	10 μm (*<0.5 μm*)/10 μm (*few nm*)
116	$LiNi_{0.333}Co_{0.333}Mn_{0.333}O_2$ **(16)**	Rich in Co but poor in Li	Rich in Li but poor in Co			11	50°C, 1500 rpm, N_2	950°C, 16 h (Li_2CO_3)	5–6 μm (*<0.5 μm*)/5–6 μm (*1 μm*)

	$LiNi_{0.40}Co_{0.20}Mn_{0.40}O_2$ **(17)** $LiNi_{0.416}Co_{0.168}Mn_{0.416}O_2$ **(18)** $LiNi_{0.45}Co_{0.10}Mn_{0.45}O_2$ **(19)**								5–6 μm (*<0.5 μm*)/5–6 μm (*1 μm*) 5–6 μm (*<0.5 μm*)/5–6 μm (*1 μm*) 3–4μm (<0.5μm)/3–4 μm (1 μm)
117	$LiNi_{0.85}Co_{0.12}Mn_{0.03}O_2$ **(20)**	$LiNi_{0.90}Co_{0.10}O_2$	$LiNi_{0.60}Co_{0.25}Mn_{0.15}O_2$		1.5 mol/L 3 mol/L NaOH 2 mol/L NH_4OH	11.5	50°C, 600 rpm, N_2	550°C, 4 h to 750°C, 12 h (LiOH)	5–17 μm (*few nm*)/5–17 μm (*<0.5 μm*)
118	$LiNi_{0.80}Co_{0.10}Mn_{0.10}O_2$ **(21)**	$LiNi_{0.60}Co_{0.20}Mn_{0.20}O_2$	$LiNiO_2$		1.5 mol/L 4 mol/L NaOH 2 mol/L NH_4OH	11.5	50°C, 750 rpm, N_2	550°C, 4 h to 750°C, 12 h (LiOH)	8–12 μm (*few nm*)/8–12 μm (*<0.5 μm*)
119	$LiNi_{0.60}Co_{0.20}Mn_{0.20}O_2$ **(13)**	$LiNi_{0.83}Co_{0.17}O_2$	$LiNi_{0.22}Co_{0.17}Mn_{0.60}O_2$	10	4 mol/L NaOH		N_2	550°C, 4 h to 820°C, 12 h under O_2 (LiOH)	10 μm (*<1 μm*)/10 μm (*~1 μm*)
120	$LiNi_{0.70}Co_{0.10}Mn_{0.20}O_2$ **(22)**	$LiNiO_2$	$LiNi_{0.51}Co_{0.16}Mn_{0.33}O_2$	10	1.5 mol/L 4 mol/L NaOH		N_2	550°C, 4 h under air to 810°C, 12 h under O_2 (LiOH)	10 μm (*0.5 μm*)/10 μm (*0.5 μm*)
121*	$LiNi_{0.60}Co_{0.20}Mn_{0.20}O_2$ **(13)**	$LiNi_{0.78}Mn_{0.02}Co_{0.20}O_2$	$LiNi_{0.40}Co_{0.20}Mn_{0.38}O_2$	10	2 mol/L 10 mol/L NaOH 2 mol/L NH_4OH	11.6	50°C, 800 rpm, N_2	800°C, 16 h (LiOH)	10 μm (*<0.5 μm*)/10 μm (*1 μm*)

(*Continued*)

Table 1. (*Continued*)

Reference	Average composition (reported as a number in Figure 13)	Bulk composition	Surface composition	Tank volume (L)	Concentrations of the transition metal solution and the basic solutions	pH	Stirring conditions (temperature, speed, atmosphere)	Calcination: temperature, time, (lithium precursor)	Size of the aggregates (*primary particles*) in the intermediate precursor/oxide
122	$LiNi_{0.50}Co_{0.20}Mn_{0.30}O_2$ **(23)**	$LiNi_{0.60}Co_{0.20}Mn_{0.20}O_2$	$LiNi_{0.44}Co_{0.20}Mn_{0.36}O_2$	170	2.5 mol/L 10 mol/L NaOH 1.5 mol/L NH_4OH	11	50°C, 500 rpm, N_2	750°C to 900°C (Li_2CO_3)	10 μm (*0.5 μm*)/10 μm (*0.5 μm*)
123	$LiNi_{0.50}Co_{0.20}Mn_{0.30}O_2$ **(23)**	$LiNi_{0.60}Co_{0.20}Mn_{0.20}O_2$	$LiNi_{0.42}Co_{0.20}Mn_{0.38}O_2$		2 mol/L 1.5 mol/L NH_4OH	~11.5	600 rpm	900°C, 2 h to 800°C, 10 h (Li_2CO_3)	10 μm (*0.5 μm*)
124	$LiNi_{0.62}Co_{0.14}Mn_{0.24}O_2$ **(24)**	$LiNi_{0.80}Co_{0.20}O_2$	$LiNi_{0.20}Mn_{0.80}O_2$		2 mol/L 2 mol/L NaOH 0.5 mol/L NH_4OH	11	N_2	800°C, 15 h (LiOH)	3 μm (*~0.5 μm*)
125	$LiNi_{0.73}Co_{0.12}Mn_{0.15}O_2$ **(25)**	$LiNi_{0.8}Co_{0.1}Mn_{0.1}O_2$	$LiNi_{0.4}Co_{0.2}Mn_{0.4}O_2$			11	N_2	480°C, 5 h to 750°C, 20 h (LiOH)	12 μm (*100–200 nm*)

Some conditions are missing because they were not reported by the authors themselves.

*Feeding rates of the precursors 0.8 L/h for $(Ni,Co)SO_4$ and 0.4 L/h for $(Ni,Co,Mn)SO_4$.

Table 2. Summary of the conditions used to prepare concentration gradients of layered oxides rich in nickel, as reported in patents available in English

Reference	Average composition (reported as a number in Figure 13)	Bulk composition	Surface composition	Tank volume (L)	Concentrations of the transition metal solution and the basic solutions	Feeding rate of solutions	pH	Stirring conditions (temperature, speed, atmosphere)	Calcination: temperature, time (lithium precursor)	Size of the aggregates (*primary particles*) in the intermediate precursor/oxide
126	$LiNi_{0.64}Co_{0.24}Mn_{0.12}O_2$ **(26)** $LiNi_{0.71}Co_{0.13}Mn_{0.16}O_2$ **(27)** $LiNi_{0.70}Co_{0.08}Mn_{0.22}O_2$ **(28)**	$LiNi_{0.80}Co_{0.13}Mn_{0.07}O_2$ $LiNi_{0.80}Co_{0.10}Mn_{0.10}O_2$ $LiNi_{0.80}Co_{0.10}Mn_{0.10}O_2$ $LiNi_{0.80}Co_{0.13}Mn_{0.05}Al_{0.05}O_2$	$LiNi_{0.40}Co_{0.40}Mn_{0.20}O_2$ $LiNi_{0.50}Co_{0.20}Mn_{0.30}O_2$ $LiNi_{0.475}Co_{0.05}Mn_{0.475}O_2$ $LiNi_{0.40}Co_{0.30}Mn_{0.15}Al_{0.15}O_2$	4	2.4 mol/L	0.3L/h (sulfates) 0.03 L/h (NH_4OH)	11	50°C, 1000 rpm, N_2	500°C, 10 h to 780°C, 20 h (LiOH)	10–30 μm/10–30 μm
127		$LiNi_{0.85}Co_{0.10}Mn_{0.05}O_2$ $LiNi_{0.85}Co_{0.10}Mn_{0.25}O_2$ $LiNi_{0.85}Co_{0.10}Mn_{0.05}O_2$ $LiNi_{0.78}Co_{0.10}Mn_{0.02}O_2$ $LiNi_{0.80}Co_{0.10}Mn_{0.10}O_2$ $LiNi_{0.85}Co_{0.10}Mn_{0.05}O_2$	$LiNi_{0.60}Co_{0.1}Mn_{0.30}O_2$ $LiNi_{0.70}Co_{0.08}Mn_{0.22}O_2$ $LiNi_{0.70}Co_{0.05}Mn_{0.25}O_2$ $LiNi_{0.80}Co_{0.05}Mn_{0.15}O_2$ $LiNi_{0.70}Co_{0.05}Mn_{0.25}O_2$ $LiNi_{0.63}Co_{0.10}Mn_{0.27}O_2$	4	2.4 mol/L 4.8 mol/L NaOH 3.6 mol/L NH_4OH	0.3 L/h (sulfates) 0.03 L/h (NH_4OH)	11	50°C, 1000 rpm, N_2	280°C, 10 h to 750°C, 15 h ($LiNO_3$)	12 μm
128	$LiNi_{0.80}Co_{0.07}Mn_{0.13}O_2$ **(29)**	$LiNi_{0.80}Co_{0.10}Mn_{0.10}O_2$	$LiNi_{0.80}Mn_{0.20}O_2$	4	2.4 mol/L 4.8 mol/L NaOH 3.6 mol/L NH_4OH	0.3 L/h (sulfates) 0.03 L/h (NH_4OH)	11	50°C, 1000 rpm, N_2	280°C, 10 h to 750°C, 15 h ($LiNO_3$)	12 μm (*few nm*)/12 μm (*~1 μm*)

(Continued)

Table 2. (*Continued*)

Reference	Average composition (reported as a number in Figure 13)	Bulk composition	Surface co Mposition	Tank volume (L)	Concentrations of the transition metal solution and the basic solutions	Feeding rate of solutions	pH	Stirring conditions (temperature, speed, atmosphere)	Calcination: temperature, time (lithium precursor)	Size of the aggregates (*primary particles*) in the intermediate precursor/oxide
128	$LiNi_{0.80}Co_{0.07}Mn_{0.13}O_2$ **(29)**	$LiNi_{0.80}Co_{0.10}Mn_{0.10}O_2$	$LiNi_{0.80}Mn_{0.20}O_2$	4	2.4 mol/L 4.8 mol/L NaOH 3.6 mol/L NH_4OH	0.3 L/h (sulfates) 0.03 L/h (NH_4OH)	11	50°C, 1000 rpm, N_2	280°C, 10 h to 750°C, 15 h ($LiNO_3$)	12 μm (*few nm*)/12 μm (*~1 μm*)
	$LiNi_{0.75}Co_{0.07}Mn_{0.18}O_2$ **(30)**	$LiNi_{0.75}Co_{0.125}Mn_{0.125}O_2$	$LiNi_{0.75}Mn_{0.25}O_2$							
	$LiNi_{0.70}Co_{0.09}Mn_{0.21}O_2$ **(31)**	$LiNi_{0.70}Co_{0.15}Mn_{0.15}O_2$	$LiNi_{0.70}Mn_{0.30}O_2$	8	10 mol/L NaOH 4.8 mol/L NH_4OH	0.2 L/h (sulfates) 0.1 L/h (NH_4OH)		50°C, 450 rpm, N_2		
	$LiNi_{0.65}Co_{0.10}Mn_{0.25}O_2$ **(32)**	$LiNi_{0.65}Co_{0.175}Mn_{0.175}O_2$	$LiNi_{0.65}Mn_{0.35}O_2$							
	$LiNi_{0.70}Co_{0.09}Mn_{0.21}O_2$ **(31)**	$LiNi_{0.70}Co_{0.15}Mn_{0.15}O_2$	$LiNi_{0.70}Mn_{0.30}O_2$							
	$LiNi_{0.72}Co_{0.10}Mn_{0.18}O_2$ **(33)**	$LiNi_{0.90}Co_{0.10}O_2$	$LiNi_{0.70}Co_{0.10}Mn_{0.30}O_2$	4	2.4 mol/L 4.8 mol/L NaOH 3.6 mol/L NH_4OH	0.3 L/h (sulfates) 0.03 L/h (NH_4OH)		50°C, 1000 rpm, N_2		
129		$LiNi_{0.88}Co_{0.05}Mn_{0.07}O_2$	$LiNi_{0.52}Co_{0.16}Mn_{0.32}O_2$	4		0.3 L/h (sulfates) 0.3 L/h (NH_4OH)		50°C, 1000 rpm, N_2	450°C, 5 h to 820°C, 10 h (LiOH)	
		$LiNi_{0.88}Co_{0.05}Mn_{0.07}O_2$	$LiNi_{0.60}Co_{0.12}Mn_{0.28}O_2$							
130		$LiNi_{0.65}Co_{0.1}Mn_{0.25}O_2$	$LiNi_{0.40}Co_{0.10}Mn_{0.50}O_2$	70	2.5 mol/L	6.9 L/h (sulfates) 0.48 L/h (NH_4OH)	11	50°C, 400 rpm	550°C, 10 h to 750°C, 20 h	8–13 *μm*
131	$LiNi_{0.80}Co_{0.10}Mn_{0.10}O_2$ **(21)**	$LiNi_{0.84}Co_{0.11}Mn_{0.05}O_2$	$LiNi_{0.78}Co_{0.10}Mn_{0.12}O_2$							10 *μm*

Table 3. Summary of the conditions used to prepare concentration gradients of layered oxides rich in manganese, as reported in the literature

Reference	Average composition (reported as a number in Figure 13)	Bulk composition	Surface composition	Concentrations of the transition metal sulfate solution and the basic solutions	Feeding rate of solutions (L/h)	pH	Stirring conditions (temperature, speed, atmosphere)	Calcination: temperature, time (lithium precursor)	Size of the aggregates (*primary particles*) in the intermediate precursor/ oxide
132	$Li_{1.2}(Mn_{0.62}Ni_{0.38})_{0.8}O_2$ **(34)**	$Li_{1.2}(Mn_{0.37}Ni_{0.63})_{0.8}O_2$	$Li_{1.2}(Mn_{0.62}Ni_{0.38})_{0.8}O_2$	2 mol/L 2 mol/L Na_2CO_3 0.2 mol/L NH_4OH	1	8	60°C, 500 rpm	600°C, 15 h to 900°C, 15 h ($LiCO_3$)	20 μm / 15 μm (< *0.5 μm*)
133	$Li_{1.13}(Ni_{0.233}Co_{0.233}Mn_{0.534})_{0.87}O_2$ **(35)**	$Li_{1.13}(Ni_{0.32}Co_{0.31}Mn_{0.37})_{0.87}O_2$	$Li_{1.13}(Ni_{0.27}Co_{0.25}Mn_{0.48})_{0.87}O_2$					500°C, 8 h to 900°C, 10 h (Li_2CO_3)	10 μm (*few nm*) / 10 μm (*~1μm*)
134	$Li_{1.14}(Ni_{0.25}Co_{0.15}Mn_{0.60})_{0.86}O_2$ **(36)**	$Li_{1.14}(Ni_{0.25}Co_{0.50}Mn_{0.25})_{0.86}O_2$	$Li_{1.14}(Ni_{0.25}Co_{0.10}Mn_{0.65})_{0.86}O_2$	1.5 mol/L 1.5 mol/L Na_2CO_3		7.5	55°C, 800 rpm	500°C, 8 h to 900°C, 16 h (Li_2CO_3)	8 μm (*few nm*) / 8 μm (*< 0.5μm*)
135	$Li_{1.5}Mn_{0.75}Ni_{0.20}Mg_{0.05}O_2$	$Li_{1.5}Mn_{0.75}Ni_{0.25}O_2$	$Li_{1.5}Mn_{0.75}Ni_{0.10}Mg_{0.15}O_2$	1.6 mol/L 1.6 mol/L Na_2CO_3		7.5	55°C, 600 rpm	500°C, 6 h to 850°C, 12 h (Li_2CO_3)	10 μm (*few nm*) / 10 μm (*< 0.5 μm*)

together. The solution thus obtained is then continuously stirred for few hours, before the precipitate is filtered, washed with deionized water, and dried. The mixed carbonate (or hydroxide) of transition metal ions thus obtained is treated at high temperature with a lithium precursor such as Li_2CO_3 or LiOH to form the lithiated oxides. In general, the nickel-rich layered oxides are calcined under N_2, whereas those rich in Mn are calcined under air. All the average compositions of the intermediates prepared as concentration gradients are shown in Figure 13, as crosses for the hydroxides and as circles for the carbonates, most of them being rich in Ni, with a nickel-rich core and a manganese-rich surface in order to combine within the same electrode material high capacity, cyclability, and thermal stability at the charged state of the battery. Indeed, the core rich in Ni brings the high capacity, whereas the surface or the external crown is responsible for the thermal stability. The concentration gradient in such an engineered material creates a synergy of properties.

b. *The attractive properties of the concentration gradients*

In the following, it is highlighted through a few examples taken from Sun *et al.* how beneficial the preparation of concentration gradients — or at least of core-shells — can be. The first example chosen is the layered oxide of average composition $LiNi_{0.64}Mn_{0.18}Co_{0.18}O_2$ **(1)** with the composition $LiNi_{0.8}Mn_{0.1}Co_{0.1}O_2$ (NMC 811) in the core and the composition $LiNi_{0.46}Mn_{0.31}Co_{0.23}O_2$ (NMC 532) at the surface, and with thus a decrease in the Ni content from the bulk to the surface and reverse for Co and Mn as shown in Figure 14.[103] The characterization of the extent of the gradient within the hydroxide intermediate and the final oxide is achieved through elemental quantitative analyses using electron-probe X-ray microanalysis (EPMA) of polished spherical aggregates. It is really critical as during the second step of the synthesis — that is, the sintering of the mixed carbonate or hydroxide with the lithium precursor — a high-temperature thermal treatment (above 750°C) induces interdiffusion of the transition metals within the aggregates, as already revealed by different authors.[112,117,119,136] A recent study conducted by Li *et al.* has shown that interdiffusion between Ni^{3+}/Mn^{4+} is the slowest even at high temperature *versus* Ni^{3+}/Co^{3+} and even Co^{3+}/Mn^{4+}.[137]

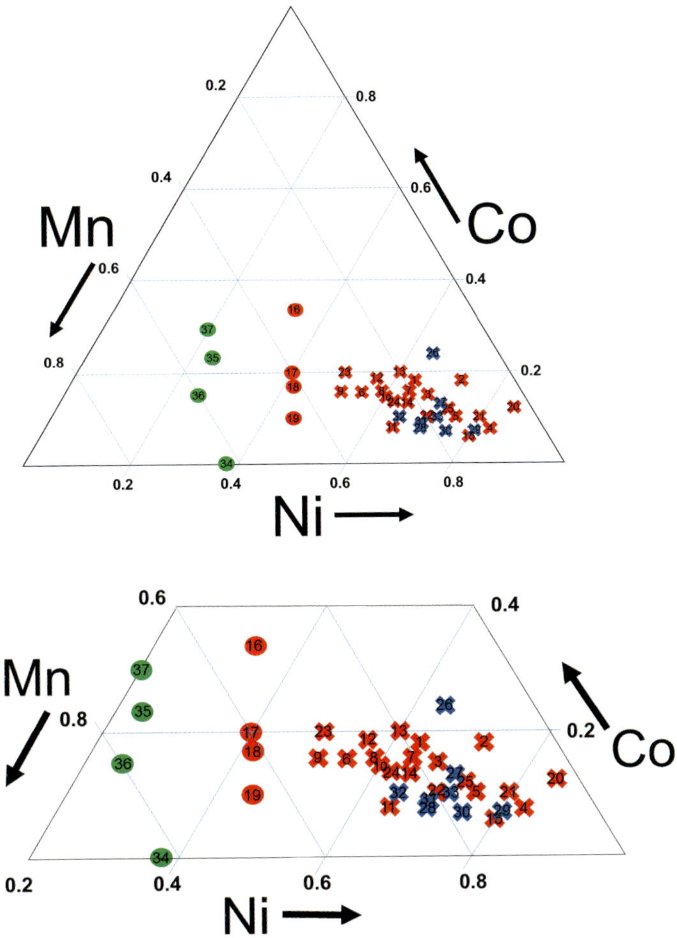

Figure 13. Phase diagram screened in the literature for the intermediate precursors rich in Ni, Mn, and Co. The crosses and circles correspond to hydroxides and carbonates, respectively, whereas the red, blue, and green colors are associated to the compositions given in Tables 1–3, respectively. The black numbers in the symbols correspond to the ones reported in the first columns of Tables 1–3.

In order to asset the beneficial effect of the concentration gradients, the electrochemical behavior of different materials were compared in full cells, at 55°C, a critical temperature to check for the stability and cyclability of the material (Figure 15): the compositions of the bulk

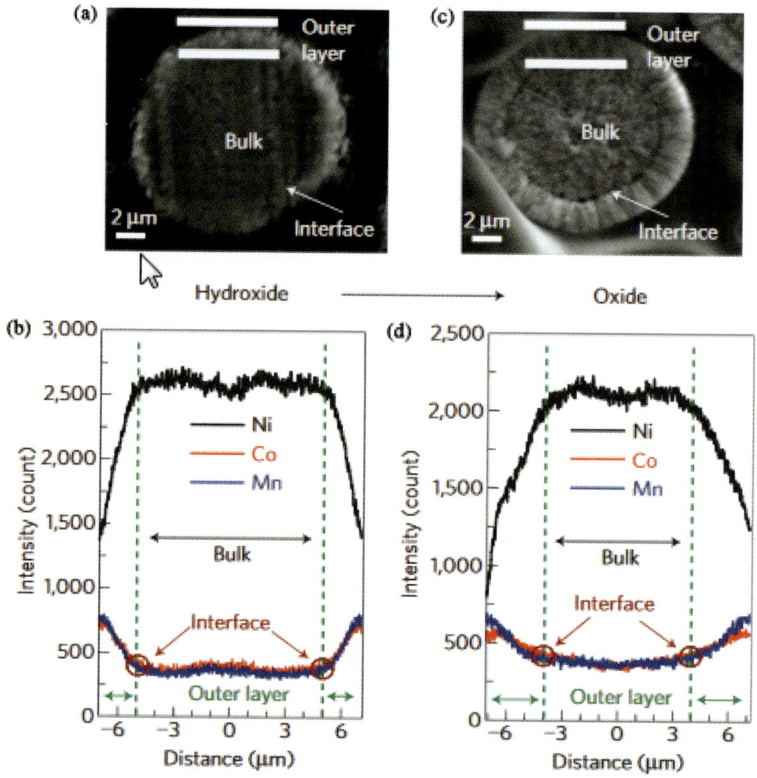

Figure 14. Scanning electron microscopy (SEM) analyses of polished surfaces of spherical aggregates and their elemental chemical analyses using EPMA: SEM photographs of the precursor hydroxide (a) and oxide (c), respectively, and the corresponding EPMA line scans in (b) and (d), respectively. In that example, the concentration gradient is limited to an outer layer of 1 µm in thickness, whose concentrations in Ni, Mn and Co change gradually as clearly shown in (b) and in (d). (Adapted from Ref. 103).

($LiNi_{0.8}Mn_{0.1}Co_{0.1}O_2$), of the surface ($LiNi_{0.46}Mn_{0.31}Co_{0.23}O_2$), and of the corresponding concentration gradient ($LiNi_{0.64}Mn_{0.18}Co_{0.18}O_2$). The FCG combines a good reversible capacity and cyclability, as expected from the composition of the core and a good thermal stability in its Li-poor deintercalated composition (i.e., the charged state of the battery) (Figure 16) as expected considering the composition enriched in Mn^{4+} at the surface of the aggregates. Indeed, the Mn-richer layered oxides are the most stable with oxygen release in their oxidized state, and they thus minimize all the degradation reactions with the electrolyte.[103]

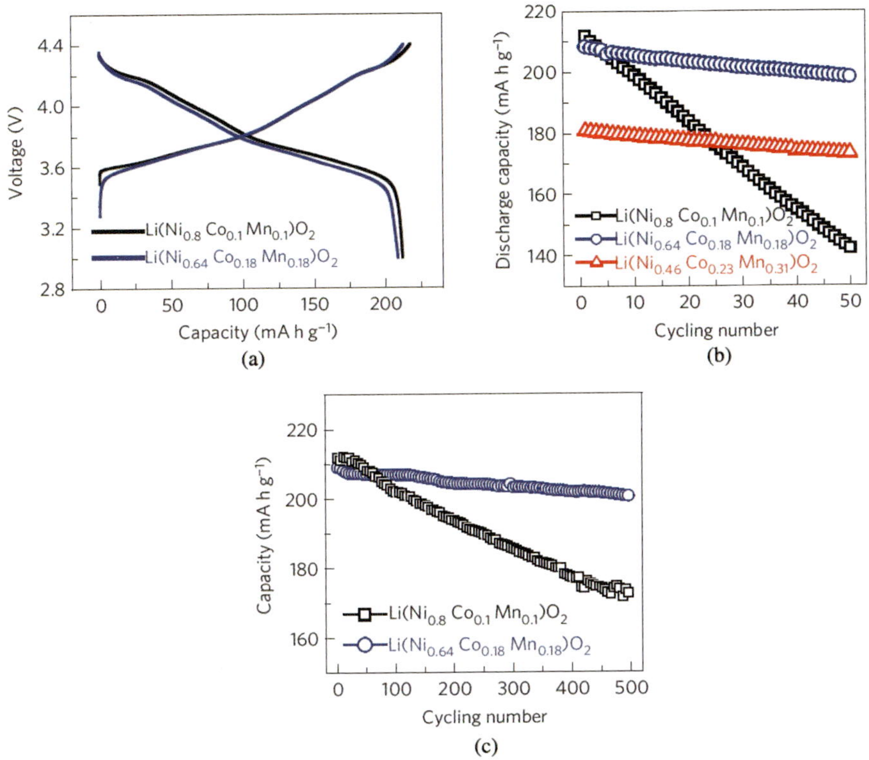

Figure 15. (a) Very similar electrochemical charge and discharge curves obtained during the first cycle *versus* Li^+/Li, at the rate of *C*/2 and at 55°C, for $LiNi_{0.8}Co_{0.1}Mn_{0.1}O_2$ (the material in the bulk) and for the concentration gradient of average composition $LiNi_{0.64}Co_{0.18}Mn_{0.18}O_2$. (b) Changes in the reversible capacity observed on cycling of $LiNi_{0.8}Co_{0.1}Mn_{0.1}O_2$ (the composition within the bulk), $LiNi_{0.46}Co_{0.23}Mn_{0.31}O_2$ (the composition at the surface), and $LiNi_{0.64}Co_{0.18}Mn_{0.18}O_2$ (the corresponding concentration gradient) in lithium batteries, between 3 and 4.4 V, at the rate of *C*/2 and at 55°C. (c) Changes in the reversible capacity obtained at the rate of 1*C* (higher rate) in pouch-cells with mesoporous graphite at the negative electrode, and either $LiNi_{0.8}Co_{0.1}Mn_{0.1}O_2$ (the composition of the bulk) or the concentration gradient at the positive electrode. The potential range is screened between 3 and 4.2 V *versus* Li^+/Li. The electrolyte used is 1 M $LiPF_6$ in a mixture of ethylene and diethyl carbonates (1:1 v/v). (Adapted from Ref. 103).

In fact, it is possible to tailor the gradient within a given composition domain limited by two extreme compositions, for the bulk and for the surface, and to tailor the properties. Indeed, in between, an infinity of average compositions exist and can be developed as concentration gradients. Recently, Lee

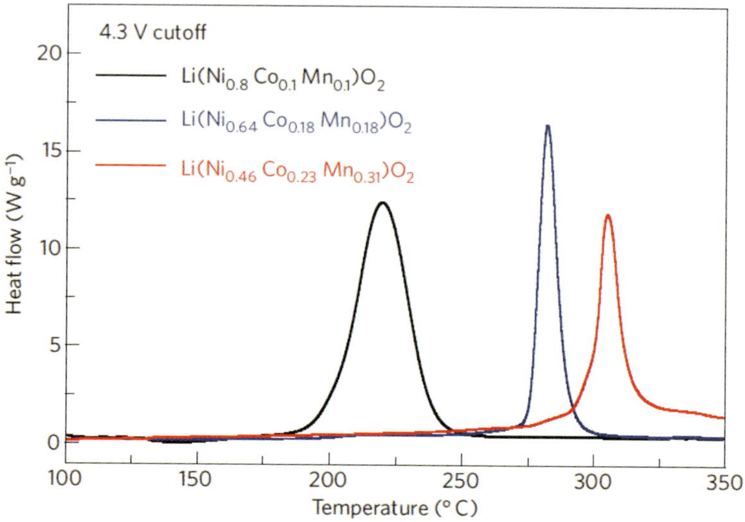

Figure 16. Comparison of the thermal stability of the deintercalated materials recovered at 4.3 V *versus* Li^+/Li, using differential scanning calorimetry analyses. The higher the heat flow the more exothermic the degradation reaction of the material with the electrolyte [1 M $LiPF_6$ EC/DEC (1:1)]. The materials are $Li_{1-\delta}Ni_{0.8}Co_{0.1}Mn_{0.1}O_2$ (the composition of the bulk), $Li_{1-\delta}Ni_{0.46}Co_{0.28}Mn_{0.31}O_2$ (the composition at the surface), and $Li_{1-\delta}Ni_{0.64}Co_{0.18}Mn_{0.18}O_2$ (the corresponding concentration gradient). (Adapted from Ref. 103).

et al.[112] for instance compared a series of concentration gradients such as $LiNi_{0.6-x}Co_{0.15+x}Mn_{0.25}O_2$ [$x = 0$ **(10)**, 0.01 **(8)**, and 0.04 **(12)**]. They have shown that the larger the contribution of the Mn-rich layered compositions, the worst are the performance at high rates and at low temperatures, as expected, as the presence of Mn^{4+} in the host structure is detrimental to transport properties. Nevertheless, the formation of a surface richer in Mn contributes to chemical and thermal stability of the deintercalated layered oxides obtained in the charged state of the batteries. As always, it appears that the optimized material is the result of combinations, which fit with the application targeted: a long range cycling high energy material will not be the best material to develop for power-demanding and low-temperature applications.

Only few strategies of material engineering have been applied to Li and Mn-rich layered oxides, all of them with the presence of the Mn-richer composition at the surface of the composites. Yang *et al.*[134] for

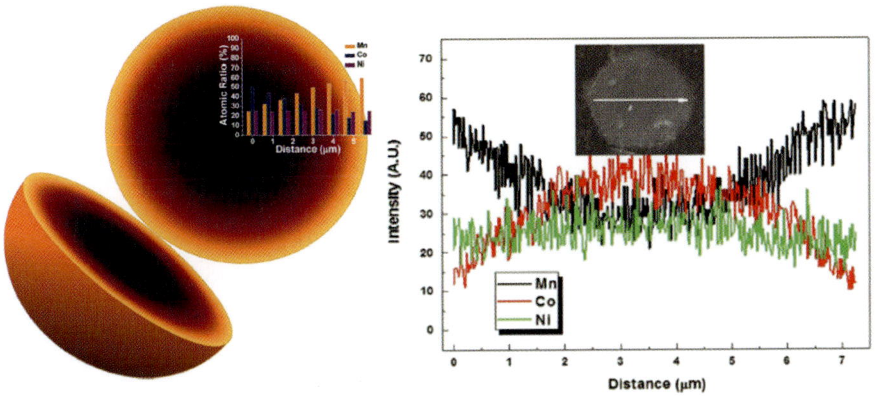

(a) (b)

Figure 17. (a) Schematic representation of a spherical aggregate for a concentration gradient Li-rich layered oxide characterized by an average composition of $Li_{1.14}Ni_{0.25}Mn_{0.60}Co_{0.15}O_2$. The theoretical change in the transition metal content is also given for Mn, Co, and Ni. (b) EDX analyses of a polished agglomerate, which shows a decreasing concentration of Co from the core to the surface, and reverse for Mn, the composition of Ni remaining constant. (Adapted from Ref. 134).

instance studied the concentration gradient of average composition $Li_{1.14}(Ni_{0.25}Co_{0.15}Mn_{0.60})_{0.86}O_2$ **(36)**, which is as shown in Figure 17 richer in Co in the bulk and richer in Mn at the surface. The changes in the transition metal content are continuous over the entire radii of the aggregates, that is, over 3.5 μm. This composite material delivers improved and very attractive performances *versus* those obtained from the corresponding homogeneous composition [i.e., a reversible capacity of ~200 mAh/g (2–4.6 V, *C*/2), and a capacity retention of 93.8% after 200 cycles *versus* a capacity decreasing from 200 mAh/g and a capacity retention of 66.3% for the homogeneous composition]. The composite behaves as if the Mn-rich shell could effectively protect the Ni and Co-rich core from degradation reactions with the electrolyte. The mechanisms involved are not obvious since the aggregates are porous, with thus the possible penetration of the electrolyte within the pores and grain boundaries. The characterization of this porosity and of the wettability by the electrolyte is not that easy even if it would be interesting to come to that understanding. As revealed in Figure 18, the voltage signature is more stable on cycling for

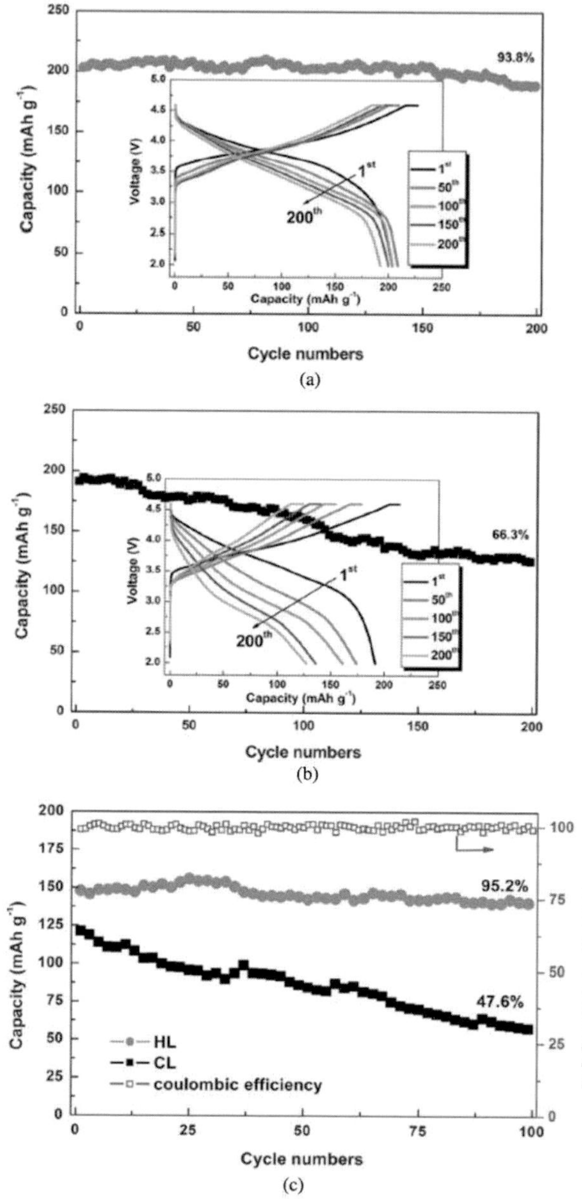

Figure 18. Performances on cycling of the material $Li_{1.14}Ni_{0.25}Mn_{0.60}Co_{0.15}O_2$ used as positive electrode in lithium cells. The potential window is 2–4.6 V and the electrolyte 1 M $LiPF_6$ in a mixture of ethylene and dimethyl carbonates (1:1 v/v). (a) $Li_{1.14}Ni_{0.25}Mn_{0.60}Co_{0.15}O_2$ is the concentration gradient (CL) characterized in Figure 17 and the cycling rate is *C*/2. (b) $Li_{1.14}Ni_{0.25}Mn_{0.60}Co_{0.15}O_2$ is the corresponding homogeneous material (HL) and the cycling rate is *C*/2. (c) The performances of the two materials are compared at a higher rate of 5*C*. (Adapted from Ref. 134).

the composite than for the homogenous composition, as if the irreversible loss of oxidized oxygen at the surface of the primary particles was mitigated. That result is unexpected considering the preferential presence of the Mn-richer composition at the surface of the material, except if by engineering this composite microstructure of the aggregates (porosity, morphology of the primary particles, etc.) was different enough to minimize this instability.

IV. Conclusions

The next step appears even much more challenging than that overpass during the last 20 years, to develop successfully the promising Li-rich layered oxides to propose the next generation of higher energy density positive electrode materials for lithium-ion batteries. To date, researches have been based on modifying the composition, structure, and morphology, but were limited to deliver successfully less than 70% of the theoretical capacity in most cases for layered oxides (i.e., less than 180 mAh/g). Struggling to reach higher reversible capacity means that the chemical and thermal instabilities of layered oxides will be exacerbated, for the layered oxides involving only the transition metal ions into the redox processes, and even more those involving the oxygen anion.

Solutions exist and are prospected, but their developments are challenging as researches are moving to material engineering with the preparation of composite materials to stabilize metastable materials at their surface and to create synergies between properties. Coatings, core-shell, and concentration gradient complex engineered materials are part of the answer, but preferential growth of oriented surface, controlled porosity, and densification of the aggregates are another part of the answer. The preparation of these materials is challenging as it requires to control the composition and structure at any stage from the core to the bulk and at the particle and aggregates level.

V. References

1. Van der Ven, A.; Ceder, G. *J. Power Sources* **2001**, *97–98*, 529–531.
2. Levasseur, S.; *et al. Chem. Mater.* **2003**, *15* (1), 348–354.
3. Kalluri, S.; *et al. Adv. Energy Mater.* **2017**, *7* (1), n/a-n/a.

4. Myung, S.-T.; Kumagai, N.; Komaba, S.; Chung, H.-T. *Solid State Ion.* **2001**, *139* (1–2), 47–56.
5. Levasseur, S.; Ménétrier, M.; Delmas, C. *J. Power Sources* **2002**, *112* (2), 419–427.
6. Verdier, S.; El Ouatani, L.; Dedryvère, R.; Bonhomme, F.; Biensan, P.; Gonbeau, D. *J. Electrochem. Soc.* **2007**, *154* (12), A1088.
7. Jung, Y. S.; *et al. Adv. Energy Mater.* **2013**, *3* (2), 213–219.
8. Andre, D.; *et al. J. Mater. Chem. A* **2015**, *3* (13), 6709–6732.
9. Delmas, C.; Croguennec, L. *MRS Bull.* **2002**, *27* (8), 608–612.
10. Peres, J. P.; *et al. J. Phys. Chem. Solids* **1996**, *57* (6), 1057–1060.
11. Guilmard, M.; Croguennec, L.; Delmas, C. *Chem. Mater.* **2003**, *15* (23), 4484–4493.
12. Ammundsen, B.; *et al. J. Electrochem. Soc.* **2002**, *149* (4), A431.
13. Rossen, E.; Jones, C. D. W.; Dahn, J. R. *Solid State Ion.* **1992**, *57* (3), 311–318.
14. Ohzuku, T.; Makimura, Y. *Chem. Lett.* **2001**, *30* (7), 642–643.
15. Ohzuku, T.; Makimura, Y. *Chem. Lett.* **2001**, *30* (8), 744–745.
16. Noh, H.-J.; Youn, S.; Yoon, C. S.; Sun, Y.-K. *J. Power Sources* **2013**, *233*, 121–130.
17. Croguennec, L.; Palacin, M. R. *J. Am. Chem. Soc.* **2015**, *137* (9), 3140–3156.
18. Boulineau, A.; Croguennec, L.; Delmas, C.; Weill, F. *Solid State Ion.* **2010**, *180* (40), 1652–1659.
19. Pasero, D.; McLaren, V.; de Souza, S.; West, A. R. *Chem. Mater.* **2005**, *17* (2), 345–348.
20. Boulineau, A.; Croguennec, L.; Delmas, C., Weill, F. *Chem. Mater.* **2009**, *21* (18), 4216–4222.
21. Croy, J. R.; Park, J. S.; Dogan, F.; Johnson, C. S.; Key, B.; Balasubramanian, M. *Chem. Mater.* **2014**, *26* (24), 7091–7098.
22. Robertson, A. D.; Bruce, P. G. *Chem. Mater.* **2003**, *15* (10), 1984–1992.
23. Yu, D. Y. W.; Yanagida, K.; Kato, Y.; Nakamura, H. *J. Electrochem. Soc.* **2009**, *156* (6), A417–A424.
24. Xiao, R.; Li, H.; Chen, L. *Chem. Mater.* **2012**, *24* (21), 4242–4251.
25. Jarvis, K. A.; Deng, Z.; Allard, L. F.; Manthiram, A.; Ferreira, P. J. *Chem. Mater.* **2011**, *23* (16), 3614–3621.
26. Bareño, J.; *et al. Chem. Mater.* **2011**, *23* (8), 2039–2050.
27. Koga, H.; *et al. J. Phys. Chem. C* **2012**, *116* (25), 13497–13506.
28. McCalla, E.; Rowe, A. W.; Shunmugasundaram, R.; Dahn, J. R. *Chem. Mater.*, **2013**, *25* (6), 989–999.
29. McCalla, E.; Rowe, A. W.; Brown, C. R.; Hacquebard, L. R. P.; Dahn, J. R. *J. Electrochem. Soc.* **2013**, *160* (8), A1134–A1138.
30. McCalla, E.; Lowartz, C. M.; Brown, C. R.; Dahn, J. R. *Chem. Mater.* **2013**, *25* (6), 912–918.
31. McCalla, E.; Li, J.; Rowe, A. W.; Dahn, J. R. *J. Electrochem. Soc.* **2014**, *161* (4), A606–A613.
32. Long, B. R.; *et al. Chem. Mater.* **2014**, *26* (11), 3565–3572.

33. Thackeray, M. M.; Kang, S.-H.; Johnson, C. S.; Vaughey, J. T.; Benedek, R.; Hackney, S. A. *J. Mater. Chem.* **2007**, *17* (30), 3112–3125.
34. Lu, Z.; Dahn, J. R. *J. Electrochem. Soc.* **2002**, *149* (7), A815.
35. Weill, F.; Tran, N.; Croguennec, L.; Delmas, C. *J. Power Sources*, **2007**, *172* (2), 893–900.
36. Koga, H.; *et al. J. Phys. Chem. C* **2012**, *116*, 13497–13503.
37. Cabana, J.; *et al. J. Mater. Res.* **2010**, *25* (8), 1601–1616.
38. Armstrong, R.; *et al. J. Am. Chem. Soc.* **2006**, *128* (26), 8694–8698.
39. Koga, H.; Croguennec, L.; Ménétrier, M.; Mannessiez, P.; Weill, F.; Delmas, C. *J. Power Sources* **2013**, *236*, 250–258.
40. Koga, H.; *et al. J. Electrochem. Soc.* **2013**, *160* (6), A786–A792.
41. Koga, H.; *et al. J. Phys. Chem. C* **2014**, *118* (11), 5700–5709.
42. Oishi, M.; *et al. J. Power Sources* **2015**, *276*, 89–94.
43. Xu, B.; Fell, C. R.; Chi, M.; Meng, Y. S. *Energy Environ. Sci.* **2011**, *4* (6), 2223–2233.
44. Qian, D.; Xu, B.; Chi, M.; Meng, Y. S. *Phys. Chem. Chem. Phys.* **2014**, *16* (28), 14665–14668.
45. Genevois, C.; Koga, H.; Croguennec, L.; Ménétrier, M.; Delmas, C.; Weill, F. *J. Phys. Chem. C* **2015**, *119* (1), 75–83.
46. Boulineau, A.; Simonin, L.; Colin, J.; Bourbon, C.; Patoux, S. *Nano Lett.* **2013**, *13*, 3857–3863.
47. Boulineau, A.; Simonin, L.; Colin, J.-F.; Canévet, E.; Daniel, L.; Patoux, S. *Chem. Mater.* **2012**, *24* (18), 3558–3566.
48. Strehle, B.; *et al. J. Electrochem. Soc.* **2017**, *164* (2), A400–A406.
49. Croy, J. R.; Kim, D.; Balasubramanian, M.; Gallagher, K.; Kang, S.-H.; Thackeray, M. M. *J. Electrochem. Soc.* **2012**, *159* (6), A781–A790.
50. Croy, J. R.; *et al. J. Phys. Chem. C* **2013**, *117* (13), 6525–6536.
51. Croy, J. R.; Gallagher, K. G.; Balasubramanian, M.; Long, B. R.; Thackeray, M. M. *J. Electrochem. Soc.* **2013**, *161* (3), A318–A325.
52. Peralta, D.; *et al. J. Power Sources* **2015**, *280*, 687–694.
53. Mohanty, D.; *et al. Chem. Mater.* **2014**, *26* (21), 6272–6280.
54. Dogan, F.; *et al. J. Am. Chem. Soc.* **2015**, *137* (6), 2328–2335.
55. Li, J.; Shunmugasundaram, R.; Doig, R.; Dahn, J. R. *Chem. Mater.* **2016**, *28* (1), 162–171.
56. Oh, P.; *et al. Nano Lett.* **2014**, *14* (10), 5965–5972.
57. Sathiya, M.; *et al. Nat. Mater.* **2014**, *14* (2), 230–238.
58. Zheng, J.; *et al. Nano Lett.* **2014**, *14* (5), 2628–2635.
59. Sathiya, M.; *et al. Chem. Mater.* **2013**, *25* (7), 1121–1131.
60. Armstrong, A. R.; Bruce, P. G. *Nature* **1996**, *381* (6582), 499–500.
61. Capitaine, F.; Gravereau, P.; Delmas, C. *Solid State Ion.* **1996**, *89* (3), 197–202.
62. Croguennec, L.; Deniard, P.; Brec, R. *J. Electrochem. Soc.* **1997**, *144* (10), 3323–3330.
63. Simonin, L.; *et al. J. Mater. Chem.* **2012**, *22* (22), 11316.

64. Yabuuchi, N.; Yoshii, K.; Myung, S.-T.; Nakai, I.; Komaba, S. *J. Am. Chem. Soc.* **2011**, *133* (12), 4404–4419.
65. Bettge, M.; *et al. J. Power Sources* **2013**, *233*, 346–357.
66. Sathiya, M.; *et al. Nat. Mater.* **2013**, *12* (9), 827–835.
67. Sathiya, M.; *et al. Nat. Mater.* **2014**, *14* (2), 230–238.
68. Xie, Y.; Saubanère, M.; Doublet, M.-L. *Energy Environ. Sci.* **2017**, *10* (1), 266–274.
69. Luo, K.; *et al. Nat. Chem.* **2016**, *8* (7), 684–691.
70. Seo, D.-H.; Lee, J.; Urban, A.; Malik, R.; Kang, S.; Ceder, G. *Nat. Chem.* **2016**, *8* (7), 692–697.
71. Delmas C. *Nat. Chem.* **2016**, *8* (7), 641–643.
72. Lee, J.; Urban, A.; Li, X.; Su, D.; Hautier, G.; Ceder, G. *Science* **2014**, *343* (6170), 519–522.
73. Chen, R.; *et al. Phys. Chem. Chem. Phys.* **2015**, *17* (26), 17288–17295.
74. Wang, R.; *et al. Electrochem. Commun.* **2015**, *60*, 70–73.
75. Yabuuchi, N.; Takeuchi, M.; Komaba, S.; Ichikawa, S.; Ozaki, T.; Inamasu, T. *Chem. Commun.* **2016**, *52* (10), 2051–2054.
76. Yabuuchi, N.; *et al. Nat. Commun.* **2016**, *7*, 13814.
77. Yabuuchi, N.; Tahara, Y.; Komaba, S.; Kitada, S.; Kajiya, Y. *Chem. Mater.* **2016**, *28* (2), 416–419.
78. Pearce, P. E.; *et al. Nat. Mater.* **2017**, *16* (5), 580–586.
79. Freire, M.; *et al. Nat. Mater.* **2015**, *15* (2), 173–177.
80. Fergus, J. W. *J. Power Sources* **2010**, *195* (4), 939–954.
81. Albertus, P.; Christensen, J.; Newman, J. *J. Electrochem. Soc.* **2009**, *156* (7), A606.
82. Manthiram, A.; Choi, W. *Electrochem. Solid State Lett.* **2007**, *10* (9), A228.
83. Nam, K.-W.; Yoon, W.-S.; Shin, H.; Chung, K. Y.; Choi, S.; Yang, X.-Q. *J. Power Sources* **2009**, *192* (2), 652–659.
84. Numata, T.; Amemiya, C.; Kumeuchi, T.; Shirakata, M.; Yonezawa, M. *J. Power Sources* **2001**, *97–98*, 358–360.
85. Gallagher, K. G.; Kang, S.-H.; Park, S. U.; Han, S. Y. *J. Power Sources* **2011**, *196* (22), 9702–9707.
86. Imachi, N.; Takano, Y.; Fujimoto, H.; Kida, Y.; Fujitani, S. *J. Electrochem. Soc.* **2007**, *154* (5), A412.
87. Whitacre, J. F.; Zaghib, K.; West, W. C.; Ratnakumar, B. V. *J. Power Sources* **2008**, *177* (2), 528–536.
88. Liu, H.; *et al. ACS Appl. Mater. Interfaces* **2015**, *7* (34), 19189–19200.
89. Araki, K.; Taguchi, N.; Sakaebe, H.; Tatsumi, K.; Ogumi, Z. *J. Power Sources* **2014**, *269*, 236–243.
90. Kobayashi, G. *et al. J. Power Sources* **2016**, *303*, 250–256.
91. Li, B.; Yu, Y.; Zhao, J. *J. Power Sources* **2015**, *275*, 64–72.
92. Shi, S. J.; *et al. Electrochimica Acta* **2013**, *88*, 671–679.
93. Liu, Y.; *et al. Ionics* **2015**, *21* (10), 2725–2733.

94. Wang, Z.; *et al. Surf. Coat. Technol.* **2013**, *235*, 570–576.
95. Huang, Y.; Chen, J.; Ni, J.; Zhou, H.; Zhang, X. *J. Power Sources* **2009**, *188* (2), 538–545.
96. Liu, X.; Liu, J.; Huang, T.; Yu, A. *Electrochim. Acta* **2013**, *109*, 52–58.
97. Camardese, J.; Abarbanel, D. W.; McCalla, E.; Dahn, J. R. *J. Electrochem. Soc.* **2014**, 161 (6), A890–A895.
98. Li, J.; Camardese, J.; Shunmugasundaram, R.; Glazier, S.; Lu, Z.; Dahn, J. R. *Chem. Mater.* **2015**, 27 (9), 3366–3377.
99. Camardese, J.; Li, J.; Abarbanel, D. W.; Wright, A. T. B.; Dahn, J. R. *J. Electrochem. Soc.* **2014**, *162* (3), A269–A277.
100. Shi, H.; *et al. J. Alloys Compd.* **2014**, *587*, 710–716.
101. Sun, Y.-K.; Myung, S.-T.; Shin, H.-S.; Bae, Y. C.; Yoon, C. S. *J. Phys. Chem. B* **2006**, *110* (13), 6810–6815.
102. Sun, Y.-K.; Myung, S.-T.; Park, B.-C.; Amine, K. *Chem. Mater.* **2006**, *18* (22), 5159–5163.
103. Sun, Y.-K.; Myung, S.-T.; Park, B.-C.; Prakash, J.; Belharouak, I.; Amine, K. *Nat. Mater.* **2009**, *8* (4), 320–324.
104. Sun, Y.-K.; *et al. Nat. Mater.***2012**, *11* (11), 942–947.
105. Sun, Y.-K.; Kim, D.-H.; Yoon, C. S.; Myung, S.-T.; Prakash, S.-T.; Amine, K. *Adv. Funct. Mater.* **2010**, *20* (3), 485–491.
106. Sun, Y.-K.; Kim, D.-H.; Jung, H.-G.; Myung, S.-T.; Amine, K. *Electrochim. Acta* **2010**, *55* (28), 8621–8627.
107. Sun, Y.-K.; Lee, B.-R.; Noh, H.-J.; Wu, H.; Myung, S.-T.; Amine, K. *J. Mater. Chem.* **2011**, *21* (27), 10108.
108. Noh, H.-J.; Ju, J.-W.; Sun, Y.-K. *ChemSusChem* **2014**, *7*, 245–252.
109. Ju, J.-W.; Lee, E.-J.; Yoon, C. S.; Myung, S.-T.; Sun, Y.-K. *J. Phys. Chem. C* **2014**, *118* (1), 175–182.
110. Yoon, S.-J.; Myung, S.-T.; Noh, H.-J.; Lu, J.; Amine, K.; Sun, Y.-K. *Chem Sus Chem* **2014**, *7* (12), 3295–3303.
111. Yoon, S.-J.; Park, K.-J.; Lim, B.-B.; Yoon, C. S.; Sun, Y.-K. *J. Electrochem. Soc.* **2015**, *162* (2), A3059–A3063.
112. Lee, E.-J.; Noh, H.-J.; Yoon, C. S.; Sun, Y.-K. *J. Power Sources* **2015**, *273*, 663–669.
113. Li, Y. *et al. Nano Energy* **2016**, *19*, 522–531.
114. Lim, B.-B.; *et al. Adv. Funct. Mater.* **2015**, *25* (29), 4673–4680.
115. Park, K.-J.; Lim, B.-B.; Choi, M.-H.; Jung, H.-G.; Sun, Y.-K.; Haro, M.; Vicente, N.; Bisquert, J.; Garcia-Belmonte, G. *J. Mater. Chem. A* **2015**, *3* (44), 22183–22190.
116. Huang, Z.; Gao, J.; He, X.; Li, J.; Jiang, C. *J. Power Sources* **2012**, *202*, 284–290.
117. Du, K.; Hua, C.; Tan, C.; Peng, Z.; Cao, Y.; Hu, G. *J. Power Sources* **2014**, *263*, 203–208.
118. Hua, C.; Du, K.; Tan, C.; Peng, Z.; Cao, Y.; Hu, G. *J. Alloys Compd.* **2014**, *614*, 264–270.

119. Liang, L.-W.; Du, K.; Peng, Z.-D.; Cao, Y.-B.; Hu, G.-R. *Chin. Chem. Lett.* **2014**, *25* (6), 883–886.
120. Liang, L.; Hu, G.; Cao, Y.; Du, K.; Peng, Z. *J. Alloys Compd.* **2015**, *635*, 92–100.
121. Hou, P. Y.; Zhang, L. Q.; Gao, X. P. *J. Mater. Chem. A* **2014**, *2* (40), 17130–17138.
122. Hou, P.; *et al. RSC Adv.* **2014**, *4* (31), 15923.
123. Song, D.; Hou, P.; Wang, X.; Shi, X.; Zhang, L. *ACS Appl. Mater. Interfaces* **2015**, *7* (23), 12864–12872.
124. Liao, J.-Y.; Manthiram, A. *J. Power Sources* **2015**, *282*, 429–436.
125. Wang, J.; *et al. Electrochimica Acta* **2015**, *174*, 1185–1191.
126. Sun, Y.-K.; Park, B.-C. Positive Active Material Comprising a Continuous Concentration Gradient of a Metal Composition for Lithium Battery, Method of Preparing the Same, and Lithium Battery Including the Same. Google Patents, 2014.
127. Sun, Y.-K.; Noh, H.-J. Anode Active Material with Whole Particle Concentration Gradient for Lithium Secondary Battery, Method for Preparing Same, and Lithium Secondary Battery Having Same. 2014.
128. Sun, Y.-K.; Noh, H.-J. Cathode Active Material with Whole Particle Concentration Gradient for Lithium Secondary Battery, Method for Preparing the Same, and Lithium Secondary Battery Having the Same. 2015.
129. Sun, Y.-K. Method of Preparing Cathode Active Material Precursor for Lithium Rechargeable Battery, Cathode Active Material Precursor for Lithium Rechargeable Battery Prepared Thereby, and Cathode Active Material Formed Using the Cathode Active Material Precursor. Google Patents, 2013.
130. Yu, J.-R.; Choi, M.-H. Method for Preparing Positive Electrode Active Material Precursor and Positive Electrode Material for Lithium Secondary Battery Having Concentration-Gradient Layer Using Batch Reactor, and Positive Electrode Active Material Precursor and Positive Electrode Material for Lithium Secondary Battery Prepared by the Method. Google Patents, 2011.
131. Hwang, D.; Lim, J.; Han, K. *Lithium Secondary Battery*. 2015.
132. Koenig, G. M.; Belharouak, I.; Deng, H.; Sun, Y.-K.; Amine, K. *Chem. Mater.* **2011**, *23* (7), 1954–1963.
133. Yang, X.; *et al. J. Power Sources* **2013**, *232*, 338–347.
134. Yang, X.; *et al. J. Mater. Chem. A* **2014**, *2* (11), 3899.
135. Yu, R.; *et al. Chem. Electro. Chem.* **2015**, *2* (9), 1346–1354.
136. Myung, S.-T.; Noh, H.-J.; Yoon, S.-J.; Lee, E.-J.; Sun, Y.-K. *J. Phys. Chem. Lett.* **2014**, *5* (4), 671–679.
137. Li, J.; Doig, R.; Camardese, J.; Plucknett, K.; Dahn, J. R. *Chem. Mater.* **2015**, *27* (22), 7765–7773.

3 Recycling of Li-Ion Batteries and New Generation Batteries

Jean Scoyer

Scaron Consulting GCV, Bornstraat 22, BE2200 Herentals, Belgium
Jean.Scoyer@eu.umicore.com

Table of Contents

I. Introduction

Although efficient recycling schemes are in place for lead–acid, Ni–Cd, and primary batteries, with recycling rates over 90% in the first case, the recycling business of Li-ion batteries is still in its infant shoes. Reason for this is twofold. First, the quantities of end-of-life (EOL) batteries are still small due to smaller production figures and much lower collection rates. Second, the recycling processes are much more complex, and most of them are still in the research or piloting phase.

This situation will dramatically evolve in the coming years due to combined effect of huge production increase and better collecting rates.

The drivers to recycling in general and of Li-ion batteries in particular are:

— The potential value of the battery. This value is of course depending on several factors as:

1) The market value of the recovered fractions,
2) The cost of the different steps of the recycling process,

3) The cost of alternative solutions, if any
4) Possible incentives/fees for the processors along the recycling chain.

— The macroeconomic considerations, like the (future) scarcity of some components, the related geopolitical considerations, and foreign currency balance of the states.
— The environment, health, and safety (EHS) aspects: impact on human health of uncontrolled disposal of heavy metals and reduction of greenhouse gases.
— The burden of place.

Based on these considerations, several countries have issued regulation on recycling of Li-ion batteries. The most advanced battery recycling legislation is the European Batteries Directive, which sets clear targets on collection and recycling rates of all types of batteries.

The drivers and legal framework to recycle Li ion will first be described. In a second part, the author's evaluation of the future recycling business at the horizons 2020 and 2025 will be presented and this for the different chemistries. In the last part, the processes developed at industrial scale as well as proposed in the literature are reviewed. Finally, the impact of the new chemistries on the economics of the recycling is discussed.

II. Drivers to Recycling

A. Market Value of the Recovered Fractions

1. *Introduction*

In order to compare different sizes and chemistries of Li ion, all data are referred to the energy content, expressed in kilowatt-hour (or its multiples, megawatt-hour and gigawatt-hour). As an example, a cylindrical format 18650, rated at 2.8 Ah has an energy content of 2.8 × 3.6 V ~ 10 Wh. A 36-Ah prismatic cell for electric vehicles (EVs) application has an energy of ~130 Wh (note that average voltage depends on the chemistry).

As the recycling business is driven by tonnages of spent batteries, and not by their initial energy, the specific energy for the different chemistries was used to convert energy data to weight data.

Unless otherwise stated, the currency unit used is USD and when needed, an exchange rate of 1.13 USD/€ was applied.

2. *Metal Prices*

The value of a spent battery depends in first instance on the chemistry used in the battery (meaning that transition metals are used in the positive electrode). Co and Ni are the most expensive metals used in Li-ion batteries. The comparative values of early 2016 are given in Table 1.

The historical evolution of the prices of Ni and Co is given in Figure 1.

For Li, the price of technical grade Li_2CO_3 is used, as this is the component that could possibly be valorized. There is no official quotation for

Table 1. Value of metals used in Li-ion batteries (early 2016)

Metal	Co	Ni	Mn	Fe	Cu	Al	Li_2CO_3
Price (USD/kg)	25	10	2	0.5	5	1.2	5
Ratio to Co	100%	40%	8%	2%	20%	5%	20%

Source: For Co, Ni, Cu, and Al: Adapted from Ref. 1.

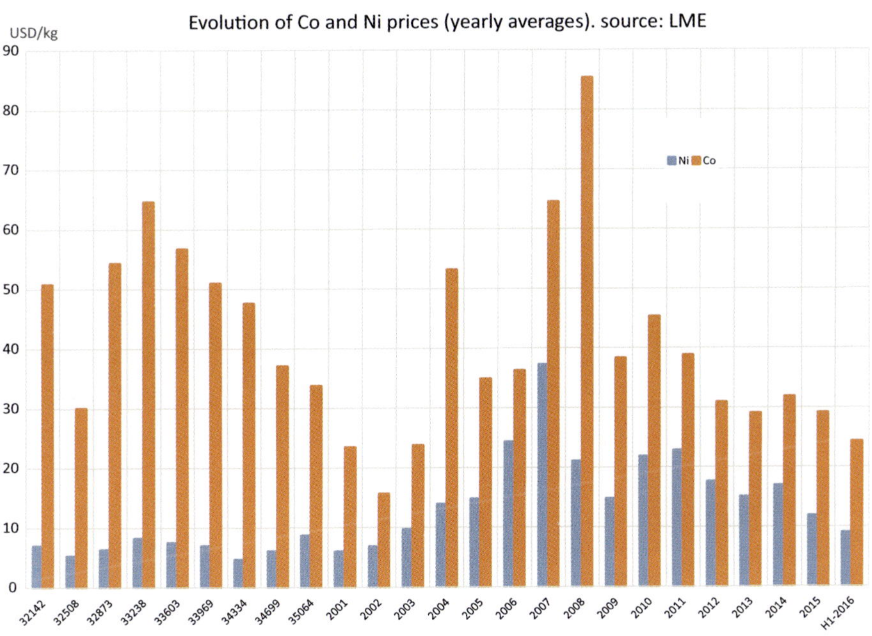

Figure 1. Historical metal prices. Adapted from Ref. 1.

Li_2CO_3; however, spot prices seem to have increased substantially between early and mid-2016. It is impossible to say if this is a mega-trend or just a temporary phenomenon due to speculation. Seen the high volatility of the metal courses, any forecast of the future price is highly speculative and out of scope of this work. Values of early 2016 will be kept throughout the chapter even for scenario developed for the longer term.

3. *Value of the Recyclable Fractions*

The price of a typical cylindrical 18650 cell in 2015 is around 170 USD/kWh.[2] The cost structure given in Figure 2 shows that materials represent 56% of the value. This means that the cost of battery materials is around 95 USD/kWh.

Table 2 gives the total value of potentially recyclable materials in a 18650 battery, based on Ni–Co–Mn mixed oxides positive electrode chemistry (NMC 1:1:1) in the market conditions of early 2016. Two approaches are considered, namely the compound recovery, where compounds (e.g.,

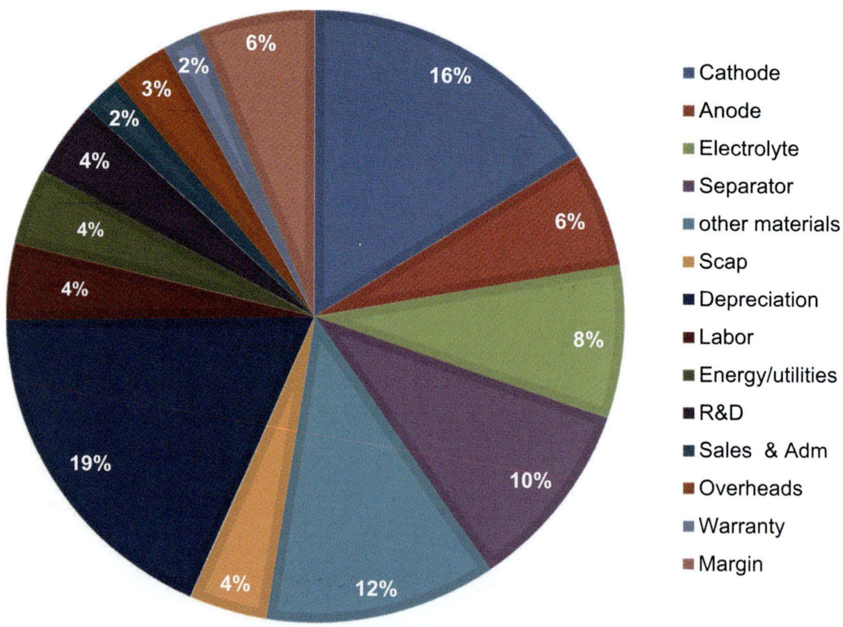

Figure 2. Cost structure of a 18650 battery in 2015. Adapted from Ref. 2.

Table 2. Value of potentially recyclable materials in a NMC 1:1:1 battery

Materials	% of Battery Price	Compound Recovery	Element Recovery
Cathode	16	16	9
Anode	6	0.5	0.1
Collectors	4	1.5	1.5
Separator	10		
Electrolyte	8	0.1	0.1
Other	8		
Scraps	4	1	0.8
% of battery cost	56	19	12
% of materials cost		35	19
Total USD/kg	21	7	4

the positive electrode active material) is recovered, purified, and reused in batteries; the element recovery, where compounds are decomposed to their elements (e.g., the metals of the positive electrode). These approaches are developed in Chapters 4 and 5 dealing with the processes.

The details of the calculation are given in Tables A1 and A2.

A similar calculation has been done for the other chemistries used at industrial scale, $LiCoO_2$ (LCO), Li(Ni–Co–Al)O_2 (NCA), $LiFePO_4$ (LFP), and $LiMn_2O_4$ (LMO) taking into account the metal prices, the energy of the batteries, and of the positive electrode material. The results, expressed in USD/kg of spent batteries are compared to NMC 1:1:1 in Table 3, details are given in Table A3.

The calculation of the potential value is based on the following approach:

1. As the value of spent batteries will be related to their weight and not to their initial energy content, all the figures expressed per kilowatt-hour were converted to weight based on their energy density. A value of 220 Wh/kg (or conversely 4.5 kg/kWh) was used for LCO and NMC, whereas 3.7, 5.0, and 5.5 kg/kWh were, respectively, taken for NCA, LFP, and LMO chemistries.

Table 3. Potential value in USD/kg of batteries of the recovered phases for different chemistries and processes

	LCO		NMC 1:1:1		NCA	
Materials	Compound	Element	Compound	Element	Compound	Element
Cathode	8.5	5.9	6.0	3.4	7.4	4.2
Anode	0.2	0.0	0.2	0.0	0.2	0.0
Collector	0.6	0.6	0.6	0.6	0.7	0.7
Separator	0.0	0.0	0.0	0.0	0.0	0.0
Electrolyte	0.0	0.0	0.0	0.0	0.0	0.0
Others	0.0	0.0	0.0	0.0	0.0	0.0
Scrap	0.7	0.7	0.5	0.3	0.8	0.6
Total USD/kg	10	7	7	4	9	6

	LFP		LMO	
Materials	Compound	Element	Compound	Element
Cathode	3.1	0.9	3.1	1.0
Anode	0.2	0.0	0.1	0.0
Collector	0.5	0.5	0.4	0.4
Separator	0.0	0.0	0.0	0.0
Electrolyte	0.0	0.0	0.0	0.0
Others	0.0	0.0	0.0	0.0
Scrap	0.3	0.1	0.2	0.1
Total USD/kg	4	2	4	2

2. By lack of better available information, the premium for positive electrode materials, on top of the metal and Li_2CO_3 price, was assumed to be identical for all chemistries. This premium was calculated from the cost structure of NMC 1:1:1. This leads, for the virgin batteries, to other total costs per kilowatt-hour and other repartition than for the developed NMC 1:1:1 case. These differences are found back when evaluating the value of the spent batteries, in particular the change of

weight of the positive electrode component per kilogram of batteries cannot be neglected.

3. The potential value is based on a 100% recovery of the related phases.
4. In the compound recovery scenario, the assumption is that the positive electrode material can be recycled at the value of fresh material. Conversely, no single attempt to reuse the graphite in new batteries was found in the literature, consequently the assumption taken is that it can be sold at 1 USD/kg as graphite for low demanding applications. For the Cu and Al current collector, only the value of the metal is recoverable as they have been shredded or molten. The separator and electrolyte cannot be recycled as such, and so loose almost all their value (electrolyte value = Li content as Li_2CO_3). Recycling of undefined "other materials," including casing, tabs, and binders, is assumed delivering no value. For the production scrap, the average value of the above percentages was taken.
5. In the element recovery scenario, only the metals and the Li_2CO_3 are valorized in the positive electrode. For the anode, the value of 0.2 USD/kg corresponds to reuse as cokes in pyrometallurgical processes. Valorization of Cu and Al is as in previous case.

4. *Remarks*

- The real value after recycling (positive or negative) will of course depend on which fractions can be recycled, with which value, which yield, and at what cost.
- Obviously, the value is higher for the LCO, leading to 3 USD/kg more potential value than for NMC in both scenarios.
- The values for NCA are close to those of NMC, the higher value for compound coming from the higher energy density of these batteries.
- For LCO, NMC, and LCA, over 80% of the value comes from the positive electrode. This justifies the accent put on recovery of this fraction in most of the processes.
- For LMO and even more for LFP-based chemistries, the value of the transition metals is very low, increasing the gap between the values of the two scenarios.

- Due to the numerous assumptions that needed to be made by the lack of accessible precise information, the values of the spent batteries need to be considered as an indication of the maximum value. Nevertheless, the relative values of the different scenarios and chemistries should be correct.
- Some batteries contain a mixture of different chemistries, for example, to optimize simultaneously the power and the energy of EV batteries.
- It is probable that the economic driver alone will be insufficient to generate a spontaneous activity, because the costs of the full recycling chain can exceed the potential revenues, certainly in the case of LFP and LMO. The legislations will thus be the driver for recycling.
- Argonne National Laboratory (ANL) has published a comparative study summarized in Figure 3.[3]

Although the absolute prices are not given and are expressed per kilogram of positive electrode material and not per kilogram of batteries, the trend is similar to this evaluation.

B. Macroeconomic Considerations

1. *Scarcity of Some Components*

The huge increase in the use of Li-ion business has raised concerns about the availability of some components, in particular for lithium and cobalt.

The issue of Li scarcity is a highly debated topic. In a recent publication, Fox-Davies reviewed the Li market.[4] They conclude that Li-ion

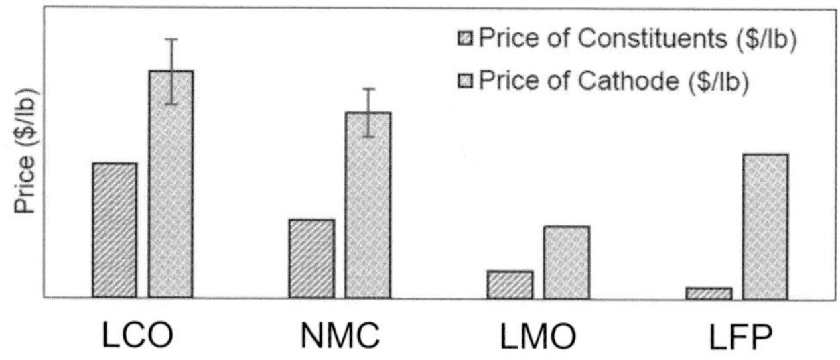

Figure 3. Price of constituents and price of positive electrode materials. Adapted from Ref. 3.

battery is the fastest growing application for Li, growing at an average annual growth rate (AAGR) of 10%, from 27% in 2012 to 40% in 2015. Consumption is expected to reach 500,000 tons (kt) equivalent Li_2CO_3 (~100 kt Li element) or 65% of the total demand in 2025. However, reserves were considered sufficient to cover the new demand.

In 2015, Oliveira compiled different recent studies on reserves and resources of Li.[5] The reserves are estimated between 20 and 70 million tons of Li element, which represent more than 100 years of consumption of 2025, in the worst case of low reserves and no recycling.

Nevertheless, promotors of an alternative chemistry based on Na use this argumentation to justify their approach. Anyhow, this new technology will require a long development path before taking a significant part of the rechargeable battery business and hence of the recycling of spent batteries. The biggest Li-supply challenge will be the expansion of production capacity to meet the growing demand.

With regard to the other components, ANL concluded in a study for US Department of Energy (DOE) that cobalt was the only possible problem on the longer term. Indeed cumulative demand up to 2050 would represent 9% of the estimated resources of 13 Mt.[3] This relatively high percentage is used to justify the need for replacing Co by more abundant sources, and develop efficient recycling of Co from the batteries (and other sources).

2. *Geopolitical Considerations*

The potential risk of supply constrains has pushed Europe to initiate the "Raw Materials Initiative."[6] It sets out a European long-term strategy to ensure the supply of raw materials in Europe. Several elements and minerals "for which a *threat to supply* could involve *harm to the national economy*, and for which the risk is higher than for most other raw materials *in the coming 10 year*" were classified as "critical" (see Figure 4). Some of these critical materials are essential to rechargeable battery materials: Co, natural graphite, and fluorspar (fluorspar is a source of F, essential element of the electrolyte). Mn and Ni are also considered as of high economic importance, but low supply risk. Li is considered as of intermediate economic importance and low supply risk. Most economic Li sources are

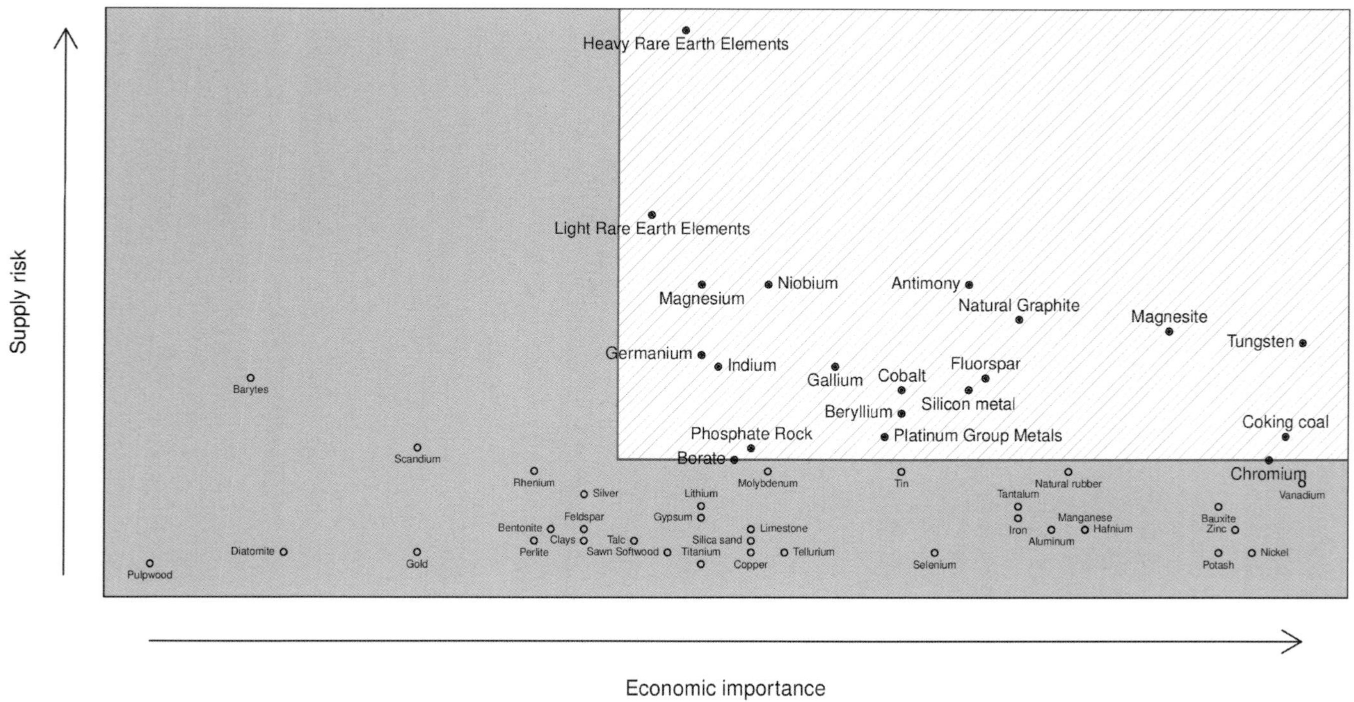

Figure 4. Critical materials as defined in the European Critical Raw Materials review, May 2014. Adapted from Ref. 6.

located in South America (80% of world Li reserves are in Chile and Bolivia). On long term, it could become competitive and of strategic importance to broaden the supply base. Reserves of Co are mainly located in Congo and Russia, and for a smaller part in Australia.

3. *Other Aspects*

At national level, recycling also contributes to job creation at all educational levels and to the improvement of the foreign currency balance, by reducing the import of feedstock and/or of energy. Besides, it offers ethical sourcing possibilities.

C. Environment, Health, and Safety

1. *Environment*

The life cycle analysis (LCA) tool is of classical use to evaluate the environmental impact of a product during its full life (from "cradle to grave" or even "cradle to cradle"). Aspects like energy consumption, greenhouse gases emission, contamination of air and water, ecotoxicity, and so on are evaluated according to standard procedures.

In the case of Li ion, this tool has been used to asses if recycling contributes to reduce the total impact of production. The ANL study in the DOE merit review 2015 shows a significant reduction of SOx emissions in the case of LCO recycling (70%–90%). Energy consumption is also significantly reduced.[3]

2. *Health and Safety*

Li ion contains a complex mixture of heavy metals and organic compounds that represent a potential chemical risk. Contact with air or water by accidental or voluntary opening of the cell can lead to fire and release of toxic gasses, for example, hydrogen fluoride (HF) from the salt in the electrolyte. As EOL batteries are not always fully discharged, shortcuts can arise from improper manipulation, with subsequent risk of fire. Finally, there is an unacceptable risk of electrocution associated with the unprofessional handling of large batteries, like those from electric cars.

Indeed battery packs of EV operate at voltages above 300 V, and can contain up to 90 kWh in the case of for pure EV.

D. Volume: Micro and Macro Scale

1. *Present and Past Situation*

Today, recycling of rechargeable batteries is very limited, not because of lack of recycling capacity, but because of low collection rate. According to figures from European Battery Recycling Association (EBRA), around 3500 tons of secondary Li-ion batteries were collected in 2012, whereas the volume of these batteries put on the market 3 years before was about 125 kt worldwide or approximately 30 kt for Europe (which represented around 25% of the world economy).[7,8] This means that only around 10% of Li-ion batteries were collected and recycled in Europe, the others being hoarded, dumped with household waste, or illegally exported. Recycling percentages in other parts of the world are probably even lower.

As the quantity of household waste increased significantly by the growing consumption lifestyle, its management becomes a huge problem for the communities. Dumping grounds gets completely full. In many countries, land is too scarce to use it as dumping ground and new dumping grounds are forbidden. Existing dumps are cleaned and sanitized and in the best scenario, materials and energy are recovered. The presence of toxic elements from batteries makes the rehabilitation of the site even more challenging. The alternative of calcination of waste had to face severe air contamination by dioxins and other toxic gases, partially due to the presence of Li-ion batteries in their feed. This leads to the necessity of huge investments for gas cleaning and even closure of some facilities.

Finally, many electronic devices exported to some countries of Africa and Asia, represent in fact an illegal export of waste as they are not suited for second-hand use as claimed in the shipment documents. They are then landfilled or recycled but following inferior standards and poor EHS consideration.[9] The international community becomes more and more aware about this unacceptable disposal of waste, and United States and European Union authorities announced increased efforts to counter these malpractices.[10,11]

2. *Future Situation*

The recycling of batteries will significantly increase in the future, so that the volume ending in waste management will be reduced, regardless of the big increase of produced batteries. Reasons, therefore, are mainly the higher percentage of batteries from EV that will not be hoarded, combined with a better awareness of the customers and more stringent regulations.

The author's scenario of recycling rates at the horizons 2020 and 2025 will be further developed in Chapter 3.

E. Legal Framework

The legislation that has been put in place in several counties is mainly motivated by the EHS risks of improper disposal of EOL batteries. The European Waste List Decision 2000/532/EC classifies the electrolyte from batteries as "hazardous waste."[12] This combination of hazardous chemicals, fire hazard, and electric shock risks has led to self-regulation by the industry and development of legal frameworks to collect and recycle batteries.

Worldwide, Europe has the most detailed legislative framework for collection and recycling of spent batteries. A number of regulatory instruments are already applicable, such as the Ecodesign Directive, the Directive on End-of-Life Vehicles, the Directive on Waste Electrical and Electronic Equipment, the Batteries Directive, and the Waste Framework Directive. Detailed information and interpretation of the legal framework in different countries can be found on the sites of recycling organization, EBRA, EUBAT, Recharge, RBRC (for North America), as well as on the Web site of Umicore UBR.[13]

Europe sets collection targets and recycling efficiency criteria for all types of portable and industrial batteries. The European Battery Directive 2006/66/EC, by September 26, 2012, has been transposed into national laws and is being implemented.[14] According to this directive, member states shall achieve a collection rate for waste *portable* batteries of 25% compared to the placing on the market (moving average other 3 years). By September 2016, the next collection target is fixed at 45%.

Batteries from EV fall under the Regulation (EU) 493/2012, stating that "*recycling of 50% by average weight should be attained for other waste batteries and accumulators.*" The directive on EOL vehicles imposes a 90% recycling of the weight of EOL cars. For all types of EV, this implies that batteries need to be recycled to cope with this percentage.

In Japan, general recycling philosophy is formulated in the Law for Promotion of Effective Utilization of Resources, and based on this law, manufacturers and importers of portable rechargeable batteries are requested to promote collection and recycling.[15–17]

In the United States, several states have voted regulations forbidding the landfilling of waste batteries and promoting their recycling.[18]

In December 2015, the "circular economy package" was adopted by the European Commission (EC). It has to be seen from the perspective of growing world population, and especially the middle class. The need for raw materials in the "technosphere" will grow tremendously, whereas the access to raw materials becomes more and more difficult. With this package, the EC wants to become more resource efficient, by working on product design (life cycling thinking and material efficiency from the design phase), reuse (longer life cycles, easier to replace worn parts, refurbishment, and remanufacturing), and recycling.[19]

III. Business of Battery Recycling

A scenario has been developed to estimate the quantities of batteries that will be amenable to recycling in the future, with detailed calculation for 2020 and 2025. Calculation for the other years can easily be done by interpolation of the values of the reference years (growth of the specific application, improvement of the recycling rates).

There are typically three types of recycling streams:

— Production scraps due to misfits, out-of-spec products, destructive testing

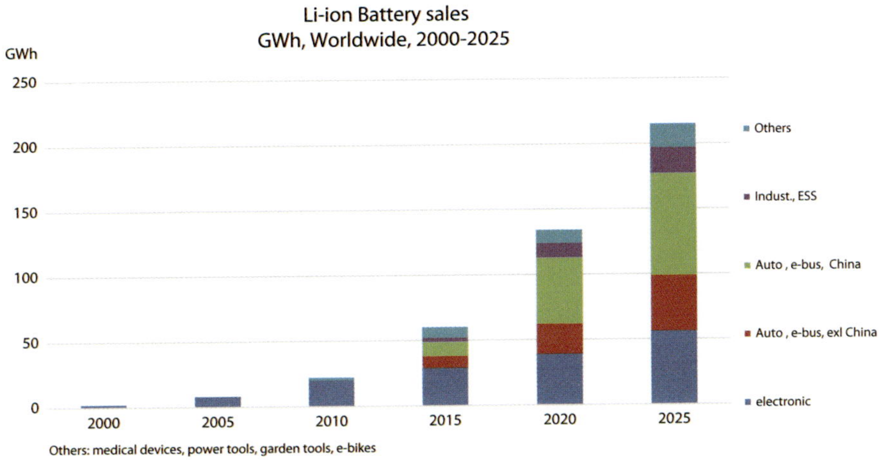

Figure 5. Total production of Li-ion batteries per application. Adapted from Ref. 2.

— Refurbishing before second life or reuse
— EOL of the battery pack or the device containing the battery

The model is based on inputs from market forecast combined with own estimation of lifetime and recycling rates.

A. Market Forecast

The market information were taken from the presentation "The rechargeable Battery Market and Main trends 2015–2025," by Avicenne Energy in March 2016.[2] Figure 5 gives the forecast of production per application, growing from 60 to 217 GWh in 2025.The forecast of chemistries (expressed in kt of positive electrode materials) is given in Figure 6.

B. Cycle Life and Recycling Rates

The cycle life and recycling rate depend on the application in which the batteries are used. The author's own evaluation of these two factors is based on several information. The scenario is summarized in Table 4. For details, see Tables A4 and A5.

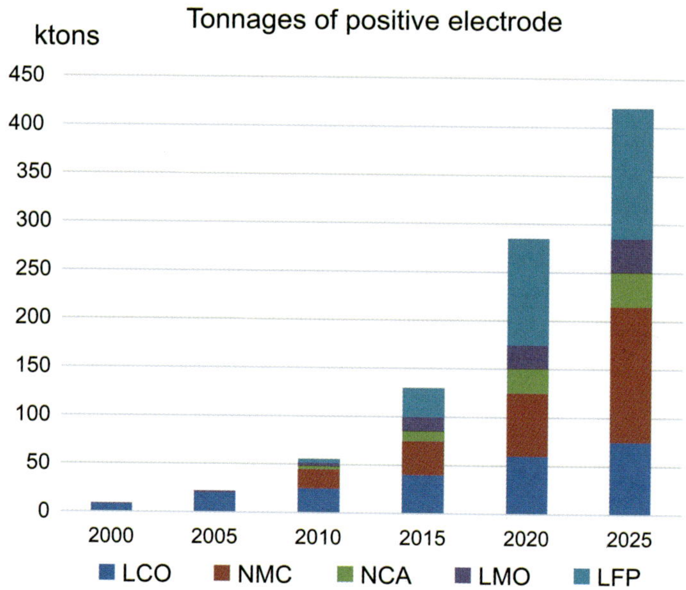

Figure 6. Repartition by positive electrode chemistry. Adapted from Ref. 2.

Table 4. Expected recycling rates and cycle life for different applications

	Horizon 2020		Horizon 2025	
Application	Recycling (%)	Years	%	Years
Phones	24	2	39	3
Computers/tablets	24	3	39	4
xEV	90	8	90	8
Industrial and ESS	90	15	90	15
Power tools/E bikes	51	3	61	3

E bike, electronic bike.

1. *Cellular Phones*

The assumptions made are the following:

— As the battery technology is mature, production scrap in 2020 and 2025 is reduced to 3% and 2%, respectively, from the 4% in 2015.[2]

— At the horizon 2020, 20% of the phones will have reached their EOL after 2 years. The recycling rate will increase to 20% from less than 15% today. The remaining 80% will get a second life after eventual refurbishing. If the refurbishing is done by specialized companies, it will include the replacement of the battery. In this case, a recycling rate of 50% of the old batteries can be expected. This will happen in 50% of the cases (case A). For the rest, if refurbishing does not include battery replacement or is done in countries where no efficient recycling programs are put in place, for example, Africa, recycling of the battery (case B) will not be done. This leads to an overall recycling of batteries of 24%, 2 years after production. The second wave after 4, 6, . . . can be neglected.
— At the horizon 2025, increase of the first life to 3 years, due to a reduction of the renewal push to acquire new technologies. Conversely, the recycling rate is increased to 25% at EOL and 60% and 25% in cases A and B, respectively. This will lead to overall recycling rates of 39%, but after 3 years.

2. *Laptops and Tablets*

For laptops and tablets, the same approach as for phones is used but with a first life of 3 and 4 years in 2020 and 2025, respectively.

3. *Portable Electronics*

The total for portable electronics is a weighted average of phones and computers/tablets.

4. *Cars*

For cars, the batteries are normally guarantied for a first life of 8 years. As residual capacity should still be 80%, some car manufacturers evaluate a second life for domestic use, coupled to renewable energies. This new application is estimated to only 20% of the smaller segment of pure EV/PHEV, and thus, this possibility has been neglected in the calculations. A 10% recycle after 1 year is also included, based on 5% misfits and 5%

total loss of cars by accidents. This ratio is reduced to 8% in 2025, due to more mature battery technology.

As battery recycling is mandatory in most countries, 95% of the car batteries will effectively be recycled in the model.

5. *Industrial and ESS*

These applications only represent 5% today, but grow faster than others and will represent 10% of the battery market in 2025. It is clear that the recycling rate for these big battery aggregates will be comparable or higher than for the cars, but also that their lifetime is very long. For ESS, a 20-year lifetime is normally mentioned. With an estimated 15-year average lifetime, their proportion in the recycling business will still be marginal at the envisaged horizons. Production scraps of 3% (2020) and 2% (2025) have been used.

6. *Other Applications*

Other applications include portable and garden tools and electric bikes (E bikes) and scooters. These applications represent 7% today and are expected to grow to 7% and 10% battery market in 2020 and 2025, respectively. A lifetime of 3 years and a recycling of 50% in between electronic and EV figures are used in the model. Scrap production is taken as 3% and 2%.

C. Quantity of Batteries to Be Recycled

1. *Methodology*

The quantity of batteries to be recycled in a particular year is based on a back calculation of the productions x years before the reference year (x being the average life cycle in the application) multiplied by the recycling yield from Table 4. These figures are then compared to the actual production forecast for the same application in that year. A more precise calculation should be based on a histogram of the life cycle instead of the average.

It is important to stress that this percentage is not equivalent to the recycling rates as usually defined, among others in the regulation, but is, in fact, discounted with the growth of the battery business during the average lifetime of the battery, and this for the different applications. This growth was deduced from the information by Avicenne.[2] A high recycling rate in a high-growing market and long life, as is the case for batteries in electric vehicles leads to 21% recycling contribution, whereas a low recycling rate in a slow-growing market and short lifetime as for portable applications leads to a similar rate (24%). In particular cases, where the application is diminishing, the discounted recycling rate can be superior to 100% (e.g., for Ni–Cd).

For all types of electric vehicles (xEV), the production between China and non-China is taken separately, as the AAGR and chemistries are different.

2. *Overall Results*

Table 5 gives an estimation of the quantities of batteries to be recycled at the horizons 2020 and 2025, according to the methodology explained before.

Recycling could cover 18% of the needs in 2020, growing to 37% in 2025. Expressed in energy, it corresponds to respectively 26 and 81 GWh. The market of x EV in China plays a major role in the total consumption and accounts for almost 50% of the recycling market at the horizon 2025.

Table 5. Recycling scenario per application at the horizons 2020 and 2025

Horizon	2020			2025		
Application	% of Total Market	Discounted Recycling Rate (%)	% Recycled of Total Market	% of Total Market	Recycling Rate (%)	% Recycled of Total Market
Portable electronic	29	24	7	26	33	8
xEV ex China	20	21	4	19	48	9
xEV in China	36	11	4	37	45	16
Industrial and ESS	7	13	1	9	12	1
Power tools/E bikes	7	30	2	9	40	3
Total	100		18	100		37
Total in GWh	137		26	217		81

3. *Breakout per Chemistry*

In the next step, the model was refined to determine which mix of chemistries would be offered to recycling. As an example, the calculation for the LCO chemistry is given in next paragraph.

At the Horizon 2020, a consumption of 60 kt LCO positive electrode material is forecasted (see Figure 6 from Avicenne).[2] This corresponds to ~32 GWh of batteries. The model estimates that 83% is used in portable electronic, representing two-thirds of the demand for this application, the remaining 17% being used in power tools and E bikes.

The recycled quantities of batteries with this chemistry can thus be estimated at

$$32\ \text{GWh} \times 83\% \times 24\% \sim 6\ \text{GWh}$$
$$32\ \text{GWh} \times 17\% \times 30\% = 2\ \text{GWh}$$

Table 6 and Figure 7 summarize the results for all chemistries at the horizons 2020 and 2025. The details of the calculation are given in Tables A6–A10, where tonnages per applications and per chemistries were reconciliated as best as possible, and where average recycling rate per chemistry were calculated.

To convert the gigawatt-hour to tonnage of batteries, the forecasted improvements in energy density of the diverse chemistries at the average year of production were used. As an example for NMC, the energy density of 220 Wh/kg (or inversely 4. 5 kg/kWh) will improve to 250 and 300 in 2020 and 2025, respectively. With an average lifetime of ~4 years (see above), the NMC batteries recycled in 2020 still weight 4.5 kg/kWh, whereas in 2025, the weight will be reduced to 4.2 kg/kWh.

It must, however, be clear that recycling will not necessary lead to reuse of these quantities to make fresh batteries, due to technical and economic issues described elsewhere in this chapter. The purpose of the calculation is to evaluate the quantities of batteries of the different types that will be amenable to a recycling process at the chosen horizons.

It is important to note that some recycled batteries will contain a mixture of different technologies, so that the tonnages are just indicative of the size of the business.

Table 6. Recycling business per chemistry

	2020			2025		
Horizon	GWh	%	kt batt	GWh	%	kt batt
LCO	8	25	36	13	33	49
NMC	7	22	37	26	35	102
NCA	2	21	12	9	46	40
LMO	2	18	13	7	40	42
LFP	6	12	32	26	39	134
Total	26	18	130	81	39	~350

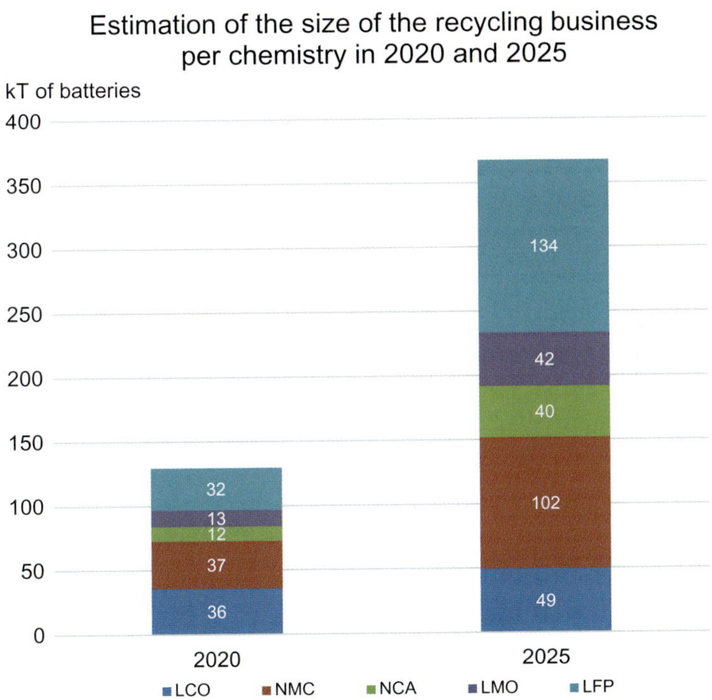

Figure 7. Recycled batteries per chemistry.

IV. Recycling Processes

A. Introduction

The processes for reprocessing of batteries and packs typically comprise a dismantling followed by three families of recycling processes: physical separations, hydrometallurgy, and pyrometallurgy. The overall processes can include a combination of these separation technologies.

The processes for the recycling of Li-ion batteries are not specific to this type of material and most of the time are derived from existing processes for the recycling of primary Li batteries, Ni metal hydride (NiMH) batteries or even other type of residues.

Two concepts of recycling are in competition. In the so-called "compound recovery," the processes aim at recovering the compounds in the battery (e.g., the cathode active material). In the "element recovery" concept, compounds are decomposed during the processes and brought back to their elements (e.g., metals).

B. Dismantling

It is generally admitted that dismantling of the heavy battery packs from car make sense, so as to remove the bulky external constituents, such as casing, electric connections, cooling system, and battery management electronics. Indeed, dedicated recycling processes are already in place for these components. Specific recycling techniques can then be applied to the cells or modules (a series of interconnected cells).

Today, this dismantling is done manually and is a costly and dangerous operation.[20] It is expected that type recognition, automation, and robotization will be applied in the future, when quantities will justify the development work.

C. Physical Methods

The first step is often a shredding followed by fine crushing. Fine crushing is needed to allow the liberation of the different fractions. To avoid risks of fire and/or evolution of HF, these operations must be done in controlled atmosphere or after inertization of the battery. Several methods have been developed and patented for this purpose (see further).

The next step is the separation of the different fractions by one or more classical operations like sieving, magnetic separation, froth flotation, eddy currents, and density differences. The different fractions are sold or further purified if required.

Compound recovery can only be achieved by physical methods, but this requires a perfect upfront separation of the different chemistries. This could prove difficult for the EOL batteries, but could apply to well-defined production scrap. It can lead to higher value of the recovered fractions, when sophisticated methods are used (see Section B. "OnTo Technology Process"), but the key concern is the adequacy of their chemistries and physical characteristics to the evolving and very stringent specification of the battery component industry.

One should, however, mention that the recovered components could also be reused in less demanding application, where value would be lower, but still higher than that of the metal value recovered in the other processes.

Basic physical separation can be done at small scale and are quite cheap; however, in the case of Li ion, the sophistication required for EHS concerns adds substantial cost to the processes.

Recovery yields of valuable fractions are seldom mentioned, but physical separation are never clear-cut, certainly when relative pure fractions are targeted.

D. Hydrometallurgy

Crushing and grinding and in some case calcination are needed prior dissolution. Co, Ni, and eventually Mn can be separated under elemental form or as their salts by classical processes of the Ni–Co hydrometallurgy. These processes are based on consecutive unit operations such as:

— Reductive leaching in acidic medium, mainly H_2SO_4 or HCl, leaving the graphite fraction in the residue.
— Cu removal by cementation, electrowinning, or precipitation as sulfide.
— Precipitation of Fe, Al, and other impurities, based on a combination of neutralization and oxidation at different pHs. Al and Fe recovery from the precipitation processes is not economically justified.
— Solvent extraction or ion exchange of impurities.

— Separation and recovery of Ni, Co, and Mn by solvent extraction, precipitation, crystallization, or electrowinning. These can further be transformed to new advanced materials for battery or other applications.
— Li salts can be produced on the final solution, but the economics of this recovery need to be justified.

High recovery yield of Co and Ni (over 95%) can be expected in optimal conditions, but overall yield depends on the yield of the up-front mechanical separation.

Contamination by fluor and organics can cause robustness problems and EHS issues.

For batteries based on LFP and LMO, valorization will be marginal.

The Recupyl process, based on these steps, is described in more details in Section D. "The Recupyl Process."

E. Pyrometallurgy

By melting in reductive conditions, an alloy is produced containing the Co, Ni, and Cu. Li and Al are deported to the slag phase. Fe and Mn are found in both fractions in proportions depending on the work conditions. Fluor is volatilized to the off-gasses, where it can be captured. Recovery rates are high since no mechanical separation is needed.

The alloy can further be handled by classical hydrometallurgical processes, similar to those described above, but cheaper due to the higher concentration of valuable elements and lower content of impurities.

Alternatively, a matte can also be produced. The process can be dedicated to batteries, or the batteries can be processed together with ores in a big smelter. As an example, US and Canadian Li-ion batteries collected by RBRC are processed in the Sudbury Glencore smelter.[21]

Pyrometallurgical processes are more robust, in the sense they can process a mixt of chemistries, but they are preferably applied at a big scale. For small batteries and packs up to E-bike types, and for modules of EV (i.e., a set of cells in series to reach the high voltage), direct processing can be applied, without prior discharging, inertization, nor crushing.

V. Description of Industrial Processes

A. Foreword

For obvious competition reasons, very little information can be found on costs, investments, recovery yields, environmental impact, nor valorization of the recovered fractions.

The legal aspects, that is, if the processes comply with the imposed recycling rates on weight basis, are mostly not addressed in the publications.

Therefore, the descriptions in next section are based on public information from publications, patents, and Web site of the companies, without guarantee that they fully correspond to the real industrial situation.

B. OnTo Technology Process

The process can be considered as an ultimate case of compound recovery. It is under development at the laboratory scale (1-L reactors) by Sloop with funding from the DOE.[22,23] The process aims at recycling positive electrode active material, after rejuvenation by different patented processes. As a variant, rejuvenation of full batteries is also proposed.

1. *Process Description*

As an example, the full process may include following steps (see Figure 8):

— Separate cells of a battery pack
— Discharge the cell to avoid electrocution
— Open the cell in a controlled atmosphere to avoid fire (low O_2 and moisture), for example, under N_2 or CO_2
— Passivate the negative electrode by controlled oxidation with air, or alternatively with water under liquid CO_2.
— Replenish the positive material with at least some Li by different processes:

- Solid-state synthesis, by mixing with Li_2CO_3 or other Li salt and firing

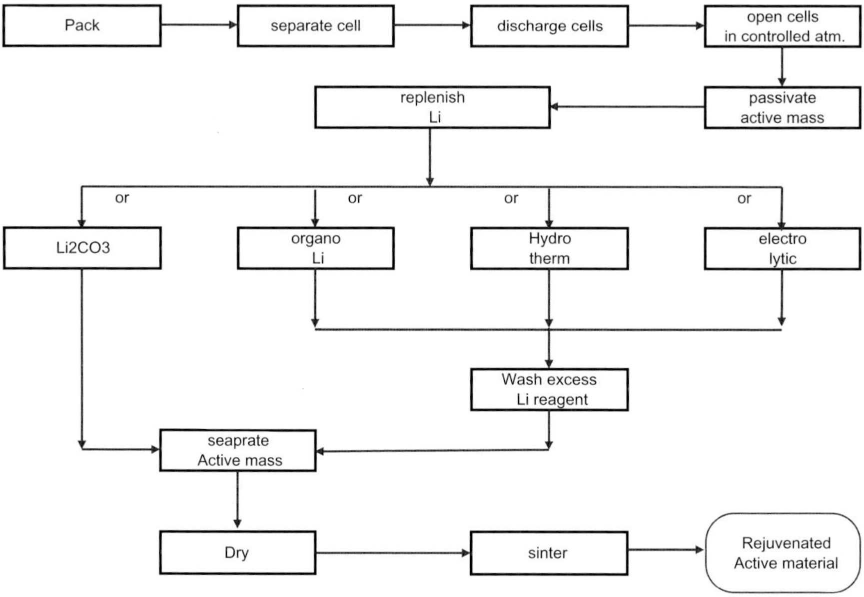

Figure 8. Schematic overview of the OnTo process. Adapted from Refs. 22 and 23.

- Hydrothermally in a solution of LiOH (e.g., for Li_{1-x} CoO_2, 4 M LiOH, 100–300°C during 12–48 h)
- By reduction with an organolithium (e.g., butyl Li)
- By electrochemical reduction

— Separate the Li-deficient positive active material, for example, $Li_{1-x}MeO_2$ or $Li_{1-x}FePO_4$. This operation can be done different methods like sieving, or filtration.
— Wash the residue with water or CO_2 to remove excess reagent.
— Dry the residue in dried air.
— Sinter the rejuvenated active material.

A variant of the process consists in applying a similar process to the positive electrode still in the battery, so as to rejuvenate the full battery:

(i) Similar to above
(ii) Similar to above
(iii) Similar to above but nondestructively
(iv) Introduction of a reductive Li-containing solution

(v) Let react to rejuvenate the positive electrode
(vi) Remove the electrolyte by supercritical CO_2 and replace by fresh electrolyte
(vii) Seal the cell

2. *Advantages of the Method*

The developers claim following advantages:

— Recovery of the most valuable compound
— Method is applicable to all battery chemistries
— Compound has similar performances as virgin material
— Firing time could be shorter than for virgin material

The performances of rejuvenated NMC and LMO show slightly lower capacities (–4%) and comparable power as virgin material.[23] The comparative electrochemical capacity of NMC is given in Figure 9.

3. *Comments*

Steps 1–4 are quite classical.

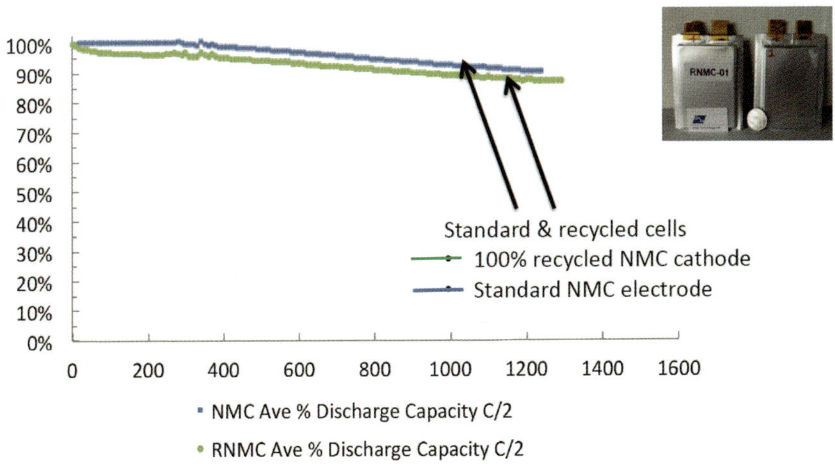

Figure 9. Performances of the rejuvenated cathode material by the OnTo process. Adapted from Ref. 23.

Steps 5–7 are key to the process; steps 5 and 7 could be inversed.

Step 6 has no details on how the positive active material is separated from the rest of the component nor how efficiently this operation can be realized. Contamination by, for example, copper current collector is critical.

Step 5i, adding the right amount of Li salt is critical to the success of the operation, as step 7 will not be applicable in this case. Finding the right ratio could be done by analysis, as stated in the patent; however, this will not work for a mixture of active mass coming from different batteries with different failure mechanisms, as will be the case in an industrial process. The average Li could be right but some particles will be overlithiated and others underlithiated.

For the other relithiation processes, step 7 could solve this problem.

Step 9, that is, sintering could be quite similar to what is done when synthesizing fresh materials.

The process has the perceived following drawbacks:

Only applicable to well-sorted battery cells, with known chemistry

Very critical separation of the positive electrode material from the rest of the components

Very critical adjustment of $LiCO_3$ addition in step 5i

No information on the recycling of the other components. Positive electrode represents only ~40% of the total weight

Industrialization to big quantities is very challenging

C. Retriev and Toxco Process

Retriev Technologies, the new name Toxco since September 1, 2013, has been recycling lithium batteries for over 20 years. Twenty million dollars were invested in the same year in a new recycling plant in Lancaster, Ohio.[24] The plant was foreseen to recycle large automotive batteries (Ni–MH and Li ion).

1. *Process Description*

The process is schematically given in Figure 10.

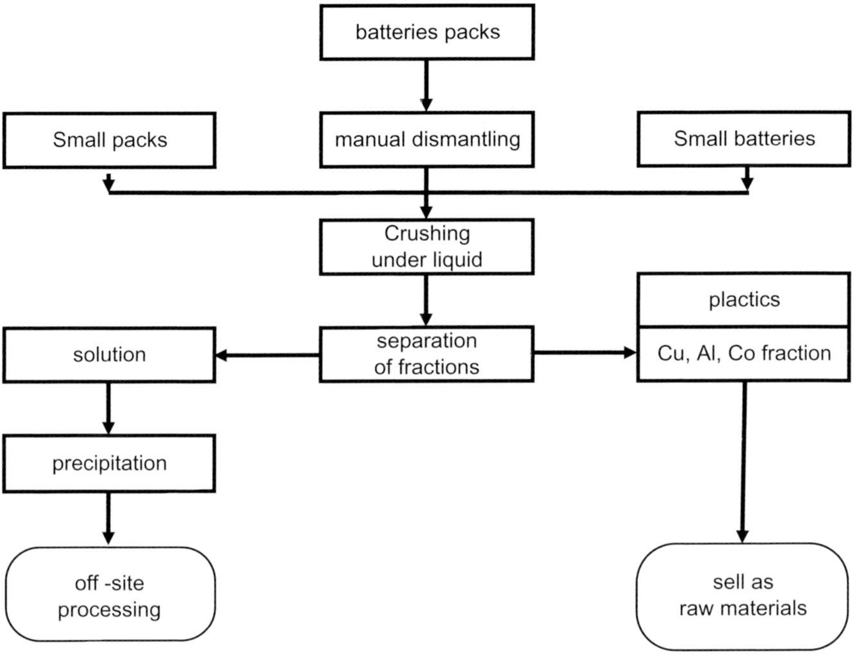

Figure 10. Schematic overview of the Retriev process. Adapted from Ref. 24.

The process consists in following consecutive steps:

— Dismantling

This operation is done manually and must be done with the utmost care, as a large battery packs contain enough energy to seriously injure those who attempt this process without the proper equipment or training.

— Crushing

This crusher operates at –200°C under liquid nitrogen to prevent fugitive emissions and to reduce the reactivity of processed batteries.

— Separation of fractions

The separation is done in aqueous alkaline conditions. Three fractions are separated: metal solids, metal-enriched liquid, and plastic fluff. The alkaline solution, containing the Li, is treated with Na_2CO_3 to recover

Li_2CO_3.The metal solid contains different amount of Al, Cu, and Co, depending on the chemistries of the processed batteries.

2. *Comments*

The further process of separation and valorization of the valuable fractions is not described. As such it is not a full recycling process, but more a preprocessing step.

D. Recupyl Process

Recupyl is a French recycling company located near Grenoble.[25]

1. *Description of the Process*

The Recupyl process for Li-ion batteries is based on hydrometallurgical treatment after inertization and crushing, and is schematically represented in Figure 11.[25,26]

It can be applied also to primary Li batteries and on mixtures of both.

The process consists in following consecutive steps:

— Inertization and crushing

The crushing is realized at room temperature under an Argon–CO_2 atmosphere, with a controlled oxygen content (typically 5000 ppm). The CO_2 and the oxygen slowly react with the Li-containing anode, so that next steps can be realized safely in air. The crushing is realized in two steps, first at low then at high speed.

— Separation of fractions

This is done by screening and magnetic separation and leads to the isolation of four fractions:

- A valuable fine fraction of mixed active material (positive electrode and carbon from negative)
- A magnetic fraction (steel from the casing)
- A dense nonmagnetic fraction (Cu and Al fractions from the current collectors)

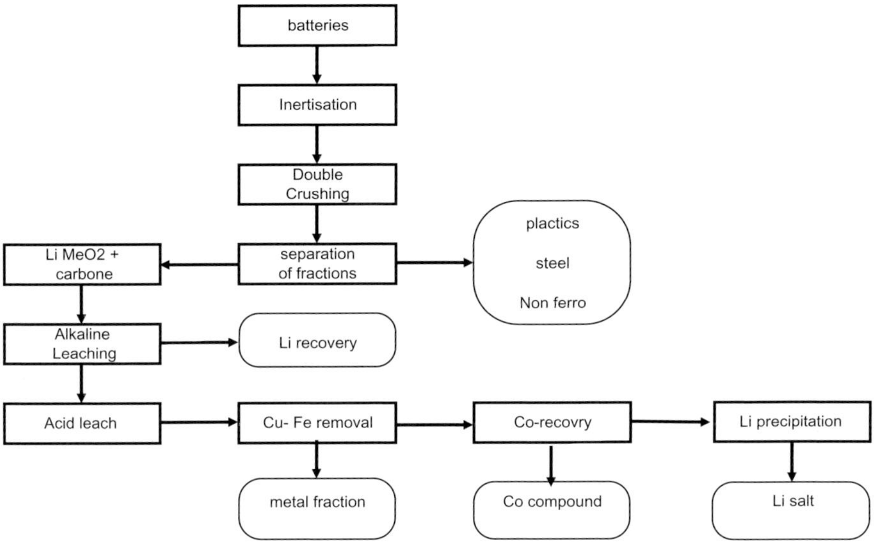

Figure 11. Schematic overview of the Recupyl process. Adapted from Refs. 25 and 26.

- A light fraction of paper and plastics (packaging and separator)

The fine fraction, after sieving at 500 μm, content less than 0.3% of Cu, and enters the hydrometallurgical process.

— Leaching of fine fraction
A first leaching is done at pH 12 with LiOH under heavy agitation to separate the active mass from residual Al of the current collector, and to dissolve the Li from the anode. The reaction leads to H_2 evolution. The reactor is blanketted with the gas from the crushing to maintain a partial pressure of O_2, and so avoid the risk of explosion. Another nasty reaction is the decomposition of the $LiPF_6$ with water, releasing HF gas according to the reaction $LiPF_6 + H_20 \rightarrow LiF + HF + POF_3$. Surprisingly, Recupyl states that this reaction does not take place in their work conditions.

— Recovery of Li
Li is precipitated as Li_2CO_3, making use of the CO_2 from the protective atmosphere of the crushing to reduce the pH to 9. As an alternative, Li is precipitated as Li_3PO_4 by addition of H_3PO_4, which leads to a much lower residual Li concentration in waste waters (400 mg/L i.o. 1.8 g/L).

— Acid leaching

The typical composition of the alkaline leach is Co: 34.5%, Li: 3.2%, Fe: 0.15%, Cu: 0.87%, C: 8.1%. It is washed and releached in sulfuric acid 2 N at 80°C. Fe shots are added as a reducing agent for the oxides of the positive electrode active mass. The leach residue contains the graphite from the anode.

— Cu and Fe purification

The residual Cu is precipitated by cementation with iron shots. Fe (and Al) is removed by hydrolysis at pH 3.85 with Na_2CO_3.

— Co–Mn recovery

Different possibilities are mentioned, depending on the feed (in particular the Mn/Co ratio).

For equal content of Co and Mn, electrolysis of Co is used. Co deposits at the negative electrode, whereas Mn precipitates at the anode as oxyhydroxide and is releached. Electrolysis is done at 55°C and 400–600 A/m².

For pure $LiCoO_2$ (LCO) feed (Mn < 2%), Co is precipitated in trivalent state by NaClO at pH 2.3–2.8.

— Recovery of Li

The Li coming from the leaching of the cathode is precipitated as Li_2CO_3 at pH 8.5 with Na_2CO_3, and further processed as in the alkaline leach stream.

2. *Comments*

The fine fraction to be leached is claimed to be free of fluor, which is surprising knowing the use of polytetrafluoroethylene (PTFE) and polyvinylidene fluoride (PDVF) as binder in the electrode formulations. The authors probably refer to soluble fluor compounds from the electrolyte. The behavior of Ni is not described. It will be codeposited in the electrolysis scenario, conversely the precipitation of trivalent Co is selective *versus* Ni, but not *versus* Mn.

E. Batrec Process

Batrec is a Swiss company, established in Wimmins.[27,28]

1. *Process Description*

The focus of their patents is on safe transportation, storage, and crushing of Li-containing batteries, by using a protective atmosphere of CO_2 at room temperature. Inertization is achieved by controlled inlet of moist air. The shredded batteries are leached in acid. No details are given on the further steps to recover the chemical substances. The process is schematically given in Figure 12.

2. *Comment*

The process focuses on primary batteries that are even more dangerous to reprocess due to the presence of very reactive Li metal negative electrode. It can, however, be used for Li-ion batteries.

As for the Toxco process described above, the Batrec process is not a complete recycling process, but must be considered as a first step of the chain.

F. Umicore Process

In 2011, the Belgian Umicore S.A. commissioned a plant for the recycling of batteries in its facilities in Hoboken, Belgium. The plant, with an investment of 25 M€ has a capacity to process 7000 t spent batteries per year.[29,30]

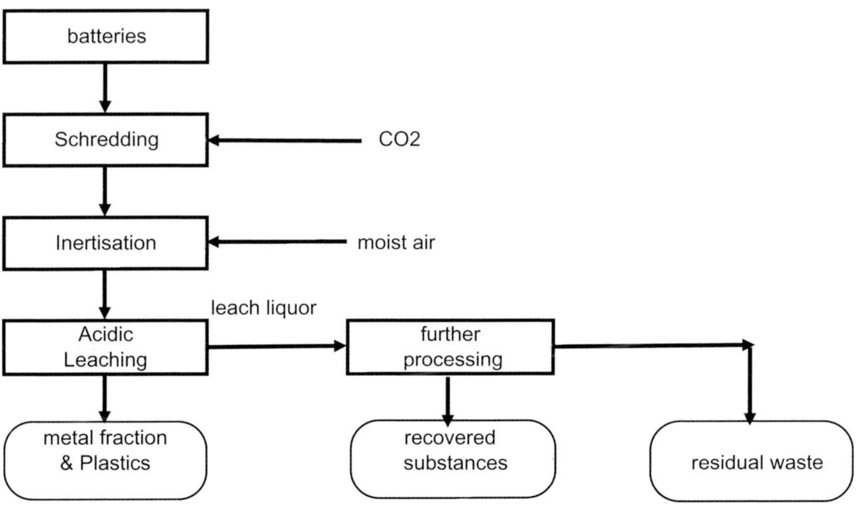

Figure 12. Schematic overview of the Batrec process. Adapted from Refs. 27 and 28.

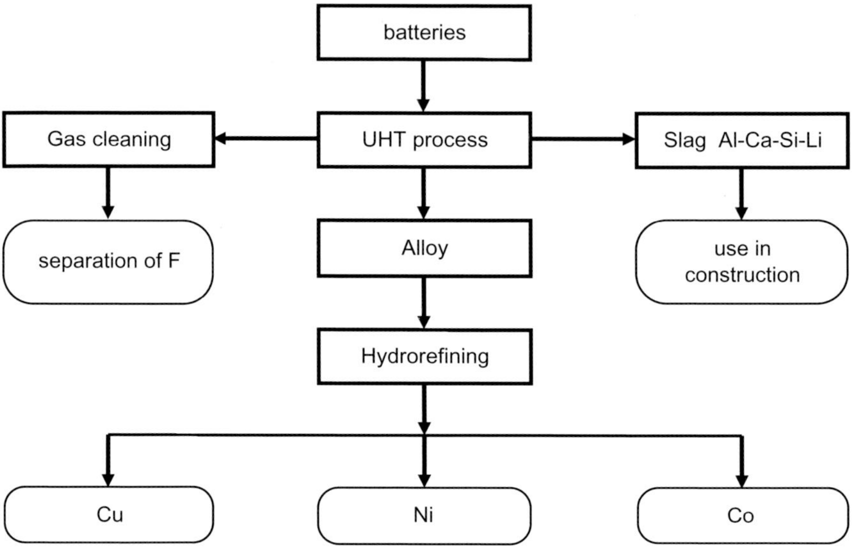

Figure 13. Schematic overview of the Umicore process. Adapted from Refs. 29 and 30.

1. *Process Description*

The process, called ultrahigh temperature (UHT), is a combined pyrometallurgical and hydrometallurgical route. The process is schematically given in Figure 13.

— Smelting

The batteries are introduced in the smelter without sorting, dismantling (except for big pack from EV), or crushing. Plastic and polymers provide the energy, whereas graphite, the C of the plastics, and the Al are used as reducing agents. The off-gases are treated to avoid the formation of any volatile organic compounds (VOCs) or dioxins, by an in-house developed process.

— Phase separation

The liquid alloy and the liquid slag are tapped separately.

The slag can be used in the cement industry.

— Treatment of the alloy
The alloy, after granulation or atomization, can be treated by hydrometallurgical processes in an existing Umicore plant, leading to

- a $NiSO_4$ solution that can be used in house as precursor for battery products like $Ni(OH)_2$ battery grade (BG) or NMC or sold for other applications and
- a $CoCl_2$ solution that is oxidized to Co_3O_4 BG, precursor for the in house synthesis of LCO, or sold for other applications.

— Recycling efficiency
As most of the materials are converted to useful products, or employed for feedstock recycling, the recycling efficiency exceeds 50% of the weight, so complying with the EU regulations. Overall, recovery yields of Co, Ni, and Cu by the combined pyro and hydroprocesses exceed 95%.

2. *Comments*

Li recovery from the slag is feasible but does not appear justified at the historic market conditions.[31]

G. Sony-Sumitomo Process

1. *Process Description*

The Sony-Sumitomo process presented in Figure 14 is a combination of pyro preprocessing and hydro treatment.[28]

— Calcination
The batteries are first calcined at 1000°C, which will open the cells and burn the flammable components as separator, solvent, and plastic casing.

— Magnetic separation
The calcined residue is separated in different fractions by magnetic and Eddy currents processes.

The metallic fraction contains the steel from the casing as well as Cu and Al from the current collectors.

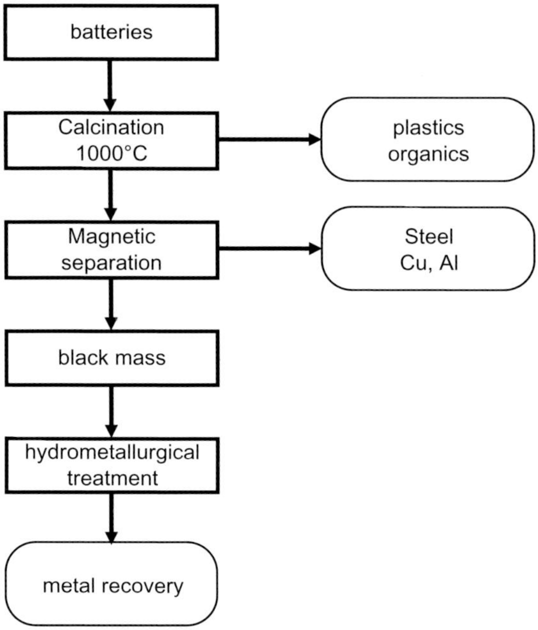

Figure 14. Schematic overview of the Sony-Sumitomo process. Adapted from Ref. 28.

The remaining fraction contains the electrode active materials, also called black mass.

— Hydrometallurgical treatment
The residue is treated by hydrometallurgical processes in Sumitomo, for the recovery of Co (and other transition metals). The Co can be converted to pure Co oxide that can be reused for the synthesis of new positive electrode materials.

2. *Comments*

Li is not recovered.

H. Review of Other Processes and Literature

Several review articles have been published the last years 2013–2016.[26,28,32–35] The great majority of the described processes are published by the scientific community and based on lab scale experiments.

In 2013, Chagnes and Pospiech published an extensive review on the hydrometallurgical route and listed over 20 processes only for the leaching, including mineral and organic acid and even bio-leaching, and described over 10 processes for the solvent extraction and separation of the impurities and valuable metals.[32]

In 2014, Meshram *et al.* published a comprehensive reviewed on the extraction of Li from primary and secondary sources, with emphasis on the hydrometallurgy, as Li could be easier to recycle via this route. He describes over 15 academic results of leaching in H_2SO_4 and HCl alone.[28] Other media as HNO_3, and organic acids are also described. Many variants of separation of valuable elements by SX processes are described in details.[28]

In 2014 and 2015, Zeng, from Tsinghua University, published critical reviews on recycling spent batteries with 86 references.[33,34] In 2015, Ekberg and Petranikova published an extensive review of Li battery recycling, including industrial as well as academic processes.[26] In 2016, a similar study was published by Ordonez *et al.*[35]

VI. New Chemistries

The only significant alternative to Li ion in 2016 is the polymer battery used in the blue car. The chemistry is based on LFP positive electrode and Li metal negative electrode. Intense research activity is ongoing worldwide to develop alternative chemistries. For more information on these chemistries, please refer to the previous chapters of this book. It should be stressed that most of these technologies are still in low technology readiness level (TRL) and market introduction is still far away.

Several road maps have been developed for the launching of advanced Li-ion or alternative technologies. Figure 15 is the vision presented by Avicenne.[2]

Furthermore, the ramp-up for a new technology is always quite slow. Li ion, introduced in 1995, represented only 2 GWh after 5 years and 8 GWh 10 years after, while after 20 year, it represents 60 GWh. This means that the recycling business for these technologies at EOL will not become significant before 2040. Nevertheless, the recycling possibilities of these new technologies will be briefly discussed.

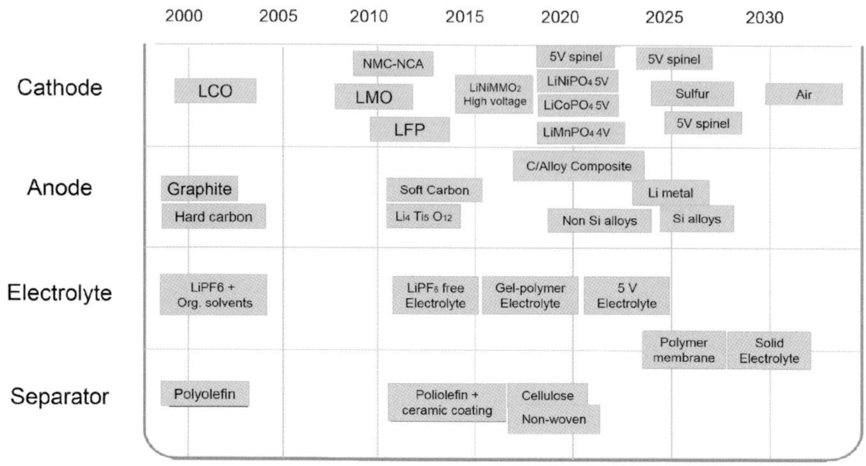

Figure 15. Road map of new applications. Adapted from Ref. 2.

Blue car

Valuable components are the Li metal negative electrode, the LFP positive electrode and the current collectors. Little information is provided on the recycling of the battery, but it can probably be realized in a similar way as for primary batteries.[36]

Advanced Li ion

In advanced Li ion, the increase of massic energy density will be realized by a combination of higher voltage, and higher capacities (in Ah/kg) of the positive electrode and negative electrode. This has already been taken into consideration in the calculation of the tonnages to be recycled.

NMC: The chemistries evolve from 33% Co in 1:1:1 to less than 20% in 6:2:2 reducing the potential value for recycling.

HNS and HLM contain low Co (5%–10%). Recycling techniques whatsoever will be similar to those developed for Li ion, while potential value of the recycled stream will decrease due to combined effect of higher energy content of the active materials and lower Co content. The weight reduction will not compensate the lower metal value per kilowatt-hour.

Negative electrodes

New negative electrodes could have double to triple capacity *versus* graphite; their tonnages will be halved or even reduced to one-third. Based on Si–C composites, the element content will even be reduced per kilogram of negative electrode, as bulk Si is cheaper than graphite. Based on Sn alloys, the value *per kilogram* could be maintained, depending on the content of the more expensive Sn.

Current collectors
Tonnage of Cu and Al are approximately 1 and 0.3 kg/kWh, respectively. With a loading of the positive electrode probably unchanged, the weight of Cu and Al will reduce accordingly to 0.9 and 0.25 kg/kWh, respectively. The thickness of the foils is expected to have reached a technical minimum.

Na ion
Also in this case, the recycling processes will be identical to Li ion. Metal value of Na (as Na_2CO_3) is much lower than that of Li_2CO_3, so Na metal value can be considered as lost.

Metal value of positive electrode and negative electrode will depend on the chemistries that are still largely undefined early 2016. Tonnages per megawatt will, however, be somewhat higher.

One of the claimed advantages of Na ion is the possible use of Al foil i.o. Cu foil at the negative electrode. This will reduce the metal value of the collectors with ~0.5 USD/kg batteries.

Li–air and Li–S
The value of spent batteries based on these technologies will come from their Li metal negative electrode. They should be treated in a similar way as primary Li batteries.

VII. Conclusions

The drivers to recycle Li ion are of economic and kt nature. Worldwide, new legislations are put in place to enforce recycling rates, Europe being a leader in the field.

The author's developed model leads to an estimation of the total recycling business equivalent to 18% of the total battery production in 2020, growing to 37% in 2025. Recycling will thus substantially contribute to reduce the stress on primary feedstock. Expressed in tons of material to be processed, the recycling will have to treat 130,000 metric tons (kt) of batteries in 2020 and ~350 kt in 2025.

The size of the market, expressed in value of the in theory recoverable components, is estimated at 560 million USD per year (MUSD/Y) in 2020 and 1.3 billion USD per year in 2025 (BUSD/Y). For some components, the economics of recycling will not be sufficiently attractive (in particular Li_2CO_3), so the business size must be considered as a maximum. For all chemistries, more than 75% of the value comes from the positive electrode (commonly called cathode), which justifies why most of the processes focus on this part of the batteries.

Although information on processing costs are not disclosed, they could be higher than the recoverable value, certainly in the case of Mn and Fe chemistries, where the maximum value does not exceed 2 USD/kg. Consequently, the business will be driven by regulation, more than by microeconomics.

The alternative concept of recovering and reuse the positive electrode component instead of its constituents would deliver an extra value of 2–3 USD/kg battery. However, recycling costs might be higher and the suitability of the recovered active material for reuse in new battery still needs to be validated.

The industrial processes are based on either mechanical separation, hydrometallurgy, pyrometallurgy, or often a combination of them. Mechanical separations can be realized at smaller scale, but are less selective. The sophistication of the processes required for the safety and a sufficient selectivity increases the costs of otherwise cheap processes. Hydrometallurgy can recover pure metals, but need a dangerous upfront crushing. Pyrometallurgy can process batteries and small pack without pretreatment, but can best be performed at a large scale.

At the chosen horizons, the impact of new chemistries on the recycling market will still be marginal.

VIII. Appendices

The tables in this chapter give the details of the calculations that lead to the table with same numbering in previous chapters. Tables 1 and 5 did not require extra details.

Table A1. Compound recovery model for NMC 1:1:1 chemistry

Battery			Compound Recovery		
Cost Structure	%	$/kWh	$/kwh	% Theory Recovered	% of Recovered
Cathode	16	27.2	27.2	16	84
Anode	6	10.2			
• Graphite			1.0	0.6	3
Current collectors		0,0			0
• Cathode	3	4.9	2.4	1.4	7
• Anode	1	1.2	0.2	0.1	1
Separator	10	17		0	0
		0			0
Electrolyte	8	13.6			0
• Li_2CO_3		0	0.1	0.1	0
Others	8	14.3			0
		0			0
Scrap	4	6.8	2.3	1.4	7
		0			0
Total	56	95.2	32.2	20	100
Total USD/kg battery		21	7		

Table A2. Element recovery model for NMC 1:1:1 chemistry

Battery			Element Recovery						
Cost Structure	%	$/kWh		$/kg	kg/kWh	$/kWh	% of Batteries	% Material Cost	% Theory Recovered
Cathode	16	27.2	Cathode	16.6	1.6	27.2	16		
		0	• Ni	10	0.33	3.3	2	3	18
		0	• Co	24	0.33	7.8	5	8	43
		0	• Mn	2	0.33	0.7	0	1	4
		0	• Li_2CO_3	6	0.60	3.6	2	4	20
		0				15.4	9.1	16	84
Anode	6	10.2	Anode				0	0	0
		0	• Carbon	0.2	0.8	0.2	0.1	0	1
Current collector		0,0	Current collector				0	0	0
• Cu	3	4.9	• Cu	5	0.5	2.4	1.4	3	13
• Al	1	1.2	• Al	1.2	0.2	0.2	0.1	0	1
Separator	10	17	Separator				0	0	0
		0	• PE/PP	0.5	0.05	0.0	0	0	0
Electrolyte	8	13.6	Electrolyte		0.4		0	0	0
• Li_2CO_3		0	• Li_2CO_3	5	0.02	0.1	0	0	1
Others	8	14.3	Others			0	0	0	0
		0					0	0	0
Scrap	4	6.8	Scrap			1.4	0.8	1	8
		0					0	0	0
Total	56	95.2				18.3	12	19	100
Total USD/kg battery		21				4			

Table A3. Impact of the chemistry on the potential value of the recycled battery

Chemistry	LCO			NMC 1:1:1			NCA		
Materials	% of Battery Price	Compound Recovery	Element Recovery	% of Battery Price	Comp Recovery	Element Recovery	% of Battery Price	Compound Recovery	Element Recovery
Cathode	22	0	15	16	16	9	16	16	9
Anode	6	0.5	0.1	6	0.5	0.1	6	0.5	0.1
Collectors	4	1.5	1.5	4	1.5	1.5	4	1.5	1.5
Separator	10	0,0	0	10	0,0	0	10	0,0	0
Electrolyte	8	0.1	0.1	8	0.1	0.1	8	0.1	0.1
Other	8			8			8		
scraps	4	2	1.7	4	1	0.8	4	1.7	1.3
% of battery cost	60	27	19	56	19	12	56	20	12
% of material cost		45	31		35	19		35	21
Total USD/kg	24	10	7	21	7	4	26	9	6

Chemistry	LFP			LMO		
Materials	% of Battery Price	Compound Recovery	Element Recovery	% of Battery Price	Compound Recovery	Element Recovery
Cathode	10	10	2.7	11	11	3.7
Anode	6	0.5	0.1	6	0.5	0.1
Collectors	4	1.5	1.5	4	1.5	1.5
Separator	11					
Electrolyte	8	0.1	0.1	9	0.1	0.1
Other	8			9		
Scraps	4	1	0.4	4	1	0.4
% of battery cost	51	13	5	53	14	6
% of material cost		25	10		26	11
Total USD/kg	16	4	1.6	15	4	2

Table A4. Details of calculation of the recycling rates per application of portable electronics and EV

Phones		Horizon 2020			Horizon 2025		
			Recycling Rate	Overall		Recycling Rate	Overall
Lifetime	Years	2			3		
EOL		20%	20%	4%	20%	25%	5%
Second life		80%			80%		
	Case A fraction	50%	50%	20%	50%	60%	24%
	Case B fraction	50%	0%	0	50%	25%	10%
Total rate				24%			39%

Computers/Tablets		Horizon 2020			Horizon 2025		
			Recycling Rate	Overall		Recycling Rate	Overall
Life time	Years	3			4		
EOL		20%	20%	4%	20%	25%	5%
Second life		80%			80%		
	Case A fraction	50%	50%	20%	50%	60%	24%
	Case B fraction	50%	0%	0	50%	25%	10%
Total rate	After 2 years			24%			39%

xEV		Horizon 2020			Horizon 2025		
			Recycling Rate	Overall		Recycling Rate	Overall
Scraps	Years	1			1		
	% of batteries		10			8	
	Recycling rate (%)		90	9		90	7
EOL	% of batteries	8	90		8	92	
	Recycling rate (%)		90	81		90	83
Second life			PM			PM	
Total rate (%)				90			90

Table A5. Details of calculation of the recycling rates per application of industrial and ESS/power tools and E bikes

Industrial and ESS		Horizon 2020			Horizon 2025		
			Recycling Rate	Overall		Recycling Rate	Overall
Scraps	Years	1			1		
	% of batteries		3			2	
	Recycling rate (%)		0.9	3		0.9	2
EOL	% of batteries	15	97		15	98	
	Recycling rate (%)		90	87		90	88
Second life			PM			PM	
Total rate (%)				90			90

Power tools/E bikes		Horizon 2020			Horizon 2025		
			Recycling Rate	Overall		Recycling Rate	Overall
Scraps	Years	1			0		
	% of batteries		3			2	
	Recycling rate (%)		0.9	3		0.9	2
EOL	% of batteries	3	97		3	98	
	Recycling rate (%)		50	49		60	59
Second life			PM			PM	
Total rate (%)				51			61

Table A6. Discounted recycling rates per application for LCO at two horizons

Year	2020					2025				
Application	GWh	Recycling (%)	% in application	% Recycling	% of application	GWh	Recycling (%)	% in application	% recycling	% of application
Portable	40	24	83	20	65	56	33	93	31	66
EV	27	21				41	48			
EV China	50	11				80	45			
Indus	10	13				20	12			
Others	12	30	17	5	45	20	40	7	3	14
Total		18	100	25			39	100	33	
Total MWh	139		32	8		217	37	40	13	

Table A7. Discounted recycling rates per application for NMC at two horizons

Year	2020					2025				
Application	GWh	Recycling (%)	% in application	% Recycling	% of application	GWh	Recycling (%)	% in application	% recycling	% of application
Portable	40	24	38	9	33	56	33	24	8	31
EV	27	21	37	8	47	41	48	25	12	45
EV China	50	11	0	0	0	80	45	9	4	8
Indus	10	13	15	2	50	20	12	21	3	77
others	12	30	10	3	28	20	40	21	8	77
Total		18	100	22			39	100	35	
Total MWh	139		34	7		217	37	73	26	

Table A8. Discounted recycling rates per application for NCA at two horizons

Year	2020					2025				
Application	GWh	Recycling (%)	% in application	% recycling	% of application	GWh	Recycling (%)	% in application	% recycling	% of application
Portable	40	24	8	2	2	56	33	5	2	2
EV	27	21	92	19	37	41	48	92	44	45
EV China	50	11		0	0	80	45		0	0%
Indus	10	13		0	0	20	12		0	0%
Others	12	30		0	0	20	40	3	0	3
Total		18	100	21			39	100	46	
Total MWh	139		11	2		217	37	20	9	

Table A9. Discounted recycling rates per application for LMO at two horizons

Year	2020					2025				
Application	GWh	Recycling (%)	% in application	% recycling	% of application	GWh	Recycling	% in application	% recycling	% of application
Portable	40	24		0	0	56	33			
EV	27	21	10	2	4	41	48	10	5	4
EV China	50	11	38	4	9	80	45	69	31	16
Indus	10	13	29	4	35	20	1	15	2	14
others	12	30	23	7	23	20	40	6	2	5
Total		18	100	18			39	100	40	
Total MWh	139		12	2		217	37	18	7	

Table A10. Discounted recycling rates per application for LFP at two horizons

Year	2020					2025				
Application	GWh	Recycling (%)	% in Application	% Recycling	% of Application	GWh	Recycling (%)	% in Application	% recycling	% of application
Portable	40	24				56	33			
EV	27	21	6	1	11	41	48	4	2	6
EV China	50	11	90	10	92	80	45	92	42	76
Indus	10	13	3	0	15	20	12	3	0	10
Others	12	30	1	0	4	20	40	0	0	
Total		18	100	12			39	99	39	
Total MWh	139		51	6		217	37	66	2	

IX. References

1. https://www.lme.com/
2. Pillot, C. The rechargeable Battery Market and Main Trends 2015–2025. *33rd International Battery Seminar & Exhibit.* **2016**.
3. Gaines, L.; Dunn, J. Lithium-Ion Battery Production and Recycling Materials Issues Project ID: ES229 Argonne National Laboratory VTO Annual Merit Review **2015**. http://energy.gov/eere/vehicles/downloads/vehicle-technologies-office-merit-review-2015-lithium-ion-battery-production
4. Fox-Davies. The Li market September 2013. http://www.globalstrategicmetalsnl.com/_content/documents/405.pdf
5. Oliveira, L.; Messagie, M.; Rangaraju, S.; Sanfelix, J.; Hernandez Rivas, M.; Van Mierlo, *J. Clean. Prod.* **2015**, *108*(Part_A), 354–362.
6. http://ec.europa.eu/growth/sectors/raw-materials/specific-interest/critical_en
7. http://www.ebra-recycling.org/sites/default/files/EBRA%20PR-%20BatteryStatistics_year2012_0.pdf
8. http://data.worldbank.org/data-catalog/GDP-ranking-table
9. http://www.ewaste2010.org/sites/default/files/Brisbane%20e-waste%202010-07.pdf
10. http://www.gao.gov/Products/GAO-10-626
11. http://www.interpol.int/Public/EnvironmentalCrime/Meetings/Ewaste2010/UNBASEL.pdf
12. http://eur-lex.europa.eu/LexUriServ/site/en/oj/2000/l_226/l_22620000906en00030024.pdf
13. http://www.rechargebatteries.org/knowledge-base/batteries/lithium-ion-cell-lilon/
14. http://eur-lex.europa.eu/LexUriServ/LexUriServ.do?uri=CELEX:32006L0066:EN:NOT
15. http://www.meti.go.jp/english/information/data/cReEffectLe.pdf
16. http://www.env.go.jp/en/laws/recycle/06.pdf
17. http://www.insg.org/presents/Mr_Kunitomo.pdf
18. http://www.call2recycle.org/laws.php?c=1&d=79&e=104&w=2&r=Y
19. http://europa.eu/rapid/press-release_IP-15-6203_en.htm
20. http://retrievtech.com/
21. RBRC http://www.call2recycle.org/physical-flowchart/
22. http://www.onto-technology.com/home
23. Sloop,S.; OnTo Technology LLC Advanced Battery recycling. http://energy.gov/eere/vehicles/downloads/vehicle-technologies-office-merit-review-2014-advanced-battery-recycling
24. Coy, T.; Toxco http://energy.gov/eere/vehicles/downloads/lithium-ion-battery-recycling-facilities-0
25. http://www.recupyl.fr/68-hydrometallurgie.html
26. Ekberg, C., Petranikova, M., 2015. Lithium Batteries Recycling. In Lithium Process Chemistry; Światowska, A.C., Ed.; Elsevier; Amsterdam, 2015; 233–267.
27. http://www.batrec.ch/en/Recycling-Services

28. Meshram, P.; Pandey, B. D.; Mankhand, T. R. *Hydrometallurgy* **2014**, *150*, 192–208.
29. Hagelüken, C; Treffer, F. Umicore *Fachtagung Ressourcenschonung München Apr. 2016* https://re-source2016.info/sites/re-source2016.info/files/documents/09_-_resource_2016_muenchen_hagelueken.pdf
30. http://www.batteryrecycling.umicore.com/UBR/
31. Elwert, T.; Strauss, K.; Schirmer, T.; Goldmann, D. *World Metall–Erzmet.* **2012**, *65*(3), 163–171.
32. Chagnes, A.; Pospiech, B. *J. Chem. Tech. Biotechnol.* **2013**, *88*(7), 1191–1199.
33. Zeng, X.; Li, J.; Singh, N. *Crit. Rev. Environ. Sci. Technol.* **2014**, *44*(10), 1129–1165.
34. Zeng, X.; Li, J.; Liu, L. *Renewable Sustainable Energy Rev.* **2015**, *52*, 1759–1767.
35. Ordonez, J.; Gago, E. J.; Girard, A. *Renewable Sustainable Energy Rev.* **2016**, *60*, 195–205.
36. http://www.bluecar.fr/les-batteries-lmp-lithium-metal-polymere

4 Na-Ion Batteries — State of the Art and Prospects

Patrik Johansson,*,§ Patrick Rozier†,§ and M. Rosa Palacín‡,§

*Department of Physics, Chalmers University of Technology, SE-412 96 Göteborg, Sweden
†University of Toulouse III Paul Sabatier, CIRIMAT CNRS UMR 5085, 118 route de Narbonne, Toulouse, Cedex 09 31062, France
‡Institut de Ciència de Materials de Barcelona, ICMAB-CSIC, Campus de la UAB, 08193 Bellaterra, Catalonia, Spain
rosa.palacin@icmab.es
§ALISTORE-ERI European Research Institute, FR CNRS 3104, 80039 Amiens Cedex, France

Table of Contents

List of Abbreviations

AN	acceptor number
BET	Brunauer–Emmett–Teller
BMIm	1-butyl 3-methyl imidazolium
BPOE	bipolar porous organic electrode
CV	cyclic voltammetry
CVD	chemical vapor deposition
DEC	diethyl carbonate
DEME	diethyl methyl methoxyethyl ammonium
DFEC	doubly fluorinated ethylene carbonate
DFT	density functional theory
DMC	dimethyl carbonate
DME	dimethoxy ethane
DN	donor number

DSC	differential scanning calorimetry
EMC	ethyl methyl carbonate
EMIm	1-ethyl 3-methyl imidazolium
EC	ethylene carbonate
ES	ethylene sulfite
ESW	electrochemical stability window
FEC	fluoro-ethylene carbonate
FSI	bis(fluorosylfonyl)imide
FTIR	fourier transform infrared
GIC	graphite intercalation compound
GPE	gel polymer electrolyte
HC	hard carbon
HFP	hexafluoropropylene
HOMO	highest occupied molecular orbital
HSAB	hard and soft acids and bases
IL	ionic liquid
Im	imidazolium
LIB	Li-ion battery
LUMO	lowest unoccupied molecular orbital
NASICON	Na super ionic conductors
NMR	nuclear magnetic resonance
NVP	$Na_3V_2(PO_4)_3$
NVPF	$Na_3V_2(PO_4)_2F_3$
OCV	open-circuit voltage
PAN	poly(acrylonitrile)
PBA	Prussian blue analogue
PC	propylene carbonate
PEG	poly(ethylene glycol)
PEGDME	poly(ethylene glycol) dimethyl ether
PEO	poly(ethylene oxide)
PMMA	poly(methyl methacrylate)
PVdF	poly(vinylidene fluoride)
Pyr	pyrrolidinium
RT	room temperature
SAXS	small-angle X-ray scattering
SEI	solid electrolyte interphase

SEM	scanning electron microscopy
SIB	sodium-ion battery
SIC	single-ion conductor
SN	succinonitrile
SPE	solid polymer electrolyte
SS	stainless steel
TDI	4,5-dicyano 2-(trifluoromethyl)imidazole
Tf	CF_3SO_3
TFSI	$N(SO_2CF_3)_2$
THF	tetrahydrofuran
ToF-SIMS	time of flight secondary ion mass spectrometer
VC	vinylene carbonate
XEVs	(hybrid) electric vehicles
XPS	X-ray photoelectron spectroscopy

I. Introduction

Those who are familiar with Li-ion batteries (LIBs) and seeing their imminent success in so many areas may ask themselves: Why scatter the research focus onto alternative battery technologies rather than trying to create even better LIBs? This is the question to answer before dwelling into the parts of the periodic table that theoretically can hold promise for alternative technologies — if the answer is any other than just by scientific curiosity.

With new manufacturing technologies emerging and materials evolving continuously for the benefit of the LIB technology, and thereby approaching the theoretically attainable limits, the prospects for attaining even higher performance by other battery technologies through innovative research become appealing. Another motivation for developing alternative battery technologies stems from concerns related to a future-limited lithium supply. This is associated to a more than probable enlarged and generalized implementation of the LIB technology from portable electronics to large-scale applications, such as different kind of XEVs ([hybrid] electric vehicles) and grid storage to balance the contributions from renewable sources. It is also of utmost importance to remember that a limited supply is not the same thing as simply looking at the resources (for example, elemental abundance in the crust of the Earth); the reserves, the most important measure, are a much more complex interplay of various properties related to availability, production, recycling, geographical/political constraints, and not the least cost issues. The actual Li reserves are thus dynamic, with the resources being the absolute maximum. To add to this, the use of lithium for other products/sectors, for example, glasses, ceramics, lubricants, and metallurgy, today accounting for almost 50% of the Li used globally, is foreseen to increase by another 3–4% annually the next decades.[1]

For the LIB technology, the theoretical performance in terms of energy density depends on the electrochemical capacities and operating potentials of the negative and positive (intercalation) electrodes. Thus, and most important, this is not in any way related to the capacity of lithium metal. Hence, there is no reason to beforehand assume an alternative metal-ion battery technology to by necessity exhibit lower energy densities, based on the higher redox potential for the metal as compared to lithium metal. The main issue to consider is rather that the lowest possible negative electrode potential limits are set by the standard redox potential of the metals, only somewhat higher for sodium (~330 mV) than for lithium.

The insertion electrodes used and needed for any metal-ion battery concept should be able to reversibly react with the largest possible amount of ions at suitable potentials, that is, close to those of the pure metals for the negative electrode, to realize high total cell voltage, and exhibit fast metal diffusion to enable high power rates (Figure 1). Moreover, they need to be stable in presence of the electrolyte, which although may decompose in contact with the highly oxidizing/reducing electrodes is expected to form a solid stable passivation layer, often denoted the solid electrolyte interphase (SEI) to enable successful cell operation. The theoretical capacity of an intercalation compound depends on the total number of the pair electron/guest ion that can be transferred to the host structure and does not depend on the nature of the guest ion itself. The maximum practical reversible capacity is then controlled by the number of vacant sites accessible in the host and can be dependent on the structural changes occurring during the intercalation process. The chemical similarity between Li and Na and early studies on intercalation of alkali metals

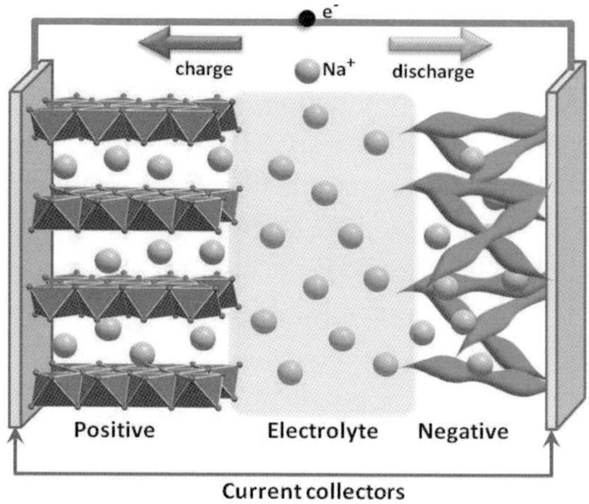

Figure 1. Schematic representation of a sodium-ion battery.

should warrant fast progress for sodium-ion batteries (SIBs) if the know-how from the LIB field could be directly implemented. From a practical point of view, this is largely the case, although some critical differences have to be considered, as will be discussed later in this chapter. Besides the fact that sodium mineral resources are "unlimited," attainable at low cost, and geographically distributed, SIBs exhibit an additional practical benefit: low-weight and less expensive aluminum current collectors can be used for both positive and negative electrodes (as sodium, in contrast to lithium, does not form alloys with aluminum).

Back in the 1980s and early 1990s, the progress in SIBs paralleled that of LIBs and several electrochemical cells were assembled using first TiS_2 and later on Na_xCoO_2 as positive electrode materials and liquid or polymer electrolytes. The first full SIB cells, using $Na_{15}Pb_4$ as negative electrodes, were also reported at the time.[2,3] Unfortunately, this research path was largely abandoned due to the advent and success of the LIB technology for portable electronics and its prospects of much more generalized fields of application. The interest in SIBs started to reemerge by 2010 and is now progressing at rapid pace; several hundreds of scientific papers are published each year, resulting in a significant number of new both positive and negative electrode materials presented (Figure 2), as well as electrolytes, and also several review papers focusing either on the

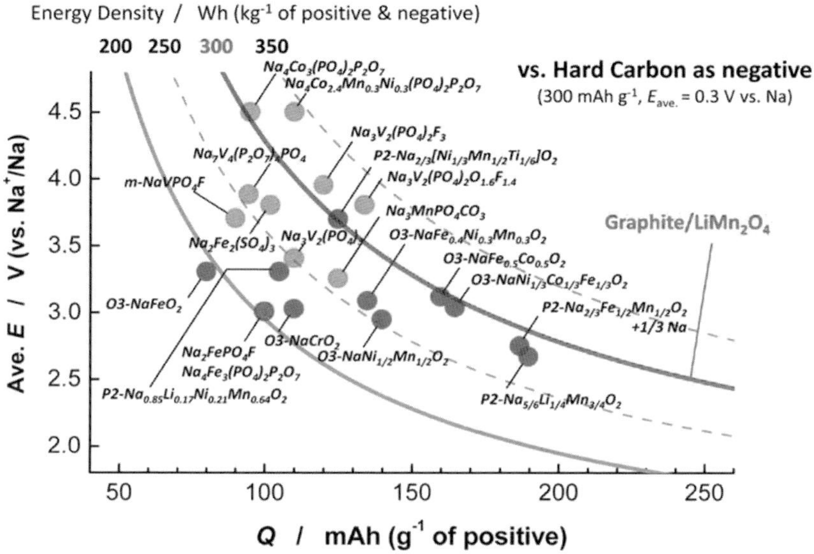

(a)

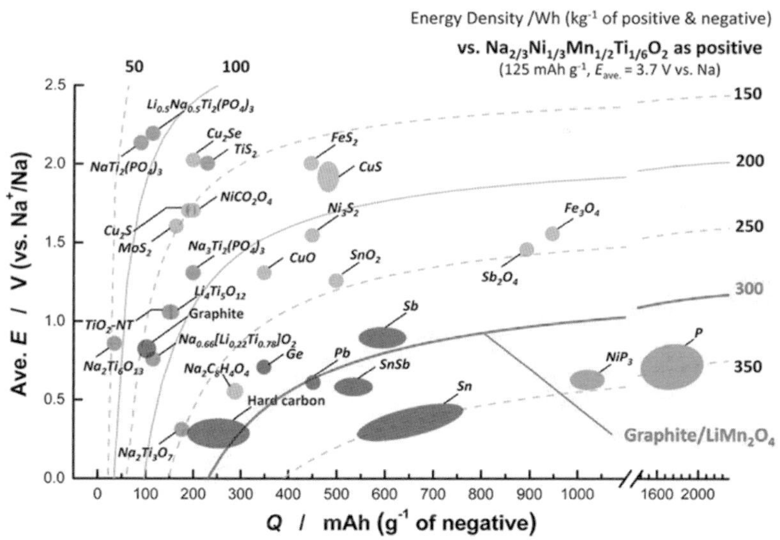

(b)

Figure 2. Average voltage (Ave. E) and energy density (Wh/kg) *versus* gravimetric capacity (mAh/g) for (a) positive and (b) electrode materials for SIBs. Taken from Ref. 4 with permission of ACS.

SIB technology as such or specific aspects.[4–6] This revival has also triggered investigations dealing with other Na-based battery technologies, for example, using Na metal anodes in room temperature (RT) Na/S or Na/air battery concepts, much following the progresses in (and accompanying problems of) the Li/S and the Li/air battery technologies, and the former in part using knowledge from the already established high-temperature Na/S battery technology.[7,8] Yet another alternative Na-based technology uses aqueous electrolytes for the benefit of low cost,[9] but at the expense of energy density by a low cell voltage. Neither of these technologies will be treated herein.

This chapter surveys the state-of-the-art SIB technology starting with separate sections devoted to SIB positive (high potential) and negative (low potential) electrode and electrolyte materials, where after a summary of reported SIB cell performance, and finally a perspective on future SIB development is given.

II. Positive Electrode Materials

The search for positive electrode materials for SIBs started back in the 1980s in parallel to the development for LIBs, underlining the expected similarities between the two alkali metals. Though the commercialization of the LIB technology largely obscured the research dedicated to the SIB for more than 30 years, the renewed interest has prompted an increasingly growing number of reports. Without surprise, most of the families of compounds investigated are analogs to those already studied for LIBs, and can roughly be classified in two families: (1) layered oxides — Na_xMO_2 with M including at least one transition metal and (2) polyanionic-based 3D structures $Na_xM_y(XO_4)_z$ — again with M including at least one transition metal and X being mainly P or S. Both families can also include other ions such as fluorine and each has its own major feature(s) — for the layered compounds, it is the versatility of compositions, whereas for the polyanionic materials, it is the stability of the structural framework and the high operating voltages attainable. In the quest of lower-cost and more sustainable materials, a third "metal-free" family — organic materials — is, while still in its infancy subject of increasing interest, paralleling the efforts of organic LIB research. The following sections aim to provide the most important characteristics of SIB-positive electrode materials and describe the main achievements made.

A. Layered Oxides

The term "layered oxides" denotes compounds with the general formula A_xMO_2 (A = alkaline metal; M = transition metal) adopting structures related to α-$NaFeO_2$.[10] The specificity of this family resides in the fact that, to release the strain induced by the large size difference between A and M, these elements tend to order to form alternating A and M planes parallel to the anion's compact planes (Figure 3a). This ordering is described using a hexagonal cell (space group *R*-3*m*) with the *c* parameter corresponding to the direction perpendicular to the planes and the *a* parameter defining the in-plane dimension. Usually, layered oxides are described as built from MO_6 octahedra sharing edges to form MO_2 layers stacked along the *c* direction, the alkaline species being located in the interlayer space (Figure 3b). To determine whether a certain A_xMO_2 compound will adopt a layered structure or not, one of the gauges is the difference between the ionic radii of the A and M cations. The larger the difference the easier the ordering is. Given the larger radius of Na^+ (1.02 Å) as compared to Li^+ (0.76 Å), ordered Na_xMO_2 phases are more easily achieved than Li_xMO_2. For example, the similarity in radii of Mn^{3+} in high spin

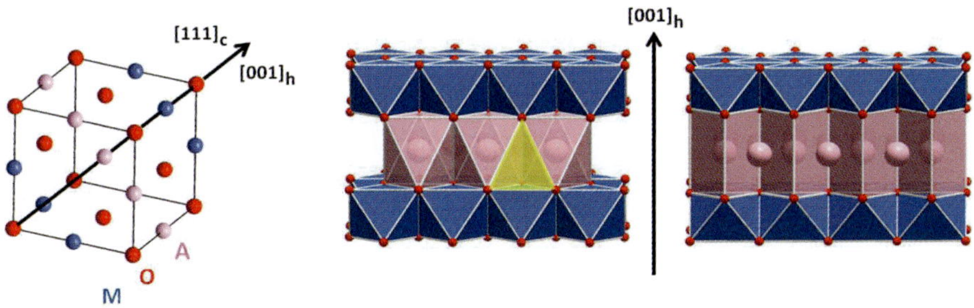

(a) (b) (c)

Figure 3. Schematic description of layered structures: (a) ordering of the occupancy of A and M species in the octahedral sites between arrays of anions and the correspondence between directions in the cubic and hexagonal unit cells. Representation of the MO_2 layers (blue) stacked to form (b) O-type structure (octahedral sites on purple and tetrahedral in yellow) and (c) P-type structure (prismatic sites in purple).

configuration (0.65 Å) and Li^+ (0.76 Å) prevents stabilization of layered $LiMnO_2$ while a stable layered structure is achieved for $NaMnO_2$.[11,12] This opens for the possibility to synthesize layered compounds with a wider range of compositions with Na than for Li. Even though many A_xMO_2 compounds adopt a structure identical to α-$NaFeO_2$, others present similar MO_2 layers and stacking sequences, but with the alkaline species located in tetrahedral rather than octahedral sites. Moreover, some compounds present similar MO_2 layers, but with a different stacking sequence defining sites with a trigonal prismatic coordination in the interlayer space rather than octahedral or tetrahedral (Figure 3c). The nature of the alkaline species governs the preferences in site occupation resulting in mainly octahedral and tetrahedral sites for Li and octahedral or prismatic sites for Na. To take into account all these possibilities, the term "layered structure" is extended to include all compounds characterized by the stacking of MO_2 layers whatever the stacking sequence and the coordination adopted by the alkaline species in the interlayer space. To rationalize this, Delmas *et al.* proposed a classification of the different polymorphs that can be exhibited using a (1) capital letter indicating the coordination of the interlayer alkali species — O for octahedral, T for tetrahedral, and P for prismatic and (2) number denoting the number of MO_2 layers in the unit cell.[13] Using this classification, α-$NaFeO_2$ is O3-$NaFeO_2$ to indicate that the Na ions are located in octahedral coordination and that 3 MO_2 layers are needed to recover the periodicity of the stacking.

Using this classification, the $NaMO_2$ layered structures tend to mainly adopt the O3-type (Figure 3b) and the P2-type (Figure 3c) structures, even though some other stacking sequences leading to coexistence of different coordinations can sometimes be observed. Closely related to their difference in structure, the two polymorphs present additional differences both in terms of accessible compositions and expected behavior.

The O3-Na_xMO_2 phases are isostructural to O3-Li_xMO_2 except for the increased interlayer space resulting from the larger radius of Na. The Na ions are located in the octahedral sites sharing faces with empty tetrahedral sites in the MO_2 layers and edges with both NaO_6 and MO_6 octahedra. This occupancy leads to interatomic Na–Na and Na–M distances large enough (> 3 Å) to allow the full occupancy of accessible sites and to reach a final composition of Na_1MO_2.

In contrast, the P2-A_xMO_2, which has never been observed for A = Li as Li^+ is not stable in prismatic coordination, is very common for A = Na. Here, the Na^+ ions are located in two types of prismatic sites sharing triangular bases with empty tetrahedral or occupied octahedral sites in the MO_2 layer. Despite shorter Na–M distances for the latter (2.8 Å) than for the former (3.15 Å), implying larger repulsion, no clear evidence of a preferential occupancy has ever been reported. These prismatic sites share their rectangular faces and the short resulting Na–Na distance (1.7 Å) induces high repulsive effect preventing their simultaneous occupancy. The maximal composition considering half occupancy of each site would also here be Na_1MO_2, but most reported P-type compounds exhibit an upper composition limit close to $x = 2/3$ with no clear reason for the Na deficiency.

The synthesis of Na_xMO_2 layered oxides is easy, using conventional ceramic routes, if M is a single transition metal. If, however, different M species are to be included, irrespective of whether or not they are electrochemically active, the synthesis becomes more complex and often soft chemistry routes are needed to allow homogeneous mixing of all M elements *via* dissolution of reactants in an appropriate solvent. Even though soft chemistry routes are well-known to allow synthesis of metastable compounds, the most thermodynamically stable (O3 and P2) polymorphs are typically obtained as almost all the reported synthesis protocols end with a high-temperature treatment. Experimentally, the tendency to achieve either the P- or O-type structure is driven by the Na/M ratio. For Na contents below 2/3, the P2 polymorph is stabilized, while increased Na contents leads to the O3 polymorph. This limit cannot be considered definite as with an adequate selection of M it can be extended up to 0.85 (see below). In addition, once a P2-type structure is achieved using conventional synthesis routes, an extra treatment, more or less energetic such as ball-milling with metallic Na or electrochemical insertion, allows for extra sodiation, leading to almost stoichiometric P2-$NaMO_2$.[14]

The structural differences, mainly in terms of Na coordination, also lead to different electrochemical behavior. O-type structures are characterized by the simultaneous presence of octahedral and tetrahedral sites in the Na^+ layers both accessible to transition metals. The irreversible migration of M in these sites during the electrochemical processes can lead to progressive capacity fading. For example, O3-$NaFeO_2$, which is electrochemically active, presents a theoretical capacity of 242 mAh/g. However, oxidizing this compound above 3.5 V leads to an irreversible migration of Fe^{4+} into the Na^+ layers and a rapid capacity loss upon cycling while limiting the cutoff voltage to 3.4 V as a remedy enables only 80 mAh/g reversible capacity.[15,16] For P-type structures, as the prismatic sites of the Na layers are highly unfavorable for transition metals, this migration cannot occur, thus allowing better capacity retention upon cycling. Another difference resides in the Na^+ diffusion, which in O-type structure occurs through the narrow tetrahedral sites (Figure 3a), thus becoming kinetically limited. For the P2-type structure, the Na^+ ions can easily diffuse through the large rectangular faces shared between two adjacent prismatic sites (Figure 3b), leading to high ionic conductivities and contributing to enhanced electrochemical performance.[17] On the basis of the above, the electrochemical behavior of P-type compounds is often stated to be superior to O-type compounds. This is illustrated in Figure 4 where the electrochemical performance of the two polymorphs of $Na_x[Mn_{1/2}Fe_{1/2}]O_2$, as determined in half-cells *versus* metallic Na, is compared.

Such studies use *half-cells* (with sodium metal counter electrodes) rather than *full cells* (with a SIB negative electrode material such as hard carbon [HC]), which enables to compensate the Na deficiency of

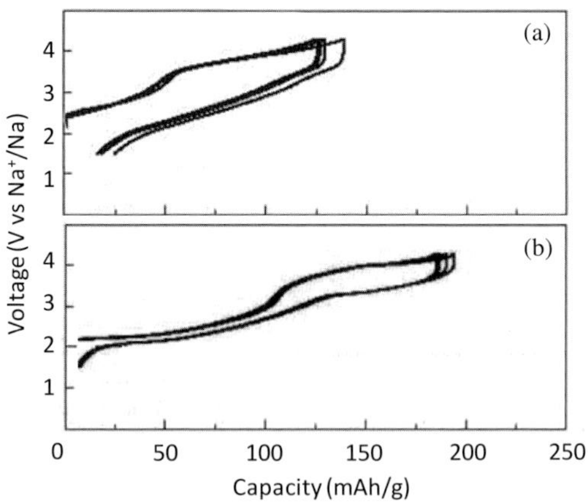

Figure 4. Voltage profiles (collected after the second cycle) of (a) O3-$Na_x[Fe_{1/2}Mn_{1/2}]O_2$ and (b) P2-Nax$[Fe_{1/2}Mn_{1/2}]O_2$. Adapted from Ref. 18.

P2-$Na_x[Mn_{1/2}Fe_{1/2}]O_2$ and after one full charge–discharge cycle, both compounds can be considered to be fully sodiated and their behavior directly compared. Comparing the data collected after the second cycle, the O3-type polymorph shows a reversible capacity of only 120 mAh/g and large polarization in agreement with the kinetic limitations, whereas the P2-type polymorph exhibits a reversible capacity of 190 mAh/g with smaller polarization.[18] Nonetheless, despite better electrochemical performance for P-type as compared to O-type, the Na deficiency needed to stabilize the P-type polymorph is the main drawback for usage in full SIB cell configuration, as without any additional source of sodium, only two-thirds of the theoretical capacity can be reached: ca. 150 mAh/g.

All the discussions above deal with the pristine compounds assuming that the structure type is maintained upon cycling. However, changes in the sodium content and the oxidation state of the transition metal during the electrochemical processes can drive structural changes. For example, gliding of MO_2 layers does not require large energy and is sometimes observed, which induces a conversion from O-type to P-type and vice versa, and can in fact lead to an unexpected better electrochemical behavior of O3-type as compared to P2-type pristine materials. This has been observed for Na_xCoO_2 where during oxidation, the lower performance O3-type $NaCoO_2$ converts to P3-type with enhanced ionic conductivity, whereas P2-type $Na_{2/3}CoO_2$ converts to lower performance O2-type when oxidized to high voltages.[19–21] Thus, the P-type polymorph will always present better electrochemical performance than the O-type, but the polymorph adopted by the pristine material should not be considered as the real criterion as structural changes can, and often will, occur. The limited knowledge about the parameters governing these conversions prevents any reliable predictions on interesting compositions with respect to performance. Nevertheless, the large radius difference between Na^+ and transition metal cations enables the synthesis of layered compounds with a large variety of compositions and allows to determine the main characteristics associated to selected elements, using know-how acquired in the LIB field, and to take advantage of synergetic effects to reach better performance.

Starting with single M element systems, $NaMO_2$ layered compounds have been reported for M = Fe, Mn, Co, Ni, Cr, and V, all of them electrochemically active, but with significant differences. Both $NaCrO_2$ and

$NaNiO_2$ exhibit reversible capacities of 120 mAh/g, but also significant capacity fading, due to several irreversible structure changes and the formation of inactive phases, often associated with the migration of oxidized transition metal ions.[22–24] $NaVO_2$, which can be achieved in both O3- and P2-type polymorphs and shows a stepped voltage profile attributed to the existence of several intermediate phases resulting from the ordering of remaining Na^+ ions in the interlayer space, but has a too low operating voltage to be of interest to SIBs.[25–27] O3-$NaFeO_2$ exhibits a theoretical capacity close to 242 mAh/g, but is in practice limited to 80–100 mAh/g due to irreversible structural changes taking place above 3.4 V related to migration of Fe^{4+} into the tetrahedral sites in the Na layers.[15,16] The Na_xCoO_2 compounds have been studied since 1980, and can be synthesized in several P and O polymorphs, but all of them exhibit low capacities: 70–120 mAh/g in a voltage range narrowed down to prevent the structural changes occurring when operating at higher potentials.[19,28–31] Moving to Mn, $NaMnO_2$ exhibits a larger theoretical capacity than $NaCoO_2$ at a lower cost. In addition to the O3- and P2-type structures, it can be also achieved in two other polymorphs: O2-type and Birnessite-type.[32,33] All are electrochemically active and exhibit high practical capacities: 180 and 219 mAh/g for the O3-type and the Birnessite-type, respectively.[32,34] However, strong Na–Na repulsion and Na–M interactions, associated with the Jahn–Teller effect of Mn^{3+}, induce large distortions and strains and thereby structural degradation and poor cycling performance.[32,34,35] Thus, no compounds containing a single transition metal compound exhibit an electrochemical performance suitable for practical application. Yet, the results derived from their study have enabled rationalization of composition-dependent trends. In turn, this has prompted, as for LIBs, focused research to benefit from synergetic effects of the ionic diffusivity of P-type structures in $NaCoO_2$, the high specific energy of Na_xMnO_2, and the high redox potential of Na_xFeO_2, and design materials with overall improved electrochemical performance.

Among these, P2-$Na_{2/3}[Mn_{1/2}Fe_{1/2}]O_2$, enlisting only low-cost elements, has been reported to exhibit a capacity of 190 mAh/g at an average potential of 2.7 V and thereby an energy density of 520 Wh/kg, quite comparable to the 530 Wh/kg of $LiFePO_4$.[18] However, oxidation at high potential induces conversion to a mixed OP4-type structure with alternating interlayers with prismatic and octahedral–tetrahedral sites. Unfortunately, on further reduction and subsequent insertion/extraction processes, the Fe^{4+} ions migrate to the newly created tetrahedral sites, hindering the Na^+ ion diffusion leading to poor cycling performance with 20% capacity fading over 30 cycles.[36] To prevent this, a Fe-poor P2-$Na_{2/3}[Fe_{1/3}Mn_{2/3}]O_2$ compound was prepared that exhibits enhanced cyclability without too much penalty on the capacity that is close to 190 mAh/g during the first few cycles and stabilizes at 152 mAh/g after 40 cycles.[37]

To totally avoid the risk of migration of Fe^{4+}, a replacement of Fe by Ni has been envisaged — as the phenomenon has not been reported to occur with Ni. Furthermore, when using Ni^{2+} rather than Fe^{3+}, oxidation of the Jahn–Teller active Mn^{3+} to nonactive Mn^{4+} is needed to balance the charge. This prevents the large Jahn–Teller distortion of the structure and the capacity loss, because the electrochemical inactivity of Mn^{4+} is compensated for by the introduction of two redox couples, Ni^{3+}/Ni^{2+} and Ni^{4+}/Ni^{3+}. Indeed, O3-$Na_1[Ni_{1/2}Mn_{1/2}]O_2$ with active Ni^{2+} and inactive Mn^{4+} exhibits a capacity of 125 mAh/g between 2.2 and 3.8 V, which can be further increased up to 185 mAh/g if the cutoff potential is increased to 4.5 V, but this is at the expense of a dramatic decrease in capacity retention: from 80% after 50 cycles to 60% after 20 cycles.[38] The oxidation at high voltages induces a loss of almost all the Na^+ ions in the interlayer space, which are no longer screening the repulsive effect between MO_2 layers, which is also enhanced due to the higher oxidation state of M. This is associated with an increase of the interlayer distance becoming large enough to allow irreversible partial intercalation of solvent from the electrolyte. To prevent this phenomenon, the previously reported negative effect of the migration of Fe^{4+} can be turned as a positive feature to anchor the MO_2 layers. This delicate balance between the positive and negative effects of Fe^{4+} migration is controlled by the amount of Fe: $Na[Ni_{0.4}Mn_{0.4}Fe_{0.2}]O_2$ delivers an initial capacity of 131 mAh/g and maintains 95% capacity after 30 cycles, whereas $Na[Ni_{0.25}Mn_{0.25}Fe_{0.5}]O_2$ shows a reversible capacity of 140 mAh/g and 90% is retained after 50 cycles.[39,40]

On the other hand, compounds containing Co in combination with low-cost Fe have also been investigated, the former being experimentally known to enhance the electronic conductivity and the Na^+ diffusion.[21] $Na[Fe_{1-x}Co_x]O_2$ with different Fe/Co ratios show similar reversible capacities of ca. 160 mAh/g and high rate capabilities, 100 mAh/g at 30°C.[41] The effect of Co on the performance of Fe-free layered oxides has also been evaluated; O3-$Na[Ni_{1/3}Mn_{1/3}Co_{1/3}]O_2$ exhibits a reversible capacity of 120 mAh/g and impressive cycling performance, as almost no capacity fading is observed after 50 cycles.[42] However, as most P-type compounds, it presents a sensitivity to moisture due to Na^+ extraction, sometimes associated with proton or H_2O intercalation, implying that without specific care taken, Na-deficient phases will be obtained — with a decreased capacity that cannot be compensated for in a full SIB cell.[43]

The benefits of combining Fe, Mn, Ni, and Co have prompted synthesis of phases with a large variety of compositions. Single, binary, ternary, and even quaternary systems have been evaluated. As for the latter, O3-$Na[Mn_{1/4}Fe_{1/4}Co_{1/4}Ni_{1/4}]O_2$ exhibits a reversible capacity of 180 mAh/g at an average voltage of 3.2 V corresponding to a specific energy of 578 Wh/kg.[44] The better performance than $Na[Fe_{1/2}Co_{1/2}]O_2$ and $Na[Ni_{1/2}Mn_{1/2}]O_2$ confirms the synergetic effect. There is no doubt that other optimized compositions may allow additional performance improvements, tough maybe only to a moderate extent.

In parallel with the above, the effect of including electrochemically inactive nontransition metal elements acting only as structure stabilizers for the $NaMO_2$ phases has also been evaluated. This was first demonstrated by, unuseful due to gravimetric penalty, heavy elements such as Sb and Te. $NaNi_{2/3}Sb_{1/3}O_2$ exhibits a reversible capacity of 133 mAh/g, matching the theoretical value, 70% of which is still maintained after 500 cycles.[45] The presence of Sb and Te seems to buffer the volume changes derived from the redox processes, and the same effect, despite less pronounced, has been also observed for Ti or Zn, but the most promising results have been reported with lighter elements such as Mg and Li.[46,47] Doping $NaMnO_2$ layered compounds with a small amount of Mg induces an increase in Mn^{4+}, to balance charge, smothering the voltage profile and decreasing the polarization, both confirming the stabilization of the structure and the suppression of the structural transformations (Figure 5a,b). P2-$Na_x[Mn_{1-y}Mg_y]O_2$ shows an initial capacity of 150 mAh/g, 95% of which is retained after 30 cycles between 1.5 and 4 V.[48] The capacity is

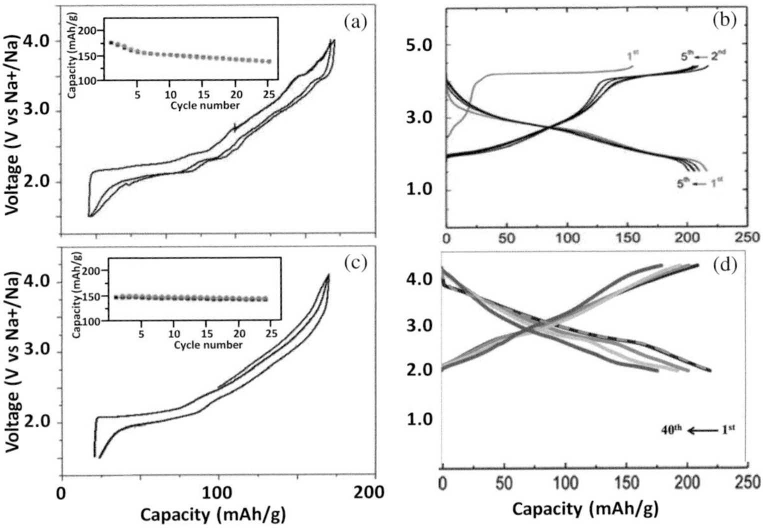

Figure 5. Voltage profiles for (a) $Na_{2/3}MnO_2$ and (b) $Na_{2/3}Mn_{0.8}Mg_{0.2}O_2$ charged up to 4.1 V (adapted from Ref. 48); (c) $Na_{2/3}Mn_{0.7}Mg_{0.3}O_2$ charged up to 4.4 V (adapted from Ref. 49) and (d) $Na_{0.95}Li_{0.15}Ni_{0.15}Mn_{0.55}Co_{0.1}O_2$ (adapted from Ref. 50).

increased up to 220 mAh/g when oxidized to 4.5 V (Figure 5c), a voltage above the theoretical maximum, thus suggesting the activation of an extra redox couple.[49] Doping with Li, for example, O3-$Na_{0.95}[Li_{0.15}Mn_{0.55}Co_{0.10}Ni_{0.15}]O_2$ (Figure 5d), leads to the same structural stabilization and smothering of the voltage profile and decreasing polarization. A reversible capacity of 200 mAh/g is observed, 85% of which is retained after 40 cycles.[50] O3-$Na_{0.78}[Li_{0.18}Ni_{0.25}Mn_{0.58}]O_2$ exhibits 240 mAh/g, which considering its average operation potential yields an energy density of 675 Wh/kg, which is among the highest value ever reported.[51] The exact role of Li is still subject to controversy, mainly because both its location in the MO_2 layer and/or the interlayer space and its behavior upon cycling are not yet clearly understood.

In addition to the enhanced stability induced by Li or Mg doping, the high voltage redox activity is associated to an increase in capacity sometimes beyond the theoretical value. To explain this extra capacity and in light of results reported for Li-rich Li_2MO_3 compounds, it has been suggested that besides the conventional cationic redox couple involving transition metals it is possible to activate the oxygen anionic redox couple.[49] This has been demonstrated using $Na_2(Ru_{1-x}Sn_x)O_3$ as a model compound where the Na^+ ions occupy the interlayer space, as commonly observed in layered $NaMO_2$ and also one-third of the sites normally occupied by M, which indicates why it can be formulated as $Na[Na_{1/3}Ru_{2/3}]O_2$ and thus usually termed as an Na-rich layered compound. Its electrochemical behavior shows that, in addition to the Ru-based redox couple, the O redox couple is activated and responsible for the observed extra capacity.[52] The mastering of the ordering of the Na–M species in the MO_2 layers seems to enhance the reversibility of this process, enabling good capacity retention upon cycling.[53]

To summarize, the versatility of the chemical composition of layered Na_xMO_2 allows tuning the electrochemical performance. Studies of compounds containing single M elements have enabled to rationalize the trends for each of the active transition metals which in turn has enabled the design of alternative phases enlisting multiple M elements to benefit from synergetic effects and layered compounds have already proven practically viable in SIB prototypes (see Section V). The combination of inactive elements, acting as structure stabilizers, and active transition metals allows reaching electrochemical performance close to the cathodes used in LIBs. Yet, there are some aspects to be improved such as the operating potential and capacity retention. Although the chemical substitution on the cationic sublattice has been widely studied, almost no work dealing with the investigation of the effect of anionic substitution has been reported. Thus, playing with partial substitution of oxygen with more electronegative elements such as fluorine can be a way to tune the operating voltage of $Na_xMO_{2-y}X_y$ layered structures. Closely related, the discovery that anionic redox couples can be activated open the way to exploration of Na-rich layered oxides. The use of the know-how already acquired for LIBs enables fast research progress, which with no doubt will lead to further improved performance.

B. Polyanionic Compounds

The structures of polyanionic compounds studied as positive electrode materials for SIBs are built from a combination of mainly tetrahedral XO_4 (X = P, S) and octahedral transition metal MO_6 units forming diverse open

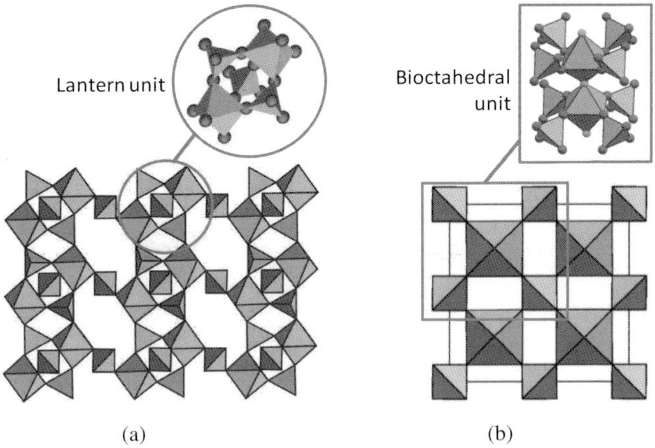

Figure 6. Schematic representation of (a) $A_3B_2(PO_4)_3$ NASICON and (b) $Na_3V_2(PO_4)_2F_3$ structure types. The inset displays the building units.

frameworks with low-energy Na^+ migration pathways (Figure 6). As for LIBs, they constitute the second most investigated family of compounds for use as positive electrode materials. Despite the weight penalty arising from the presence of the redox inactive polyanions, the strong covalent character of the X–O bond leads to several positive attributes: stable frameworks providing long-term structural stability, essential for extensive cycling and safety issues, and tunable redox voltages, by selecting the nature of X, higher than for the layered oxides.[54,55]

The interest in Na-containing polyanionic compounds started in 1976 with the discovery of the high Na^+ ion conductivity for the family of compounds with the general formula $A_3B_2(PO_4)_3$, approaching the values of β-Al_2O_3, which lead to the denomination of NASICON: Na Super Ionic CONductors.[56,57] The phases with B = Zr were first studied as solid-state electrolytes, but their potential as electrode materials was realized in the mid-1980s, mainly by Torardi and Delmas, replacing the inactive Zr element by Fe and Ti.[58–63] Since then, a lot of research has been devoted to tune the redox potential of the active element using the inductive effect proposed by Goodenough, and to tune the capacity *via* the amount of alkaline species needed to balance the transition metal oxidation state and the global charge of the polyanion.[54,55]

One of the main drawbacks of these compounds is their poor electronic conductivities due to the absence of any direct M–O–M connections, preventing overlap of orbitals. In a few cases, a direct M–O–M connection is observed, for example, in the maricite-type structure adopted by $NaFePO_4$ exhibiting a 1D chain of edge-sharing FeO_6, but still the electronic conductivity remains too low and at the same time, the loss of Na^+ ion migration pathways induces a poor ionic conductivity.[64,65] To overcome this issue, a lot of work has been devoted to mastering of particle sizes and/or coating with conductor additives — work, despite only briefly mentioned in this section, of tremendous importance to obtain good electrochemical performance, just as demonstrated for polyanionic materials such as $LiFePO_4$ for LIBs.

1. *Phosphate Based*

$NaFePO_4$, the chemical sodium analog to $LiFePO_4$, does not exhibit the same triphylite structure but adopts the thermodynamically stable maricite-type structure, wherein the Na^+ ions are isolated from each other by the PO_4 tetrahedra. This induces a large barrier for Na^+ migration leading to poor electrochemical performance.[66] To overcome this issue, $NaFePO_4$ with the triphylite type structure has been prepared by ion exchange from the stable $LiFePO_4$ using chemical or electrochemical routes. This metastable form of $NaFePO_4$ is electrochemically active, reversibly exchanging almost 1 Na without any structural change.[64,67,68] The reduction curve shows two plateaus attributed to the formation of a Na^+ ion–ordered intermediate phase $Na_{0.7}PO_4$ and the redox mechanism enlists two successive two-phase domains. This, which is not observed for $LiFePO_4$, seems to be responsible for the poor electrochemical kinetics and the resulting moderate electrochemical performance.

The NASICON-type $A_3B_2(PO_4)_3$ compounds seem more promising due to their stable 3D open framework which, built up by an arrangement of edge-sharing octahedral MO_6 sharing corners with tetrahedral XO_4, to form the so-called lantern unit $M_2(XO_4)_3$, exhibits large tunnels for fast Na^+ ion conduction (Figure 4a). Among the various NASICON compounds, $Na_3V_2(PO_4)_3$ (NVP) has received particular attention due to its

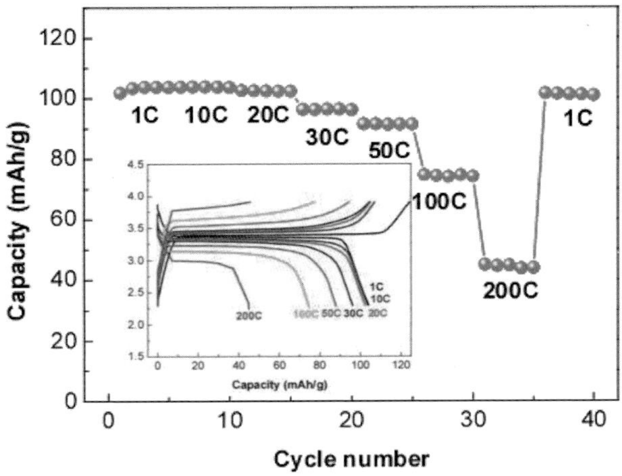

Figure 7. Electrochemical performance of carbon-coated nanoparticles of $Na_3V_2(PO_4)_3$ showing high rate performance and capacity retention. The inset displays the galvanostatic charging–discharging profiles. Adapted from Ref. 70.

high energy density, ~400 Wh/kg, and thermal stability in the oxidized state.[69] The electrochemical curve is characterized by the presence of two plateaus at 3.4 and 1.6 V corresponding to the redox couples V^{4+}/V^{3+} and V^{3+}/V^{2+}, respectively, where the latter is at a too low potential for application as a SIB positive electrode material. The use of adapted synthesis processes to decrease the particle size allows to prepare nanostructured composites with carbon coating, enabling impressive cycling stability, almost 100% capacity retention, after 1000 cycles at 1C (50% capacity retention after 1000 cycles at 100C Figure 7).[70,71]

The synthesis of fluorine-containing compounds to increase the operating voltage has also been tested. The extra Na needed to balance the charge induces a Na/M ratio > 1 and thus these compounds can be termed as Na rich. Na_2FePO_4F adopts a 2D structure built up from bioctahedral $Fe_2O_7F_2$ units made from face-sharing FeO_4F_2 octahedra connected *via* bridging F atoms to form chains and joined by PO_4 tetrahedra to form $[FePO_4F]$ infinite layers in between which the Na^+ ions are located.[72] Almost 80% of the theoretical capacity of 120 mAh/g is retained after 10 cycles with two plateaus at 2.9 and 3.06 V, only a little bit higher than in $NaFePO_4$.[73] In this case, ordering of Fe^{2+}/Fe^{3+} is observed, which induces the formation of paired dimers with a strong localization of electrons leading to poor electrochemical kinetics.[74]

The benefits of using fluorine are much more evident in NASICON-type structures containing vanadium. The first studies of $Na_3(VO_{0.8}PO_4)_2F_{1.4}$ demonstrated outstanding electrochemical performance with a reversible capacity of 120 mAh/g at 3.8 V leading to an energy density around 520 Wh/kg and very good capacity retention: 95% after 100 cycles at 1C rate.[75] This compound belongs to the family $Na_3(VO_{1-x}PO_4)_2F_{1+2x}$ and the end member with the composition $Na_3V_2(PO_4)_2F_3$ ($x = 1$) (NVPF) shows the highest theoretical energy density and has been extensively investigated.[76] Its structure is built up from VO_4F_2 octahedra sharing one fluorine to form $V_2O_8F_3$ bioctahedra units (Figure 6b). These share corners with tetrahedral PO_4, defining tunnels in which the Na^+ ions are located.[77] The electrochemical behavior (Figure 8) confirms the outstanding performance with reversible extraction of two Na^+ at an average voltage of 3.95 V leading to an energy density of ca. 510 Wh/kg.[78] The electrochemical mechanism enlists the V^{4+}/V^{3+} redox couple and corresponds to a complex sequence of structural transformations with an extremely small volume change (< 3%), which seems to be the reason behind the impressive capacity retention and rate performance reported in full SIB laboratory cells and further confirmed in upscaled 18650-sized

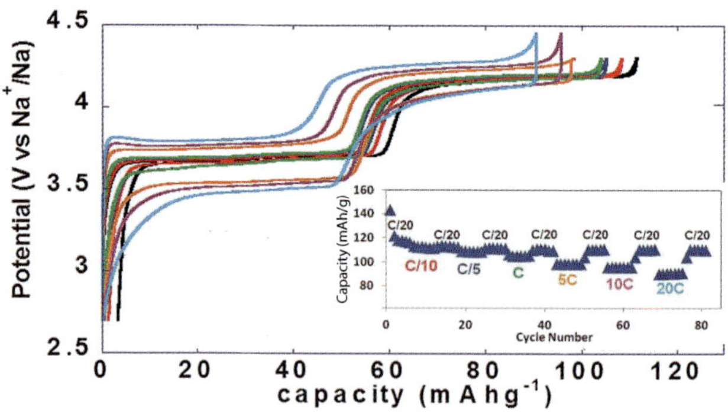

Figure 8. Voltage *versus* capacity profiles for $Na_3V_2(PO_4)_2F_3$ recorded at C rates ranging from C/20 up to 20C for both charge and discharge. The inset displays the discharge capacity *versus* cycle number at diverse rates. Taken from Ref. 80 with permission of Royal Society of Chemistry.

SIB prototypes (see Section V).[79,80] The extraction of the third Na^+ ion has never been achieved, probably due to the high redox voltage of V^{5+}/V^{4+} in these materials and a drop in the Na^+ ion diffusivity due to strong coupling of the remaining ions. Computational work suggests that the partial replacement of F by the less electronegative Cl should be able to decrease the redox voltage enough to ensure a full extraction of Na^+ ions and that the introduction of defects by aliovalent substitution in the anionic sublattice should decrease the redox voltage and increase the Na^+ ion diffusivity, with a combined effect large enough to reach energy densities close to 760 Wh/kg.[81,82]

Finally, the reports on $Li_2FeP_2O_7$ as a possible positive electrode material for LIBs pushed for many studies on the homologous $Na_2MP_2O_7$ (M = Fe, Mn, and Co) pyrophosphates. The Fe- and Co-based compounds show an average redox voltage not drastically increased by the presence of P_2O_7 groups, but $Na_2MnP_2O_7$ shows the highest ever reported Mn^{3+}/Mn^{2+} redox potential: 3.6 V *versus* Na^+/Na. This is unfortunately not enough to compensate for the weight penalty induced by the pyrophosphate group and the theoretical capacity is only 95 mAh/g for a wide potential window: 2–4 V.[83]

2. *Sulfates*

The sulfate SO_4^{2-} anion forms stronger covalent bonds than phosphate and should thus have a larger inductive effect to raise the redox potential further. A major achievement was obtained with $Na_2Fe_2(SO_4)_3$, which exhibits the highest Fe^{3+}/Fe^{2+} redox potential ever reported, 3.8 V *versus* Na^+/Na — even higher than the Fe^{4+}/Fe^{3+} redox potential observed in oxides.[84] This compound adopts the alluaudite-type structure in which two FeO_6 octahedra share edges to form Fe_2O_{10} dimers connected to each other by the SO_4 groups to form a 3D framework with suitable pathways for Na^+ diffusion. The compound exhibits a smooth voltage profile with a reversible capacity of 102 mAh/g, 85% of the theoretical capacity based on the exchange of 1 Na, leading to an energy density of 460 Wh/kg. The redox mechanism involves the formation of a solid solution with a very small volume change (1.6%), which is most likely the reason behind the high reversibility and impressive rate performance.

Fluorosulfates that adopt the tavorite-type structure also reveal interesting features compared to the phosphate analogs, notably in the increased operating voltage.[85] However, being moisture sensitive, their synthesis requires careful handling and the limited ion mobility observed along with the large volume changes during the Na extraction–insertion process result in electrochemical performance much poorer than for their Li counterparts.[86–88]

3. *Other Polyanionic Compounds*

Trying to take advantage of polyanionic compounds in stabilizing the 3D networks and/or raising the redox potential, the combination of polyanions has also been explored. For example, $Na_4M_3(PO_4)_2(P_2O_7)$ (M = Fe, Mn, Ni, and Co) adopts the same structure that can be described as built up from layers constituted by edge- and corner-sharing MO_6 octahedra and PO_4 tetrahedra. These layers are connected by P_2O_7 dimers to form a 3D network in which Na ions are located in tunnels.[89] The capacity for the Fe-based compound approaches 100 mAh/g, thus higher than for $Na_2FeP_2O_7$, at an average potential of 3.2 V *versus* Na^+/Na corresponding to the exchange of 3 Na and full activation of the Fe^{3+}/Fe^{2+} redox couple. Good capacity retention and rate performance are reported for both Na and Li cells and interpreted as a result of small volume changes (4%) on operation.[90] Other compounds with mixed polyanions PO_4/P_2O_7 and PO_4/CO_3 have also been reported, but in spite of the high redox potential most show poor reversibility.

Partial replacement of oxygen by nitrogen has also been reported to form PO_3N nitrido phosphates. The increase in the covalency related to inductive effect of N^{3-} in P–N bonds induces a higher redox potential for the active element. This has been demonstrated for the Ti^{4+}/Ti^{3+} redox couple which equals 2.1 V in NASICON-type $Na_3Ti_2(PO_4)_3$ and 2.7 V in $Na_3Ti_2(PO_4)_2(PO_3N)$.[91] This allows to take benefit of a stable structure *versus* Na exchange as shown by a smooth voltage profile characteristic of a solid solution profile with good cycling behavior, again explained by a very small volume change (< 1%). Of course only one Na can be reversibly exchanged (Ti^{4+}/Ti^{3+} redox couple), thus leading to small capacity, but this can in principle be greatly improved by selecting other multivalent elements.

Prussian blue analogues (PBAs) with the general formula $A_xMM'(CN)_6.yH_2O$ (*y* being the number of H_2O molecules needed to

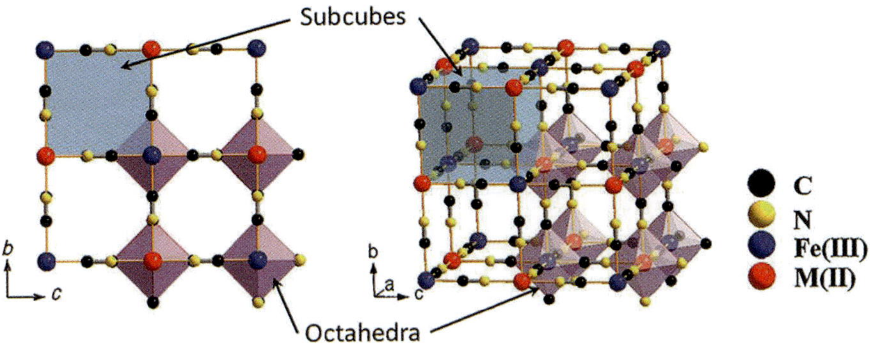

Figure 9. Framework of Prussian blue analogues showing the double-perovskite lattice of M^{2+} and Fe^{3+}, and the ordered arrangement of the CN linkers. Taken from Ref. 92 with permission of Royal Society of Chemistry.

ensure the crystallization of the compound) have recently emerged as promising positive electrode materials for both LIBs and SIBs and are now being extensively studied. The interest in PBAs lies in a widely open 3D framework (Figure 9) built up by M and M' cations bridged together *via* large cyano ligands. This network defines large channels (3.2 Å) and large interstitial sites (4.6 Å), which enable fast diffusion of Na^+ ions.[92] As a consequence of facile synthesis and low cost, the hexacyanoferrates $A_xMFe(CN)_6.yH_2O$ are the most widely investigated. Despite the theoretical ability to exchange 2 Na per formula unit, most of the prepared PBAs exhibit low Na contents, which limits the capacity, as well as a large amount of vacancies, sometimes filled with water, which contribute to poor capacity and/or low reversibility.

Improvements in the synthesis process to prevent these issues enabled preparing $Na_{1.6}Co(Fe(CN)_6)_{0.9}\cdot 2.9H_2O$, which exhibits a capacity close to the theoretical value, 132 mAh/g, at an average voltage of 3.6 V, leading to a quite high energy density of 490 Wh/kg.[93] A strategic approach to enhance the reversible capacity of these compounds involves the synthesis of reduced phases formulated as $Na_2MM'(CN)_6$ with both M and M' in the +2 oxidation state. The interest of this approach has been demonstrated as nearly stoichiometric $Na_2MnMn(CN)_6$ exhibits a high reversible capacity of 209 mAh/g, which is above the theoretical value, 170 mAh/g, which is not fully understood (Figure 10).[94]

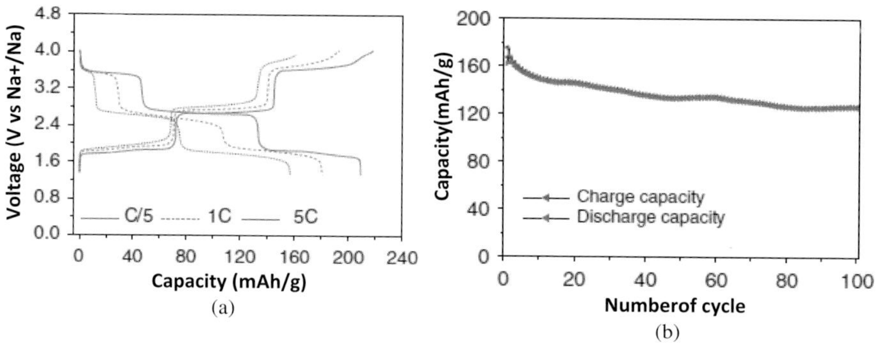

Figure 10. (a) Charge and discharge curves as a function of different C rates and (b) capacity retention at 2C for $Na_2MnMn(CN)_6$. Taken from Ref. 94 with permission of Nature.

To summarize, polyanionic compounds undoubtedly suffer from lower gravimetric capacities than layered ones. However, the stability, especially for phosphate-based compounds, and open 3D frameworks are clear benefits for long-term cycling stability and fast ion diffusion. In addition, strongly covalent polyanionic groups can be used to induce a large increase in the redox voltage of the active transition metal, especially for phosphates and sulfates, which can be further fine-tuned with electronegative anions such as fluorine. In all, this leads to high energy densities as demonstrated by the electrochemical performance of the prototypes developed using NVPF as the positive electrode material (see Section V).

C. Organic Compounds

One strategy to develop low-cost and resource-sustainable SIBs (mimicking trends in LIB research) is to shift to metal-free organic hosts that could even be synthesized from biomass. Various organic compounds and polymers have been reported as potential positive electrode materials since the mid-1980s, but their low inherent energy density hampers prospects of practical application.[95,96] The structures usually exhibit flexible frameworks allowing accommodation of Na^+ ions without any structure collapse and the redox voltage is expected to be easily tuned by chemically tailoring the functional groups or the structure of the molecules. In the absence of any metal, they react following basically two processes based on either cation or anion insertion. The former corresponds to a reversible extraction/insertion of Na^+ ions balanced by electrochemical evolution of the functional group, whereas in the latter, the electron release/uptake is balanced by uptake/release of electrolyte anions, making it independent of sodium ions. Even though the research on organic electrode materials has mostly focused on the negative side (see Section III), some positive electrode materials for organic SIBs have been reported since 2013. They usually demonstrate acceptable reversible capacities, but with low operating voltages, ca. 2.2 V, and low capacity retention due to poor electronic conductivity and dissolution of the material in the electrolyte upon cycling. Among them, π-conjugated polymers, already widely studied for application in light-emitting diodes, deserve special attention.[97,98] These materials are classified as p-type or n-type depending on their charge storage mechanism related to anion or cation insertion. Some of them, termed bipolar, can act both as p-type and n-type depending on

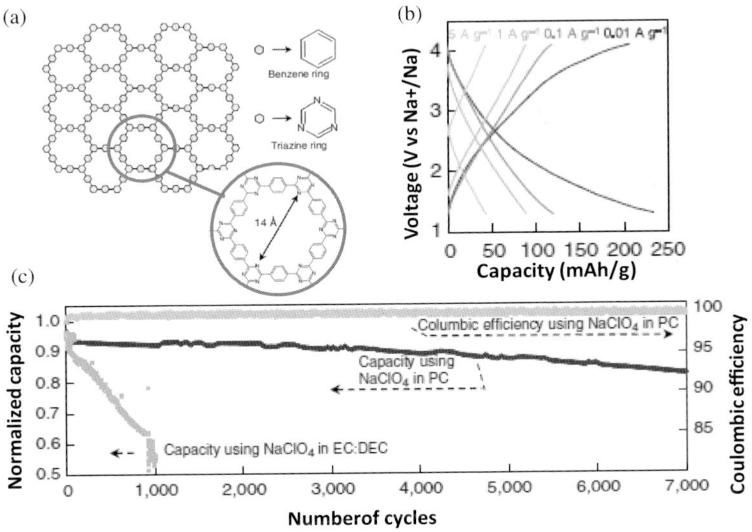

Figure 11. (a) Schematic illustration of the ideal 2D structure of a BPOE having a porous, honeycomb structure constituted by triazine rings and benzene rings with a pore diameter of 14 Å. (b) Charge–discharge curves of a BPOE using a $NaClO_4$ in PC electrolyte at various current densities. (c) Capacity retention upon cycling using two different electrolytes. Taken from Ref. 99 with permission of Nature.

the voltage range and thus lead to higher capacity. As an example, a bipolar porous organic electrode (BPOE) built up from benzene and triazine rings forming an honeycomb-like porous structure (Figure 11a) has been reported to exhibit a discharge capacity of 240 mAh/g in the 1.3–4.1 V range related to a combination of the p-type anion insertion above 2.8 V and the n-type Na^+ insertion below 2.8 V.[99]

The average voltage close to 2.8 V leads to an energy density of 500 Wh/kg (Figure 11b). The porosity enables fast ion diffusion and minor structural changes upon cycling, and therefore this compound shows high power density, 10 kW/kg, and 80% capacity retention after 7000 cycles (Figure 11c). To further increase the energy density, the redox potential can be raised *via* tuning the functional group of the organic compound that determines the lowest unoccupied molecular orbital (LUMO) energy.[100] This approach has enabled successful introduction of electronegative species such as nitro groups in polyaniline to reach a reversible capacity of 180 mAh/g at 3.2 V *versus* Na.[101] The incorporation of conductive carbon or the encapsulation of active molecules in the carbon matrix has also proved to enable enhanced both capacity retention and rate capability. For example, 9,10-anthraquinone encapsulated in CMK-3 mesoporous carbon shows an initial capacity of 214 mAh/g, the theoretical capacity is 257 mAh/g, and a capacity retention of 88% after 50 cycles. However, all largely depend on the nature and concentration of the electrolyte.[102]

In conclusion, the rapid progress in the understanding of the mechanisms involved in the organic compounds as well as the extension of the know-how acquired to tune their properties has allowed the synthesis of complex compounds exhibiting enhanced electrochemical performance. Although strategies to solve the drawbacks related to poor electronic conductivities and low redox voltages seem successful, the dissolution of the compounds into the electrolyte remains a challenge and the low energy densities an intrinsic penalty.

III. Negative Electrode Materials

In analogy with LIBs, the redox mechanisms in positive electrode materials for SIBs, the organic compounds being an exception, are all based on reversible, usually topotactic, insertion and extraction of sodium ions without major structural changes. The redox center is a transition metal and the

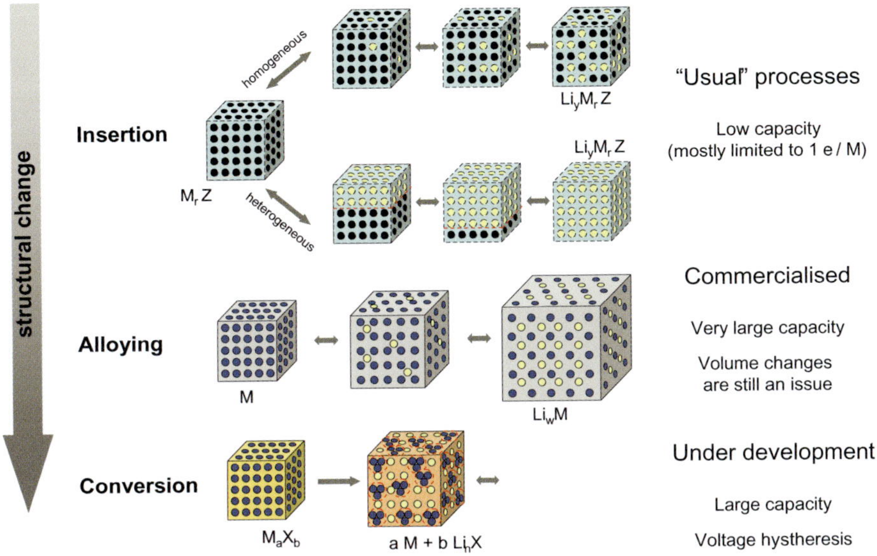

Figure 12. Schematic representation of different reaction mechanisms observed in electrode materials for lithium batteries. Black circles, voids in the crystal structure; blue circles, metal; yellow circles, lithium. Taken from Ref. 103 with permission of Royal Society of Chemistry.

electrochemical capacities are typically limited to one electron per transition metal. For negative electrode materials, the situation is, however, radically different, and, as for LIBs, alternative redox reaction mechanisms, alloying and conversion (Figure 12), can be used to enable larger electrochemical capacities. This is achieved at the expense of major structural changes caused by the large modifications in composition, which are difficult to buffer and significantly penalize cycle life.

A. Carbonaceous Materials

A significant difference between the fields of LIBs and SIBs is that while graphite is the negative electrode of choice in most commercial LIB cells (94%), it cannot be used to intercalate sodium ions to any significant amount.[104] Puzzling as this behavior seemed at first sight from the experimental point of view, as graphite is able to intercalate larger alkali ions, it was soon rationalized through structural energy calculations, which found the Na–C interaction to be anomalously weak.[105] Independent density functional theory (DFT) calculations further confirmed the instability of sodium graphite intercalated compounds (GICs).[106] Nonetheless, graphite is indeed able to insert solvated sodium ions to form ternary GICs with solvent molecules being cointercalated in the interlayer space, with the $Na(diglyme)_2C_{20}$ example exhibiting very good reversibility and a sustained reversible capacity close to 100 mAh/g for more than 2500 cycles.[107–109]

Most nongraphitic carbons are prepared by pyrolysis of different precursors at $T < 1500°C$ and exhibit structures in which most of the carbon atoms are arranged in a planar hexagonal network, but without any long-range ordering in the stacking of the layers. These structures are characterized by amorphous areas embedding and partially crosslinking, by sp^3-hybridized carbon atoms, more graphitic, layered, structure segments consisting of aligned small-dimensionality sp^2 graphene layers. If the crosslinking between the layers is weak (*soft carbons*), these layers may be mobile enough to form graphite-like crystallites upon further heat treatment at temperatures from 1500 to 3000°C. In contrast, strong crosslinking (*hard carbon, HC*) will immobilize the carbon layers and prevent any true development of a graphite structure even by treatment at high temperature.[110] The degree of crosslinking depends on the state of aggregation of the intermediate phase during pyrolysis. HCs are commonly products of solid-phase pyrolysis, whereas soft carbons are usually produced by liquid- or gas-phase pyrolysis.[111] Gases evolved during the pyrolysis process induce the development of a porous structure in the carbonaceous material, which is largely influenced by the pyrolysis conditions.[112] The lateral extent of the graphene sheets and the number of sheets stacked are typically of the order of 10–40 Å and 1–3 Å, respectively.[113] This complex structure is often understood through a simplified "falling cards model."[114]

Back in 1970, Hérold and coworkers observed that sodium vapor was retained by nongraphitic carbons through adsorption and insertion mechanisms, with the relative Na amounts depending on the origin of the carbon and the treatment temperature.[115] The commercialization of the LIB in 1991, with the first generation indeed using HC anodes, prompted a more thorough study on the relationship between the degree of Li^+ uptake and its reversibility and the carbon microstructure. It was shown that higher storage capability was related to a higher proportion of single-layer sheets

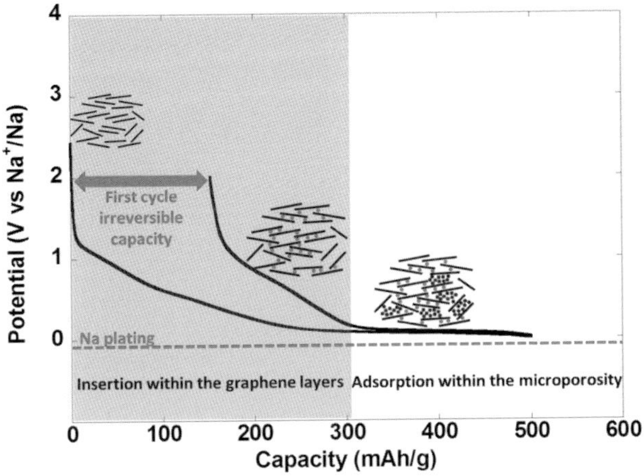

Figure 13. Typical potential *versus* capacity profile for HC when tested against sodium metal counter electrodes. The different steps of the mechanism are depicted (see text for details). Taken from Ref. 112 with permission of ECS.

and micropores.[116,117] Further studies revealed that the mechanisms for lithium and sodium insertion were similar and involve two distinct steps: (1) a sloping potential region extending to ca. 0.2 V *versus* Na and (2) a low potential plateau (Figure 13).[118–120]

The first step is related to the insertion of the alkaline ions between the layers, with the turbostratic disorder producing sites with a distribution of chemical environments resulting in a sloping profile and inducing an expansion of the interlayer distance between the graphene sheets from 3.8 to ca. 4.15 Å. The second step corresponds to the adsorption of ions in the micropores, experimentally evidenced using small-angle X-ray scattering (SAXS) experiments, and in full agreement with recent DFT calculations.[121–123] In spite of the random alignment of small-dimensional graphene layers in HC that provides a significant porosity able to accommodate lithium and thus a high capacity, graphite is generally used as the negative electrode material in commercial LIBs.[124] This stems from its higher density, which results in higher volumetric density, and its lower irreversible capacity. As mentioned earlier, graphite is, however, not able to insert "naked" sodium ions, and hence the initial research on carbonaceous negative electrode materials for SIBs focused on nongraphitic carbons.[125–128]

The first reports dealing with sodium intercalation in HCs showed, at its best, moderate performance upon cycling, but later Komaba and coworkers, focusing on screening of electrolytes, succeeded in achieving a stable capacity of ~230 mAh/g for 100 cycles using 1 M $NaClO_4$ in propylene carbonate (PC) or ethylene carbonate (EC): diethyl carbonate (DEC) electrolytes.[121] This study showed the existence of a nonuniform SEI deposited on HC, mostly consisting of inorganic compounds, as deduced from mass spectroscopy. Systematic parallel studies on SIB electrolytes allowed to conclude that electrolytes based on a mixture of EC and PC solvents, with either 1 M $NaClO_4$ or $NaPF_6$, allowed much better capacity retention for HC due to the beneficial effect of EC inducing the formation of a stable SEI layer.[129] An addition of 10% dimethyl carbonate (DMC) to the solvent mixture was found to decrease its viscosity and hence enhance the ionic conductivity, still keeping the electrode polarization low upon cycling and allowing access to the full capacity of the low potential plateau: 300 mAh/g at C/10 for more than 120 cycles.[80] This formulation has also recently shown to enable good capacity retention upon low (–15°C) and high (75°C) temperature operation, stressing the critical influence of the total cell resistivity (IR drop) on the total measured capacity values.[130] Indeed, as a significant part of the electrochemical capacity delivered by HC is achieved at very low potential, large cell polarizations related to a nonoptimized electrolyte composition may result in the loss of a significant amount of capacity (the lower cutoff potential for most tests reported in the literature is 0.02 V *versus* Na^+/Na), and divergent performance reports for HCs with similar microstructures. Moreover, nonoptimized electrolytes may induce a continuous growth of the SEI layer and hence result in significant capacity fading upon cycling. For more on the specific SIB electrolytes and the SEIs developed, refer to Section IV. In spite of such issues making it difficult to extract microstructure–performance correlations, two trends seem to be clear: the reversible capacity associated with sodium insertion between the graphene layers (the sloping region) is enhanced with increased d_{002} spacing and the capacity related to adsorption (low potential plateau) increases with decreased micropore size.[112]

The main practical drawback of using HC as the negative electrode material for SIBs is its high irreversible capacity, which translates into a loss of *cyclable* sodium in the cell. This effect, coupled to the need to build cells with an excess of negative active material, as for LIBs, to prevent safety hazards derived from sodium deposition upon overcharge or low temperature operation, would induce a severe penalty in energy density in any practical SIB. The irreversible capacity loss has typically been correlated to the electrode surface area accessible for the electrolyte.[131–133] For LIBs, and in spite of a linear relationship between Brunauer–Emmett–Teller (BET)-specific surface area and irreversible capacity loss, other parameters have been shown to be important such as the concentration of surface functional groups, typically chemisorbed oxygen–containing species, which can react with the lithium atoms.[134] An inverse relationship between porosity and reversible capacity has also been reported.[135] The

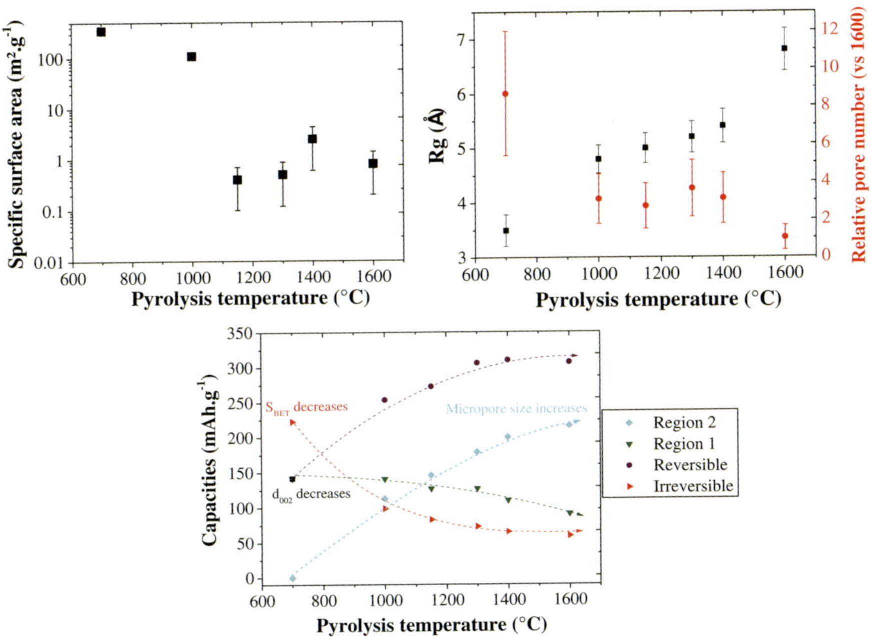

Figure 14. Plots of specific surface area, micropore radius of gyration, and relative pore number *versus* the pyrolysis temperature for carbons prepared by pyrolysis of cellulose at different temperatures. Evolution of the capacity in region 1 (sloping) and region 2 (low potential plateau) as well as the total reversible and irreversible capacities as a function of pyrolysis temperature (the evolution of the BET surface and d_{002} are also indicated). Taken from Ref. 141 with permission of Communaute Universite Grenoble Alpes.

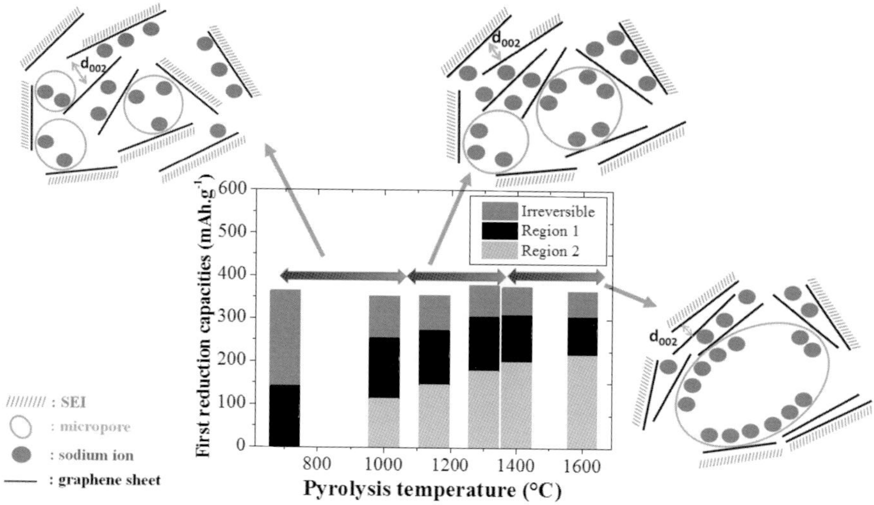

Figure 15. Schematic view of the proposed HC redox mechanism. Taken from Ref. 141 with permission of Communaute Universite Grenoble Alpes.

porosity and surface area can typically be tuned through the pyrolysis conditions (Figure 14), which also influence the surface chemistry and in turn the electrochemical performance (Figure 15). Indeed, the proportion of micropores increases with the crosslinking density in the precursor, and is modified by thermal treatment, as increasing the temperature results in an increase in the micropore size and decreasing their number, and closing of the pores above 1100–1200 °C, which all can be nicely related to the electrochemical performance.[136–139] Moreover, surface groups are also desorbed at different temperatures and d_{002} does also decrease with enhancing temperature.[140] When the pyrolysis takes place under gas flow, typically Ar, the flow rate has been found to have an effect on surface area: high rates allow flushing away the released gas, mostly CO_2, which would otherwise react with carbon (burn-off) creating porosity and enhancing surface area.[131]

In agreement with the above discussed trends, irreversible capacity losses in LIBs have been shown to be reduced by a soft carbon deposit on the surface of HC carried out by chemical vapor deposition (CVD), and similar results have recently been achieved for SIBs with alternative carbon deposition methods.[130,142,143] The highest coulombic efficiency upon the first cycle reported to date is ca. 85% for samples either commercial or prepared at laboratory scale at T ≥ 1100°C with low specific surface areas, < 10 m^2/g, which compare very well with the state of the art HC electrodes in LIBs.[121,131,144]

Finally, the highest reversible capacity, ca. 430 mAh/g at C/10, and rate capability, ca. 340 mAh/g at 2C, for HC in SIBs has been recorded at 75°C.[130] The ratio between the Na^+ insertion and adsorption capacities is similar at RT and at 75°C, ca. 47% and 53% of the total capacity being associated with insertion and adsorption, respectively, together with the significant increase in the total capacity at 75°C suggests that either the Na^+ ion diffusion within the HC particles is improved with temperature and/or the molar density of Na is increased. Indeed, the smaller reversible capacity recorded in cells using sodium metal counter electrodes as compared to those using lithium metal counter electrodes was early ascribed to the lower molar density of Na, 0.042 mol/cm^3, as compared to Li, 0.077 mol/cm^3, although it is also clearly influenced by the different standard potentials of reduction for both metals.[145] Moreover, the much higher impedance recorded for passivated Na as compared to Li metal counter electrodes seems to be related to a lower SEI stability stemming from enhanced solubility of their components, for example, sodium salts, inducing a significant additional ohmic drop in the cell, which may hinder the recovery of the capacity from the low potential plateau.[146]

Overall, in spite of the issues related to irreversible capacity losses, HC is yet the most practical negative electrode material for SIBs (see Section V). The possibility for further improvements in performance, not only by decreasing the irreversible capacity losses but also by enhancing the reversible capacities through mastering the microstructure–performance relationships, is currently very challenging.

B. Alloy-Based Materials

Alloying reactions between lithium and a large number of metallic or semimetallic elements proved feasible in electrochemical cells at RT in conventional organic electrolytes and have been widely studied since the 1970s as they enlist large electrochemical capacities, that is, a large amount of Li atoms reacting per mole of alloying material. This results, however, in large strains in the particles that promote microcrack formation and propagation and a progressive decohesion, particle shuffling, and, subsequently, severe capacity fading. This has prevented the practical utilization of alloys despite the wide spectrum of materials engineering strategies developed to limit the effects of the volume changes. The latter are based on better accommodation of volume changes by particles that are amorphous, nanosized, and less prone to break upon stress. The use of active particles embedded in a conducting matrix, that is, carbon, which could buffer the volume expansion has also been widely explored. However, these approaches do also involve drawbacks related to the use of nanosized materials, which are by no means to be disregarded such as higher surface/volume ratios that promote enhanced reactivity with the electrolyte. Another crucial factor is the stability of the SEI layer, which is much more critical for alloying materials than for conventional insertion electrodes. Indeed, it will break if its endurance limit is exceeded by the amplitude of the stresses generated and subsequently fresh, naked electrode surface will be exposed on which an additional SEI would start to grow, a process ultimately resulting in cell failure due to electrolyte consumption and loss of electrode porosity.

As mentioned in Section I, $Na_{15}Pb_4$ was the negative electrode in the very first proof-of-concept SIB reported in the literature.[2,3] No other alloy-based negative electrodes for SIBs were, however, reported before 2011 when a DFT computational study appeared comparing the volumetric energy densities for carbonaceous and alloying electrodes for both LIBs and SIBs.[147] The larger ionic radius for sodium is expected to result in much larger volumetric expansions for the formation of sodium alloys than for lithium alloys and the DFT study concluded the former unlikely to be able to compete (with Li) in terms of volumetric energy density (Figure 16).

Despite these negative expectations, research proliferated on alloying electrodes based on elements of groups 14 and 15, with relevant performance achieved for a few. Based on abundance, cost, safety, and toxicity issues, Si and Sn are the most attractive options, although promising results were achieved also with Pb, Sb, Bi, and P. Electrochemical alloying of sodium with crystalline silicon or germanium has not been reported

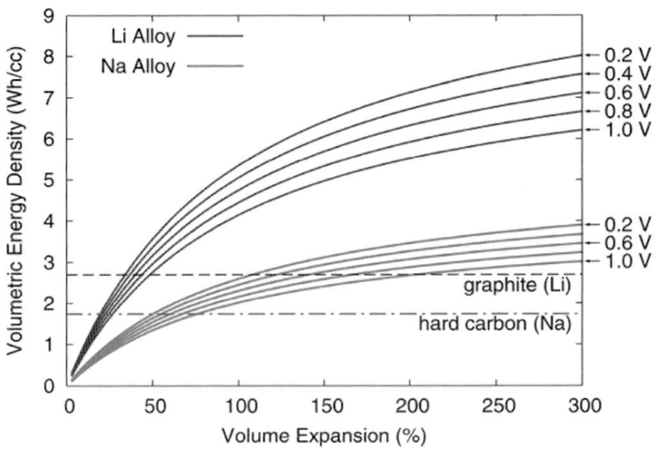

Figure 16. Volumetric energy density *versus* volume expansion for alloying electrodes for LIBs and SIBs. The volumetric energy densities of lithiated graphite and sodiated HC are included for ease of comparison, but are not functions of the volume expansion *x* axis. Taken from Ref. 147 with permission of ECS.

to date, but given the fact that some alloys are known, NaSi and NaGe, this may be ascribed to kinetic issues and hence could be solved with the use of nanosized amorphous particles, which would practically be penalized by enhanced electrolyte decomposition. Sn and Pb do form electrochemical alloys with sodium at low potentials, < 0.5 V *versus* Na^+/Na, with $Na_{15}Sn_4$ and $Na_{15}Pb_4$, respectively, as end members. For P, Sb, and Bi, the sodium uptake per mole is slightly lower and proceeds to reach Na_3X. The theoretical gravimetric and volumetric capacities are depicted in Figure 17. Besides the pure metals/semimetals, reactivity of sodium also exists with binary compounds based on the same elements such as SnO, Cu_2Sb, and TiSnSb and composites within a matrix, most commonly carbonaceous — putting even higher demands on nanosizing to compensate for inactive mass in the electrodes. The extent to which these approaches exhibit practical advantages is not clear and should be assessed in full SIB cell prototypes and further upscaled (see Section V).

The reaction of Na with Sn/Pb and Sb/Bi to finally form $Na_{15}Pb_4$/$Na_{15}Sn_4$ or Na_3Sb/Bi_3Sb involves formation of diverse intermediate alloys that may exhibit differences compared to the reported NaX binary phase

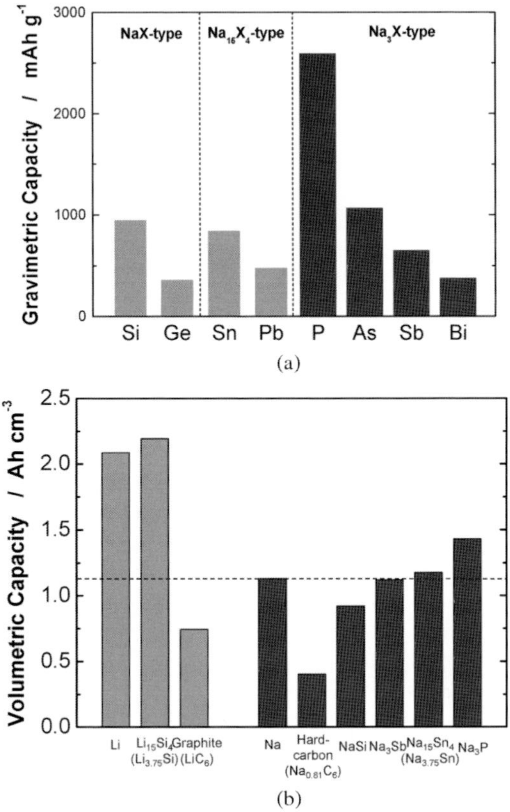

Figure 17. Theoretical (a) gravimetric and (b) volumetric capacity estimated from the most sodiated phase in the available phase diagrams for elements of groups 14 and 15. The volumetric capacities for sodium, HC, lithium, lithium silicon, and graphite are provided for comparison. Taken from Ref. 4 with permission of ACS.

diagrams, including some phases with limited long-range order.[148–150] This may explain why, for example, crystalline Sb has a better electrochemical performance against sodium than lithium counter electrodes in spite of the larger volume expansion.[151] Moreover, since the stability of the SEI plays an important role for the cyclability, changes in the electrolyte composition, including the presence/absence of additives, may trigger significant differences in coulombic efficiency and capacity retention (Figure 18). Remarkably, the formation of Na_3P proceeds with the highest capacity and

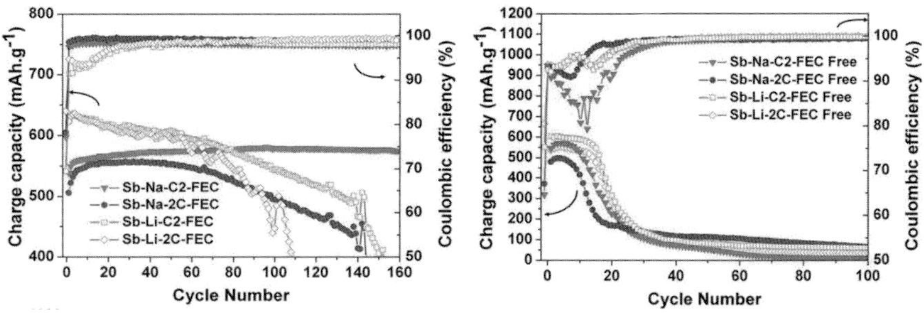

Figure 18. Capacity *versus* cycle number for Sb electrodes cycled against sodium or lithium counter electrodes at C/2 or 2C rate with (left) or without (right) the addition of 5% fluoro-ethylene carbonate FEC in the 1 M $NaClO_4$ in PC or 1 M $LiPF_6$ in EC:PC:3DMC 1% vinylene carbonate (VC) electrolytes. Taken from Ref. 151 with permission of ACS.

a small volume expansion.[147] Nonetheless, in spite of using the relatively stable red P allotrope, safety constrains may limit its practical use.

C. Other Materials

Conventional insertion–deinsertion processes at low potentials can only be expected for compounds containing early transition metals, mostly Ti and V, as conversion reactions are observed for the others. Ti (IV) oxides are interesting as negative electrodes for LIBs due to their moderate operation potential, ~1.5 V *versus* Li^+/Li, where the penalty in the energy density is expected to be balanced by the enhanced safety. Based on crystal chemistry considerations, the known TiO_2 polymorphs did not a priori seem to be interesting as electrode materials for SIBs, due to a lack of available sites large enough to accommodate sodium ions, but turned out to exhibit capacities close to 150 mAh/g between 2.5 and 1 V *versus* Na^+/Na, all related to a pseudocapacitive storage mechanism.[152] Layered $Na_2Ti_3O_7$ reversibly uptakes 2 Na^+ ions per formula unit, resulting in 177 mAh/g at a surprisingly low average potential of 0.3 V *versus* Na^+/Na. Unfortunately, this compound exhibits unsolved issues related to extensive capacity fading upon cycling, common to other titanates, which seem to originate in a loss of crystallinity and a reactivity of the reduced phase toward the electrolyte.[153]

Conversion is the term applied to define the reaction of a binary transition metal compound, M_aX_b (M = transition metal, X = O, S, N, P . . .) to yield metallic nanoparticles embedded in a matrix of Na_yX according to: $M_aX_b + (b{\cdot}n)\ Na \leftrightarrow a\ M + b\ Na_nX$ and in principle involves large capacities related to the full reduction of the metal.[154] A myriad of studies on diverse binary compounds materials, either pure or composites, exhibiting diverse particle sizes and shapes including nanoarchitectured electrodes have been published. Unfortunately, and similarly to what has been reported for LIB cells, the practical prospects of such materials are limited by low first cycle coulombic efficiencies and severe hysteresis between oxidation and reduction, which ultimately result in unacceptably low round-trip efficiencies.[155] The fact that conversion electrodes can simultaneously show large voltage hysteresis and fast kinetics indicates that the round-trip asymmetries are not due to simple ohmic or charge transfer polarization, but possibly and more likely related to an inherent difference in the reaction path between reduction and oxidation. Thus, strategies based on the reduction of diffusion lengths or the improvement of charge transfer kinetics, widely reported in the literature for materials operating through a conversion reaction mechanism, are unlikely to produce any overall decisive gain. Indeed, significant challenges lie ahead if conversion-based electrodes are to ever be implemented in energy storage devices and substitution of lithium by sodium is unlikely to bring in any significant relief.

Along the interest in organic compounds as electroactive materials for LIBs, these have also been studied for SIBs.[156] The redox mechanism is intrinsically different to that of inorganic compounds and deals with addition/withdrawal of electrons to/from electron poor/rich functional groups. Organic molecules and polymers with conjugated carbonyl groups and polymeric Schiff bases have attracted an increasing interest, for example, disodium terephthalate ($Na_2C_8H_4O_4$) that reacts with two more moles of sodium at ca. 0.45 V *versus* Na^+/Na corresponding to a capacity of 250 mAh/g, and (–N=CH–Ar–HC=N–) with Ar being an aromatic group, exhibiting 350 mAh/g between 1.5 and 0 V *versus* Na^+/Na.[157–159] It is expected that organic materials would in general exhibit easier processability, lower cost, and lower environmental footprint than their inorganic counterparts and may be more prone to multielectron reactions and higher capacities. This will probably come at the expense of poorer thermal stability, lower packing density, sluggish kinetics, and solubility issues with conventional electrolytes. Nonetheless, further investigations are required to assess the practical application prospects of such materials — and their possible market niches.

IV. Electrolytes

While the electrodes applied in the SIB technology will determine how successfully it can compete with the prevailing LIBs in terms of energy and power densities, the electrolyte chosen will ultimately determine, for example, the safety and the capacity fading. Both in terms of materials and concepts, the SIB electrolyte has not yet conformed to any standard choice, in stark contrast to the mature LIB technology where more or less an unanimously accepted "standard electrolyte" exists: the ca. 1 M $LiPF_6$ in EC/DMC + a number of special additives, the latter being each manufacturers secret recipe. While there are better and worse choices to be made for the electrolyte in terms of salts, solvents, and additives, the exact composition also interplay with the electrodes chosen. Therefore, while SIB electrolyte research benefits from the know-how gained from LIBs, the rational for precise choices of electrolyte components depends on the specifics of each cell setup, especially the electrode chemistry, and usage conditions. The following sections first give an overview of why and how the SIB electrolyte needs to act, then continue by reviewing the main achievements made within a set of electrolyte families — both without and in combination with electrode(s), and finally the ever so important SEI will be commented on from an electrolyte design and demand perspective.

A. Roles and Requirements

The role of the electrolyte is sometimes termed "passive" — as compared to the "active materials" in the electrodes. In one way nothing could be more wrong. While at a first glance any SIB electrolyte should only be responsible for containing and shuttling Na^+ ions between the electrodes and hindering electrons to do the same, it has a far more complex role. As an example, it has to provide the electrochemical stability windows (ESWs) needed to enable the full voltage range of the electrodes to be used — and this is set by the LUMO and the highest occupied molecular orbital (HOMO) of the electrolyte (components). The strongly reducing/oxidizing electrode active materials make this even more difficult why degradation reactions often take place — which can be a safety issue. On the other hand, if made in a controlled manner, this can be utilized to result in insoluble products adhering to the negative electrode surface and thereby form (parts of) the SEI. In practice, SIB cell operation, just as for LIBs, is made possible through chemical (and kinetic) passivation — extending the thermodynamic limits imposed by the HOMO and LUMO levels. The main roles are thus to conduct Na^+ ions, hinder electronic conduction, and stabilize cell operation.

The requirements are partly connected to the main roles above, but there is a much longer list — especially if one wants to not only get a functional cell at the lab scale, but also make the SIB technology a commercial reality. A high ion conductivity, often a target value of 1 mS/cm is quoted, is the property most common to first mention, which really should focus on the exact species that is to be intercalated at the electrodes — here Na^+. The overall conductivity is important not the least to avoid cell polarization and gradients and ensure appreciable power densities. Second, and perhaps this is even more important, is that placed between the two electrodes and interacting with both of them the interfaces created often dictate the true performance and hence the stability or indeed metastability of the electrolyte is crucial to control. There should ideally be no chemical reactions occurring, else there would be a continuous loss of performance during shelf life, or other changes involving the electrolyte during the SIB cell operation, something that becomes more cumbersome to attain while pushing the low (negative) and high (positive) limits of the cell voltage window further and further apart when working to enhance the energy densities by extended voltage. Therefore, most often a metastable electrolyte is chosen and then the reactions should be as controlled and predictable as possible. This not only goes for the interaction of the electrolytes with the electrodes — it must also be true for the separator, the current collectors, the packaging material, and so on.

From a practical SIB operation and manufacturing point of view, there are quite a few other additional requirements, the electrolyte should be (1) thermally stable and have a wide range without any phase transitions occurring — especially both the melting and boiling points of the electrolyte, not only the solvent(s), should be well outside the (internal) operation temperatures of the SIB, (2) all components and the final electrolyte must have low a toxicity and successfully meet also other measures of limited environmental hazard and impact, (3) given the overall argumentation for SIBs *versus* LIBs, it should preferably be based on very sustainable chemistries, using only abundant elements and processes with an as low impact as possible with respect to energy, pollution, and socio-economic issues, and (4) for a commercial success, it must reduce costs, both in terms of materials, production, and maintenance as possible.

Most of the above requirements are general, equally valid for SIBs as for LIBs, and not given in terms of specific optimal or even lower threshold values, but there is of course a need to optimize them.[5,160] The latter can be done either separately or in conjunction with each specific SIB cell chemistry and application. It is stressed, however, that no single measure can ever tell the full story. Some SIB electrolyte properties mainly originate in the salts used, others in the solvents, and yet other are more "holistic" — especially the latter kind is difficult to predict prior to actual testing in full SIB cells. While much of the know-how from more than 30 years of LIB electrolyte R&D can and is extensively applied to SIBs, there are some notable differences. First of all, while the similarity of Na^+ to Li^+ often is stressed when marketing SIBs, there are some fundamental physicochemical differences affecting the electrolytes, for example, the different charge/radius ratio, the lower Lewis acidity, and the Na salts' solubilities. However, the most striking is that the electrodes used will not be the same, and thus also the interfaces will be different. Overall, the electrolyte optimization is made by varying the constituents of the electrolyte; the Na salt(s), the matrix — often the solvent(s), the additive(s), and their concentrations/ratios. Below the current development of SIB electrolytes is summarized in paragraphs divided by the type of matrix used: organic solvents, ionic liquids (ILs), polymers, and other concepts.

B. Nonaqueous Organic Electrolytes

Just as for LIBs, the most common electrolyte concept is based on one or several Na salts dissolved in one or several organic solvents. The wide ESWs wished for turns into the stringent and absolute demand that solvents with active protons cannot be used. This most often translates into nonaqueous, but should indeed more correctly be aprotic. The very interesting recent development of superconcentrated, indeed including using even water as solvent, electrolytes — largely circumventing the issue of complying with the electrochemical stability of each electrolyte component, has not yet seen any SIB application/example.[161,162] Each of the salts and solvents applied hence need to comply with the prerequisites set above: stable — chemically and electrochemically — nontoxic, inexpensive, and so on. In addition, presence of polar groups in the solvents to dissolve sufficient amounts of Na salts easily dissolvable is a compulsory feature, even if the cation–anion interaction strength of Na^+ can be expected to be only 80% of Li^+.[163] The SIB electrolytes aiming to match the requirements above can, however, be designed many different ways, each having their own set of promises and problems. While the combinations of salt(s) and solvent(s) may differ between LIBs and SIBs, there are as of yet no specific salts or solvents developed specifically for SIBs.

The salts used are most often the same independent on the particular type of matrix chosen and in many cases analogous to those used for LIBs. The Na salts for SIB electrolytes need to be easily soluble, and together with the stability demands (ESW, chemical), there are in reality very few candidates. They are all based on inorganic anions with strong electron-withdrawing ligands to create a delocalized negative charge and thereby weakly coordinating anions — ensuring low lattice energies of the salt itself and few ion-pairs in solution. This results in the same anions as applied in LIB research; ClO_4^-, BF_4^-, PF_6^-, $CF_3SO_3^-$ (Tf), and $[N(CF_3SO_2)_2]^-$ (TFSI). As many resulting electrolyte properties are dependent on the anion rather than the cation applied, also the pros and cons are more or less the same. To briefly summarize, the ClO_4^- anion is a strong oxidant and therefore more or less banned for any practical cell development — but is extensively used for SIB studies; BF_4^- produces less conductive SIB electrolytes by virtue of a stronger cation interaction than the other anions listed; PF_6^- has very much the same role for SIBs as in LIBs — the $NaPF_6$ salt is often used, but has the same severe safety issues, it is the anion that suffers hydrolysis to possibly yield PF_5, POF_3, and HF; Tf produces less conductive electrolytes and corrodes the aluminum current collectors — the onset potential is ca. 2.7 V *versus* Li^+/Li for Tf in a LIB electrolyte; the TFSI anion finally has the same problem as Tf and furthermore NaTFSI is more expensive than other Na salts.[164] NaTFSI though, and its family member NaFSI, have some special

Table 1. Basic properties of the most commonly used Na salts for SIB electrolytes

Salts	M_w [g/mol]	T_m [°C]	σ [mS/cm] 1 M NaX in PC at 25°C
$NaClO_4$	122.4	468	6.4
$NaBF_4$	109.8	384	
$NaPF_6$	167.9	300	8.0
NaTf	172.1	248	
NaTFSI	303.1	257	6.2
NaFSI	203.3	118	

promises for IL and foremost polymer electrolytes (see below). Scrutinizing the problematic features above related to the choice of Na salt, some are indeed altered by moving from SIBs to LIBs. One example is that the sodium salts in general have higher melting points — making them easier to dry and a larger thermal stability is also expected to bring advantages in terms of safety. In Table 1, a comparison of a few of the salts for SIB electrolytes is summarized.

As is clear from the table, it cannot be the ionic conductivity that determines the choice of electrolyte salt, why this often made determination is just a very basic requirement — but clearly not enough. Instead, it is the aforementioned diverse factors that determine the Na salt to be chosen for a specific SIB design. Looking at the literature, there are some general observations to be made: the most common salt used, especially in academic studies, is surprisingly $NaClO_4$, largely due to historical and cost aspects, but apart from the problematic safety aspects, these salts are also notoriously difficult to dry — despite not being more hygroscopic than other salts.[165] Drying the salt at 80°C under vacuum overnight and making an electrolyte the $NaClO_4$ based, electrolytes can exhibit water contents > 40 ppm, compared to, for example, $NaPF_6$ based at < 10 ppm. $NaPF_6$ is indeed the second most popular salt — enabling "direct" comparisons of SIB *versus* LIB studies. Strongly emerging candidates, despite cost and Al corrosion issues, are NaTFSI and NaFSI. While NaTFSI and NaFSI perhaps cannot be applied as single salts for SIB electrolytes, but together with an additive or as cosalt, their nontoxicities, high thermal stabilities, and the resulting conductivities being higher than for NaTf-based electrolytes are all promising features. However, it seems that the initial doubts on FSI are not yet resolved; FSI was early reported for Al corrosion, this was later attributed to Cl impurities remaining from the synthesis route, but there is still stainless steel (SS) corrosion reported for FSI.[166] Comparative studies of Na salts are very few. $NaPF_6$, $NaClO_4$, NaTf, and NaTFSI were compared using a EC:DMC matrix by Bhide *et al.* and the ionic conductivity reported also as functions of Na salt concentration.[167] The maximum conductivities obtained were 6.8 mS/cm for 0.6 M ($NaPF_6$), 5.0 mS/cm for 1 M ($NaClO_4$), and 0.8 M NaTf showed a much lower conductivity. The ESWs against inert and $Na_{0.7}CoO_2$ positive electrodes were also improved and for the latter attributed to formation of a stable surface film.

To create nonaqueous organic electrolytes from the above Na salts, the choices of solvents are the same or similar for SIB and LIB electrolytes — above an EC:DMC matrix was used for the comparative study of Na salts. The solvents should (1) be polar, a dielectric constant $\varepsilon > 15$, to facilitate the salt dissociation and limit ion-pair creation, (2) have low viscosities to

Table 2. Solvents commonly used in SIB electrolytes

Solvents	T_m (°C)	T_b (°C)	T_f (°C)	η at 25°C (cP)	ε at 25°C	AN/DN
EC	36.4	248	160	1.9 (40°C)	89.78	/16.4
PC	−48.8	242	132	2.53	64.92	18.3/15.1
DMC	4.6	91	18	0.59	3.107	
DEC	−74.3	126	31	0.75	2.805	/16.0
EMC	−53	110		0.65	2.958	
DME	−58	84	0	0.46	7.18	10.9/18.6
Diethylene glycol dimethyl ether (diglyme)	−64	162	57	1.06	7.4	9.9/19.2
Triethylene glycol dimethyl ether (triglyme)	−46	216	111	3.39	7.53	10.5/14

DEC, di-ethyl carbonate; DMC, di-methyl carbonate; DME, dimethoxy ethane; EC, ethylene carbonate; EMC, ethyl methyl carbonate; PC, propylene carbonate, T_m = melting point; T_b = boiling point; T_f = flash point; η = viscosity; ε = dielectric constant; AN = acceptor number; DN = donor number.

improve ionic mobility, (3) be inert to the electrodes, (4) have wide liquid ranges — allow also low and high T operation. To comply with this list in the best way most often solvent mixtures are used — based on organic carbonates, linear and cyclic, esters, and ethers, some of which are listed in Table 2. An obvious example is to combine a low viscosity, but also low dielectric constant, solvent such as DMC with the RT solid EC — which has a dielectric constant higher than water. Simplified, this is how the MX in EC:DMC electrolytes has come to dominate LIBs and now also makes significant contributions to the field of SIBs.

From the properties listed, the Lewis acidity/basicity concept will influence the ESW and the ANs and DNs correlate with the HOMO/LUMO levels. Moreover, the acidity (basicity) will determine the solvation properties; strong (low) acidity/basicity results in easy (hard) solvation of anions/cations.[168] According to the hard and soft acids and bases (HSAB) rule of thumb, this will determine the solvent–solvent and ion–solvent interactions. Na^+ is a weaker acid than Li^+, why some differences are expected in terms of ion–solvent interactions, that is, the size and nature of the solvation shell and the solvation energy.[163,169,170]

PC was among the first solvents used in both LIB and SIB cells, mainly due to its combination of a high dielectric constant and a large liquid range. However, when HC was replaced by graphite in LIBs, PC was replaced by EC as PC was found to cointercalate between the graphene layers of the graphite causing exfoliation, while EC passivates the surface by the formation of a thin and functional SEI.[171] For SIBs, where HC currently is one of the more viable negative electrode materials, PC has remained the main solvent used and is used in ca. two-thirds of the electrolytes used in the SIB literature. Nonetheless, HC tested *versus* pure PC-based electrolytes exhibits a strong capacity fading and low coulombic efficiencies, attributed to a continuous SEI growth.[129] Other electrolytes with focus on varying the solvents were tested by Alcántara *et al.* using various disordered carbonaceous electrodes and they concluded that THF was a solvent improving performance, but THF clearly has a too narrow ESW to be practically interesting.[172] Komaba *et al.* showed by using $NaClO_4$ in a 1 M concentration in EC:DEC or PC to outperform EC:DMC or EC:EMC.[173] Using a similar strategy, Vidal-Abarca *et al.* moved away from PC to EC:DEC and observed an improved performance

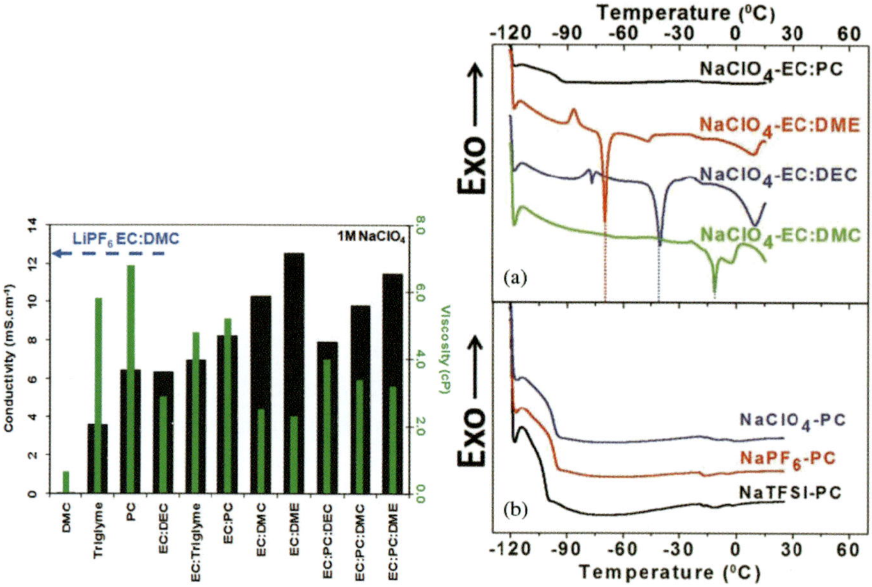

Figure 19. Conductivity (black bars and left-hand side *y* axis) and viscosity (green bars and right-hand side *y* axis) values of electrolytes consisting of 1 M $NaClO_4$ dissolved in various solvents and solvent mixtures with the ionic conductivity of 1 M $LiPF_6$ in EC:DMC given for comparison. Differential scanning calorimetry (DSC) heating curves of electrolytes after cooling the sample down to 120°C. (a) Electrolytes based on 1 M $NaClO_4$ dissolved in various solvent mixtures and (b) PC-based electrolytes with 1 M of various Na salts. Adapted from Ref. 80.

for a fluorophosphate-based cathode.[174] The former decomposed at low voltages, whereas the latter was limited by the high voltage. While the above quoted studies have contributed to the understanding of the limits of the exact applied nonaqueous organic electrolytes, they were not really systematic. However, some systematic studies were published soon after, focusing thermal behavior, ionic conductivity, viscosity, and the ESW, both using inert and HC electrodes.[80,175]

These studies indicate that a mixture of EC and PC, with either 1 M $NaClO_4$ or $NaPF_6$, provides a much better capacity retention than using PC alone (Figure 19). This was explained as due to EC inducing the formation of a more stable SEI with a lower polarization enabled achieving the full capacity of the HC negative electrode (the low potential plateau). This follows the LIB results for mixed solvent, EC:DMC or EC:PC, electrolytes. Along the same path of development, addition of 10% DMC to an EC:PC mixture, creating a ternary electrolyte, resulted in a decreased viscosity and further enhanced ionic conductivity.[80] In terms of thermal stability, electrolytes show interesting and sometimes puzzling results. By DSC, only a glass transition near −95°C was observed for the PC-based electrolytes (Figure 19b), whereas an EC:PC solvent mixture showed a single exothermic peak at ca. 250°C resulting in an impressive liquid range.[129] Moreover, by DSC it was shown that most Na salts are thermally more stable than their Li-salt analogues.

To create a functional electrolyte, target-specific additives in amounts typically less than 5% are added to the basic salt and solvent electrolyte. For SIBs, just as for LIBs, the focus of additives has been on preferred actions at the electrolyte/electrode interfaces rather than in the bulk of the electrolyte. Typically this means to modify the SEI, increase the wetting of the surface, and/or protect toward overcharging events by redox shuttles. For the SEI-forming additives, the difference between SIBs and LIBs is clear-cut, the different solubilities of the Li_2CO_3 and Na_2CO_3 carbonates as well as LiF and NaF will affect the nature of the SEI formed. For more on this topic, refer to Section F "The Solid Electrolyte Interphase." Currently, but this can be changing fast, FEC, that is, the singly fluorinated EC, is the most common SEI former/enhancer used for SIBs. Several additives were screened for SIB electrolytes in 2011 including FEC, VC, ethylene sulfite (ES), and the doubly fluorinated EC (DFEC), and results pointed to a clear advantage of FEC with respect to the others.[176] Interestingly, VC, very popular in the field of LIBs, was not found to work the same way for SIBs. However, also the effects of FEC for SIB cells with HC anodes remain controversial; a beneficial effect was reported for 1 M $NaClO_4$ in PC as the electrolytes, thus an intrinsically "bad" SEI builder, but in contrast, for electrolytes better at creating high-quality SEIs themselves, for example, EC:PC, the FEC addition was found to be detrimental to the cell performance.[129,175] These differences are not due to any differences in water content, but are true and due to the more resistive SEI built in the presence of FEC causing larger cell polarization. The benefits of FEC are then again clearly proven for negative electrode materials displaying significant volumetric expansions, that is, SnO_2, Sn, Sb, and SnSb — all operating at much higher potentials than HC and there the use of FEC seems mandatory to achieve agreeable capacity retention and in fact allows SIB cells to outperform their lithium analogues.[151,177,178] Interestingly, again though, when changing salt to NaTDI for an EC/DMC electrolyte cycled at C/10 against Sb_xO_y/C, delivering 425 mAh/g, FEC was not used (in contrast to the same setup for Li which used 5% FEC).[179]

To summarize, the main track of SIB electrolytes holds no surprises compared to the LIB development — right now the main concept used is 1 M $NaPF_6$ in different carbonate solvent mixtures — binary or ternary. The role of additives is less clear than for LIBs, as is natural at this stage of development. Today, this family of electrolytes is the only one allowing to assess the full potential any SIB electrode materials at ambient temperatures in terms of energy and power densities and is therefore used to benchmark novel electrode materials — but this might sometimes be scrambled if not an a priori understanding of the limitations of the electrolytes themselves exists.

C. Ionic Liquid Based Electrolytes

Ionic liquids (ILs), most often salts of the type [Cation][X] that are liquid below 100°C, have several properties matching the requirements for SIB electrolytes: large liquid ranges, thermal and electrochemical stabilities, and no or very low vapor pressures — hence non-flammable. The latter is often the most emphasized property of ILs, but perhaps more important are the ability to solvate salts and the intrinsic high ionic conductivities. And not the least the stability *versus* Na metal that can be obtained by proper salt + IL selection.[180] As ILs often are made from the same anions X as the NaX salts used, an IL-based SIB electrolyte can simply be denoted as $Na_xCation_{(1-x)}X$ with x usually in the range 0.1–0.25. X is most often TFSI or FSI and the cation often imidazolium (XMIm) based, despite having problems with the cathodic limit of the ESW, or pyrrolidinium (Pyr) based, with better electrochemical stability.

As compared to conventional organic solvent–based electrolytes, the main drawback of using IL-based electrolytes is the high viscosities, of the order of tens of cP at RT for the native ILs and increasing further upon doping with Na salt. This is due to the formation of stronger ion–ion interactions, which turns the mobility of the Na ions being a complex,

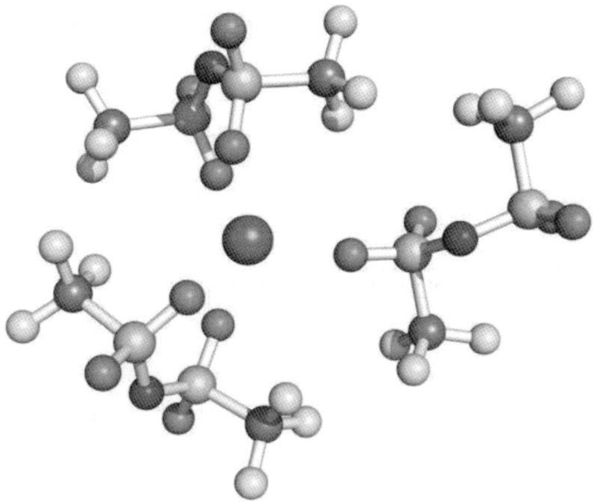

Figure 20. The energetically preferred structure for the $[Na(TFSI)_3]^{2-}$ complex from the DFT calculations. Taken from Ref. 181 with permission of Elsevier.

unresolved issue. It has been suggested that the Na^+ ions primarily are moving in doubly negatively charged complexes $[NaX_3]^{2-}$ (Figure 20).[181]

The high viscosities have implications both for ion transport (C-rates) and for manufacturing, for example, filling of a SIB pouch cell and proper wetting of the electrodes. The competing effects of ion–ion interactions being reduced and the solvent shell of the cation being larger make the issue of SIB *versus* LIB electrolyte viscosities difficult to predict.

The second main drawback is the price of ILs — this is largely ruled by the anion design and is therefore of the same order as for the salts used. The total material cost of the electrolyte will thus be high, sometimes believed to be excessively high for any practical purposes. The use of FSI-based ILs rather than TFSI-based ILs can reduce this problem somewhat as FSI has a less cumbersome and thus less expensive synthesis path. An additional drawback is the large number of studies published with not dry enough or impure ILs, which diffuse the know-how and hinder the development. As a fact, studies of IL-based electrolytes have a tendency to have an initial very heavy focus on the electrolyte material properties prior to any real SIB studies being made. Sometimes the ionic conductivity is highlighted, but as these electrolytes are composed only of ions, the ionic conductivities will always be appealing, meaning that this is a less than usual proof of any SIB promise. Therefore, assessments of the Na^+ ion transport and/or electrochemical tests/experiments involving cycling *versus* real electrode materials are even more important tools for any true appraisal of the potential as SIB electrolytes.

Moving to the major findings in the literature, for Pyr-based systems, Ding *et al.* reported NaFSI in $Pyr_{13}FSI$ (2:8 ratio) to have conductivities of 3.2 and 15.6 mS/cm at 298 and 353 K, respectively, and an ESW stretching to 5.2 V *versus* Na^+/Na.[182] A follow-up study aimed at optimizing the salt/IL ratio for a cell of Na/NaFSI in $Pyr_{13}FSI/NaCrO_2$ at a specific operating temperature, with the performance measured as capacity and rate capability.[183] More systematic studies were also made by Noor *et al.* on NaTFSI in $Pyr_{14}TFSI$, reporting ionic conductivities of 1–2 mS/cm at RT, with no deposition of Na metal on the Cu working electrode until –0.2 V *versus* Na^+/Na.[184] The latter was most probably due to the high viscosity, ca. 100 cP. More recent work by the same group dealt with the ion transport in NaTFSI in $Pyr_{13}FSI$ *via* ^{23}Na nuclear magnetic resonance (NMR) spectroscopy, but most interest attracts the stable cycling *versus* Na metal even at elevated temperatures, up to 100°C, with a stable CV of plating/stripping on the Ni working electrode.[185] This dismissed the doubts raised for this setup *via* a careful assessment of the role of water contamination at the ppm level. This again stresses IL purity and dryness, already mentioned above to obstruct progress. Very recently, NaTFSI dissolved in $Pyr_{1x}FSI$ ILs were used *versus* anatase nanosized TiO_2 electrodes to emphasize a "totally safe" SIB offering 155 mAh/g even after 80 cycles — with an initial capacity of 159 mAh/g — much ascribed to a stable SEI.[186]

The other major class of IL-based electrolytes typically uses the imidazolium ILs EMImX (1-ethyl 3-methyl imidazolium) or BMImX (1-butyl 3-methyl imidazolium) often with X = TFSI and doped with NaTFSI. In a recent communication, the equivalent FSI system, NaFSI–EMImFSI, was used for both basic properties evaluation and some cycling tests *versus* Na metal.[187] The latter again confirmed the wide ESW for the FSI anion, 5.1 V *versus* Na^+/Na. A proof-of-concept SIB symmetric cell was assembled by the group of Yamaki in 2010 using $Na_3V_2(PO_4)_3$ as both the positive and the negative electrode and a 1 M $NaBF_4$ in $EMImBF_4$ electrolyte, but with a cell voltage of merely 1.8 V — the difference between the two redox couples: V^{4+}/V^{3+} (3.4 V) and V^{3+}/V^{2+} (1.6 V) *versus* Na^+/Na.[188] An IL-based electrolyte with a mix of anions, 0.45 M NaTFSI in $Pyr_{14}FSI$, was benchmarked against an 0.5 M $NaPF_6$ in PC electrolyte and shown to outperform the latter in terms of cyclability of an $Na_{0.45}Ni_{0.22}Co_{0.11}Mn_{0.66}O_2$-positive electrode in a half-cell *versus* Na metal.[189] This was attributed to a better stability of the FSI anion at higher potentials and the formation of a suitable SEI. Another recent study on the effect of salt concentration, 0.1–1 M, of NaTFSI in $Pyr_{14}TFSI$, and rate capability, 0.05–5C, revealed a better capacity retention after 100 cycles at 0.3C and 50C (87%) as compared to 1 M $NaClO_4$ in EC:DEC (62%) for a cell setup of $Na/NaFePO_4$.[190]

The benchmarking *versus* organic electrolytes using the same setups is perhaps currently the most interesting development for IL-based electrolytes, but yet most often the conditions are a bit biased for the IL-based electrolytes, that is, the 50°C operating temperature above. Nevertheless, it shows that given specific conditions, this family of electrolytes might be viable alternatives for SIBs functionality. The issues of manufacturing and cost, however, still remain. As a prospective remedy, hybrid organic/IL-based electrolytes have been launched, primarily focusing on finding compromises/synergies in terms of SIB cell efficiency and safety, but should also cut materials costs.[191] Indeed, against a HC electrode, a better SEI than for IL-based electrolytes, more mechanically and electrochemically stable, forms, and for 0.8 M NaTFSI in $EC_{0.45}$:$PC_{0.45}$:Pyr13$TFSI_{0.10}$ electrolyte, ca. 182 mAh/g at C/10 over 40 cycles was obtained.

D. Polymer Electrolytes

Polymers are considered as matrices mainly because they offer (1) dimensional, thermal, and chemical stability, all promising better safety and (2) a flexibility that simplifies cell manufacturing. First, however, polymer electrolytes were launched due to their mechanical properties hindering the dendrite growth of Li metal anodes. The mechanical properties avoid the use of a separator in the cell. However, polymers are typically bad solvents — they have very high viscosities and low dielectric constants. Therefore, the polymers used must have strongly solvating groups such as ether oxygen atoms, carbonyl groups, and nitrile groups. The archetype solid polymer electrolyte (SPE) is based on poly(ethylene oxide) (PEO), first used in 1973 and launched for Li batteries in 1978, and is today a "component" of the Bolloré Bluecar battery used in the Autolib' cars in Paris. Quite many different SPEs have been suggested, but as of today almost all rely on ether oxygens to solvate salts, Na salts as well as Li salts. The main problems of SPEs are two: bad interfacial contact with porous electrodes and that thin film fabrication (10–20 μm) is needed due to low conductivities at RT: 10^{-5}–10^{-7} S/cm. The latter actually counteracts their safety advantage by risk of pinhole formation and/or internal short circuits. The polymer may also have limited stability at high voltages. For Li batteries using SPEs, the temperature must be set to allow for adequate ion transport, typically above 65–75°C, for Autolib' the internal operating temperature is ca. 85°C, but for any Na-based batteries, the same approach is very problematic as the melting point of Na metal is 98°C. Hence the urge to find Na-based SPEs more conductive at ambient temperatures is actually larger than for Li-based SPEs.

A functional battery with an SPE, Na/$NaClO_4$-PEO/V_2O_5, was presented as early as in 1985 by West *et al.*[192] Avoiding a Na metal anode altogether, Ma *et al.* later reported on a SIB cell built using $Na_{0.7}CoO_2$ as the positive electrode, $Na_{15}Pb_4$ as the negative electrode and $NaTf(PEO)_8$ as the SPE — operating at 60–80°C.[3] While there still are only a few full SIB studies made using SPEs, especially compared to the vast amount of studies using Li-based SPEs, there are arguably some fundamental advantages to Na-based SPEs: (1) the less strong interactions of Na^+ with the polymer host in theory allows for a faster segmental motion, and thereby cation transport and (2) the weaker ion–ion interactions NaX — important for a bad solvent such as the polymer host. But in practice, the ionic conductivities of Li- and Na-conducting SPEs are always close.[193] The current status of SPEs for Na batteries can be exemplified by the recent NaTFSI-PEO studies, with the Li analogue reported almost 20 years ago(!).[194,195] The ionic conductivity reaches 10^{-3} S/cm when above 70°C and an addition of SiO_2 ceramic nanofillers (7 nm) improved the mechanical properties and enhanced the capacity retention for a Na/$NaTFSI(PEO)_{20}$/Na symmetric cell. The use of NaTFSI is instrumental as the TFSI anion acts plasticizing reducing the concomitant increase in glass transition temperature with Na salt allowing for softer and more conductive SPEs.

Interestingly, the combined use of Na and TFSI seems to render the SPEs even "too" soft as ceramic particles are needed to create films with mechanical integrity. One alternative might be replacing NaTFSI by NaFSI, the FSI being somewhat less plastizicing.[195,196] A different approach to Na-based SPEs is the single-ion conductor (SIC) developed by Villaluenga *et al.* having the main advantage of suppression of concentration polarization.[197] Conductivities of ca. 2×10^{-5} S/cm at RT and an anodic ESW limit > 4.4 V were obtained, whereas ^{23}Na and ^{19}F NMR line-width studies confirmed the materials as SICs.

Some of the intrinsic drawbacks of SPEs can be counteracted by swelling the polymer matrix by an organic solvent to create a gel polymer electrolyte (GPE) — where the polymer provides the mechanical support and the liquid the ion transport. The solvent(s) used are typically those used for nonaqueous liquid electrolytes — and while not suitable *versus* metallic anodes GPEs can be made from many different polymers; PEO, poly(acrylonitrile) (PAN), poly (methyl methacrylate) (PMMA), poly (vinylidene fluoride) (PVdF), and so on. Indeed, the "Bellcore LIB concept" is partially based on a copolymer of PVdF with hexafluoropropylene (HFP) — PVdF–HFP. GPEs have not yet been given any major attention for SIBs — only a few publications deal with GPEs, and without any SIB cell tests. NaTf in a combined IL (EMImTf) and polymer (PVdF–HFP) matrix was suggested for SIBs targeting the problem of slow depreciation of GPE performance due to loss of solvent by evaporation, which was carefully monitored also for succinonitrile (SN) in a NaTf-PEO SPE — a system with better mechanical properties than other GPEs, due to the high viscosity of SN.[198,199] Egashira *et al.* made a ternary electrolyte of $NaBF_4$ in PEGDME adding $DEMEBF_4$ (where DEME is diethyl methyl methoxyethyl ammonium) as a cosolvent — this could almost be classified as a GPE.[200] The aim was improved cell safety and conductivities, viscosities, and thermal properties, and for a 8:1:2 PEG:$NaBF_4$:$DEMEBF_4$ ca. 1.2 mS/cm was obtained, higher than the analogous Li-based electrolyte.

E. Other Electrolyte Concepts

LIBs are currently finding a scientific revival of the all-solid-state battery concept by different new glassy and crystalline electrolytes. For SIBs, there are no such materials launched known so far — rather the classical NASICON (see Section B "Polyanionic Compounds") remains the symbol for solid-state Na conductors. As recent addition to the NASICON class of compounds, $Na_{3.1}Zr_{1.95}Mg_{0.05}Si_2PO_{12}$ was reported in 2016, showing a RT ionic conductivity of remarkable 3.5×10^{-3} S/cm, but tested for a Na salt setup rather than for.[201] Yet, for SIBs, the main current other concepts are variants/modifications of IL-based electrolytes.

As an example, glymes and a Na salt in a 1:1 ratio were used to create "solvate ILs" by creating $[Na(glyme)]^+$ cations. Using pentaglyme, G5, and NaTFSI, an ionic conductivity of 0.61 mS/cm (30°C) and an upper ESW limit of ca. 4 V were observed.[202] Subsequently, Na/[Na(G5)]TFSI/$Na_{0.44}MnO_2$ cell delivered 100 mAh/g for 50 cycles when operated at 60°C. Another, but similar track, followed by Forsyth *et al.*, is organic ionic plastic crystals, for example, $Na_xC2Pyr12_{(1-x)}TFSI$ displaying an ionic conductivity of 0.3 mS/cm at 60°C for $x = 0.4$.[203]

Eutectic salt mixtures, most simply binary NaX–MX systems, have also been pursued as SIB electrolytes. An inherent advantage with eutectic electrolytes is their fully inorganic nature, reducing side reactions with the electrodes and improving safety. The main drawbacks are cost and the operating temperature range. In 2010, the team of Hagiwara presented thermal properties of MTFSI systems, and reached 325 K for the lowest melting Na system, $Na_{0.47}Cs_{0.53}TFSI$, and the range 325–349 K for all choices of second M.[204] Studied at 423 K the former exhibits an ionic conductivity of ca. 12.1 mS/cm with an upper ESW limit of 4.9 V *versus* Na^+/Na.[205] A NaFSI–KFSI electrolyte with a conductivity of 3.3 mS/cm was used in cells with $NaCrO_2$ and Na metal and studied at 353 K — with an ESW up to 5.2 V *versus* Na^+/Na.[206,207] While no corrosion of the aluminum current collector was found, the NaK alloy is a liquid at RT for a wide range of compositions, possibly causing problems for this electrolyte/Na metal electrode combination. However, the same group also used an alloy electrode, Sn–Na and a $Na_2FeP_2O_7$ positive electrode setup.[208,209] Also ternary systems, NaFSI–KFSI–CsFSI, melting at ca. 323 K and stable up to 4.9 V *versus* Na^+/Na, have been presented.[210]

To summarize, the main electrolyte path for SIBs is liquid electrolytes based on organic solvents and runs very much in parallel to the avenues developed for LIBs. $NaPF_6$ dissolved in a mixture of alkyl carbonates is the current "state of the art." As the main handicap of LIBs due to graphite does not apply for SIBs, PC can be a key component for SIBs to a much larger extent, but EC seems compulsory for stable SEIs (more on this below). For alloy negative electrodes, EC does not seem to be enough, and use of FEC as an additive seems mandatory although the mechanism behind the observed benefits is not yet completely elucidated. Other concepts, such as IL-based electrolytes, might offer safe(r) use of HC electrodes and better capacity retention, and this at only slightly elevated temperatures, the caveat is that the ion transport details and mechanisms are not well understood. The main drawback, however, is likely cost and problems of purity and dryness. The eutectic electrolytes have the same basic promises and problems, but are further limited by the operating temperature ranges. For SIBs based on organic solvent or IL-based electrolytes, a separator is needed to contain the electrolyte and mechanically separate the electrodes. This is not needed for SPEs or a strong GPE. The separator could affect the power/energy densities, capacity retention, and safety, as observed for LIBs, but for SIBs, there are no studies on the influence of the separators available. Most SIB studies have used glass microfiber or polypropylene film separators, as for LIBs. For SPEs, there is no evidence of progress toward RT operation and the GPEs are extremely much mirroring the LIBs for materials, application, and scientific interest — but the current interest in Na-based GPEs is very limited.

F. Solid Electrolyte Interphase

The SEI, crucial for the function of the SIB, and the characterization of the SEI are both complex because the analytical conditions/sample preparation procedures can have a significant impact on the results achieved and conclusions made. Following the experience gained in the field of LIBs, the SEI composition for SIBs has also been studied by a myriad of techniques. A first comparative X-ray photoelectron spectroscopy (XPS) study combined with time of flight secondary ion mass spectrometer (TOF-SIMS) analysis of fully sodiated or lithiated HC electrodes in 1 M $MClO_4$ in PC (M = Li, Na) revealed similar compounds for both LIB and SIB cells (i.e., M_2CO_3, $ROCO_2M$, CH_2–, –CO-O–, ester linkages, . . .). The proportions differed, however, with significantly larger amounts of inorganic compounds for the SIB cells (i.e., Na_2O^+, Na_2OH^+, Na_2Cl^+, $Na_3CO_3^+$) as compared to the "organics" found for the LIB cells (i.e., $C_4H_3^+$, $C_2H_5O_3^+$, $C_2H_2O_5Li_3^+$, $Li_3CO_3^+$).[121] A significant increase in the amount of C–O environment of carbon was reported later when EC was

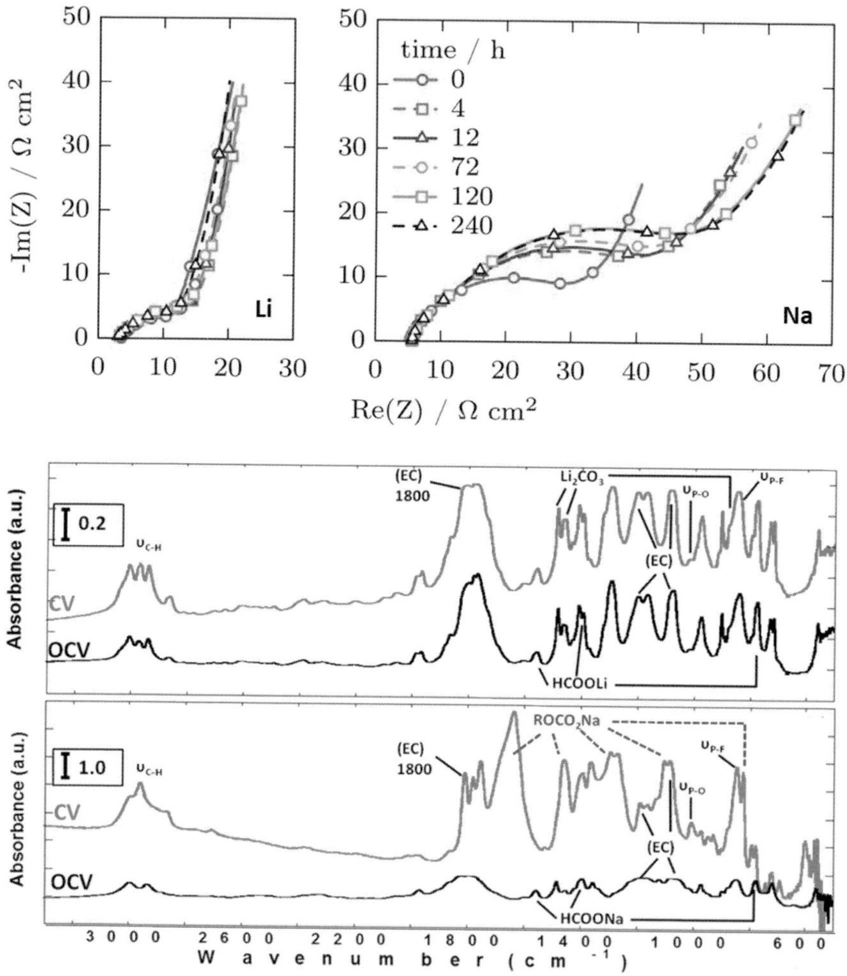

Figure 21. Impedance of HC measured in the 500 kHz–10 Hz frequency range and a three electrode configuration with Li or Na reference and counter electrodes at different times of storage in open circuit and corresponding FTIR spectra from Li or Na surfaces after 24 h storage at open-circuit voltage (OCV) or after 30 cyclic voltammetry cycles using 1 M MPF_6 in EC:DMC as electrolyte (M = Li, Na, respectively). Taken from Ref. 146 with permission of ECS.

present in the electrolyte, attributed to the ring opening of EC with formation of PEO-style oligomeric/polymeric species.[80] Alternative XPS studies using alloy or conversion electrode materials in diverse electrolytes indicated that thick layers (> 5 nm) rich in carbonates (Na_2CO_3 and $NaCO_3R$) are fully covering the electrode after full discharge (0 V *versus* Na), which are either thinner or cracked after oxidizing the electrodes up to 2 V *versus* Na.[211–214] Important differences in the morphology of the SEI formed on sodiated or lithiated HC electrodes were observed being rough and nonuniform for the former, whereas smooth and slightly thicker for the latter.[121] At this point, the legitimate question arises as to whether the decomposition of the electrolyte follows different pathways or if there are other issues, such as solubility of the products formed that can affect the SEI composition and hence also stability. Indeed, recent studies using symmetric Li/Li and Na/Na cells allow to infer significantly larger cell overpotentials for the latter related to an enhanced interfacial resistance (Figure 21).[146] This was correlated to evolution of the SEI layer upon cycling for the case of sodium, as deduced from Fourier transform infrared (FTIR) spectroscopy and scanning electron microscopy (SEM). In addition, dissolution of SEI components can result in a cross-contamination between electrodes with soluble products resulting from electrolyte decomposition at the negative electrode migrating to the positive electrode and affecting the observed electrochemical behavior. These findings call for a systematic investigation of the SEIs of SIBs, with special emphasis given to stability and correlation with the possible use of electrolyte additives, as discussed in previous sections.

V. Cell Performance

In spite of LIBs and SIBs not enlisting metal anodes, most of the tests on new electrode materials within the battery research community are carried out in *half-cells* using either lithium or sodium metal anodes. These are mostly intended to assess the electrochemical performance of a given compound as either positive or negative electrode material in a simple way, avoiding the assembly of *full cells* in which electrode balancing can severely affect performance. Indeed, if the only source of cyclable sodium is the positive electrode, it should be present in excess to compensate for losses related to the formation of the SEI, typically not known a priori.

Moreover, practical SIB cells are typically built with excess negative electrodes to avoid risks of metal plating (and dendritic growth), especially upon fast charging at low temperatures. The outcomes of *half-cell* tests are typically the electrochemical capacity and operation potential values at different rates, but the reliability of extrapolating *half-cell* testing results to potential performance in full-cells relies on the stability of the SEI. This fact was assessed for LIBs, and the SIB research, which initially ran in parallel to the avenues developed for LIBs, assumed similar stabilities resulting from similar electrolyte compositions, such as $NaPF_6$ dissolved in a mixture of alkyl carbonates. Yet, the above-mentioned doubts about the stability of the SEI call for caution when interpreting *half-cell* results and makes mandatory either to address these issues through the use of suitable electrolyte additives and/or the test in *full cell* configuration, paying due attention to electrode balancing. Moreover, figures of merit for performance rely not only on the active electrode materials but also on electrode formulation, for example, type and amount of carbon and binder, and architecture, for example, thickness, porosity, and active material loading. As a comparative note, most commercial LIB designs contain ca. 50 wt% inactive materials.

Besides the early proofs of concept, around 1990, which exhibited rather poor performance in terms of capacity retention, no other reports dealing with full SIB cell assembly appeared until a decade later with the assembly of a 3.7 V SIB cell using $NaVPO_4F$ and HC as electrode materials.[3,215] This was most certainly a side result from intense research on phosphates for LIBs at the time.[216] The electrolyte used was 1 M $NaClO_4$ in EC:DMC and the cell, tested at RT at C/10, showed a reversible capacity of 80 mAh/g on the second cycle, which faded to 50% on the 30th cycle. Nonetheless, as mentioned before, research on SIB was scarce at the time until momentum was regained around 2010, and the cycling performance of a HC|$Na[Ni_{0.5}Mn_{0.5}]O_2$ cell using 1 M of different salts, $NaClO_4$, $NaPF_6$, or NaTFSI, in PC as electrolytes was reported.[121] All cells showed initial capacities above 200 mAh/(g carbon) at 1C rate with an operation voltage of 3 V, but drastic capacity fading was observed with $NaClO_4$ in the electrolyte, while those containing $NaPF_6$ or NaTFSI retained 70% of the initial capacity after 50 cycles. Concomitantly, a stable capacity of 100 mAh/(g cathode active material) for 150 cycles at 0.5C

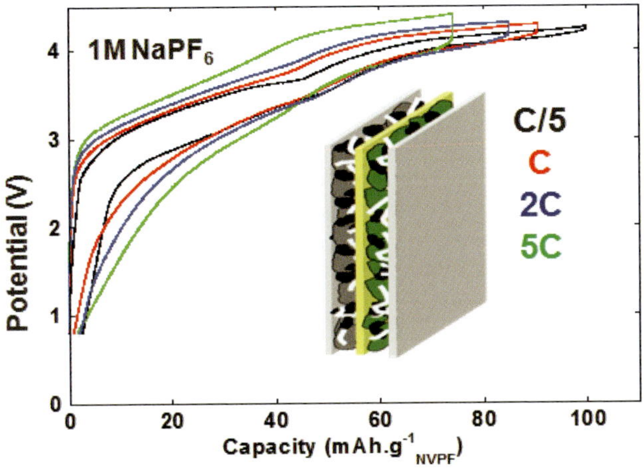

Figure 22. Voltage *versus* capacity profiles for HC//NVPF full Na-ion cells cycled in 1 M $NaPF_6$ in $EC_{0.45}$:$PC_{0.45}$:$DMC_{0.1}$ at different rates. Taken from Ref. 80 with permission of Royal Society of Chemistry.

rate was reported for a $Na_yC|Na_{1-y}[Ni_{1/3}Fe_{1/3}Mn_{1/3}]O_2$ ($y = 0.46$) cell with an operating voltage of 2.75 V using 1 M $NaClO_4$ in PC as the electrolyte.[217] Rate capability tests performed between 0.1 and 1C yielded reversible capacities ranging from 130 to 94 mAh/g. Full HC||NVPF SIB cells have been also assembled using 1 M $NaClO_4$ or $NaPF_6$ in $EC_{0.45}$:$PC_{0.45}$:$DMC_{0.10}$ as the electrolyte, which exhibit average potentials of 3.75 V and theoretical energy densities comparable to that of graphite|$LiFePO_4$ LIB cells. A stable capacity of 97 mAh/(g positive active material) for 120 cycles at C/5 and a Coulombic efficiency > 98.5% and good performance at high rates were shown (Figure 22).[80]

Full SIB cells have also been built using graphite as the negative electrode and NVP as positive electrode and 1 M $NaCF_3SO_3$ in tetraglyme as electrolyte, which with a moderate operating potential exhibited good cyclability.[218] The practical bottlenecks associated to alloying, conversion, or organic materials described above have not been an prohibiting obstacle to build laboratory full SIB cells using such materials, but these were in most cases *precycled* against sodium counter electrodes to form SEIs to avoid the issues derived from consumption of cyclable sodium from the positive electrode in the full-cell configuration. This approach, while useful at the laboratory level to assess performance, is of course not viable from any practical, commercialization, point of view. As an example, the conversion Fe_3O_4 negative electrode has been coupled to $Na[Ni_{0.25}Fe_{0.5}Mn_{0.25}]O_2$, a layered positive electrode, using $NaClO_4$ in a mixture of FEC and ethyl methyl sulfone as the electrolyte.[219] The cell operated reversibly around 2.4 V and delivered a capacity of about 130 mAh/(g cathode active material) with 76.1% capacity retention after 150 cycles and a Coulombic efficiency approaching 100%. A full Pb||NVPF cell has also been reported with 1 M $NaPF_6$ in diglyme as the electrolyte, using highly loaded negative electrodes, and exhibited a capacity of 464 mAh/(g Pb) after 50 cycles.[220] Full SIB cells have also been reported using organic negative electrodes, mono- or di-sodium terephthalate, and $Na_{0.75}Mn_{0.7}Ni_{0.23}O_2$ as the positive electrode using 1 M $NaPF_6$ in EC:EMC as the electrolyte, delivering an operating voltage of 3.6 V and ca. 268 mAh/(g cathode active material) discharge capacity after 50 cycles.[221] Finally, the operating conditions of the full SIB cells can also be changed; a 1.5 mAh cell using HC and $NaCrO_2$ and Na[FSI]-[N-methyl-N-propylpirrolidinium] [FSI], with FSI being bis(fluorosylfonyl) imide, was shown to exhibit an efficiency of 97.5% and C/5 rate, but only when running at 90°C due to the electrolyte, and this cell concept was later further upscaled to 27 Ah.[222]

Since technological parameters are important in order to achieve any practical energy density, insight on the effects of electrode balancing, charge and discharge cutoffs, and rates is of utmost importance. The trends derived, even if obtained from the lab scale, can be most useful in evaluating the real prospects of entering the market. An interesting comparison has been performed in cells using HC negative electrodes (Figure 23), either a layered $Na_{2/3}(Fe_{1/2}Mn_{1/2})O_2$ or a polyanionic NVPF as positive electrode, and a 1 M $NaPF_6$ in EC:DMC electrolyte and combining extrapolations from *half-cell* tests to experimental data achieved in full SIB cells, resulting in 208 Wh/kg for $Na_{2/3}(Fe_{1/2}Mn_{1/2})O_2$ and 237 Wh/kg for NVPF for 4.2 V cut-off voltage).[223]

Upscaling from lab scale to larger cells is a necessary step to benchmark different battery chemistries against the state of the art technologies, and most specifically SIBs *versus* LIBs, the main contender for most application prospects. Indeed, despite the environmental issues and the lower

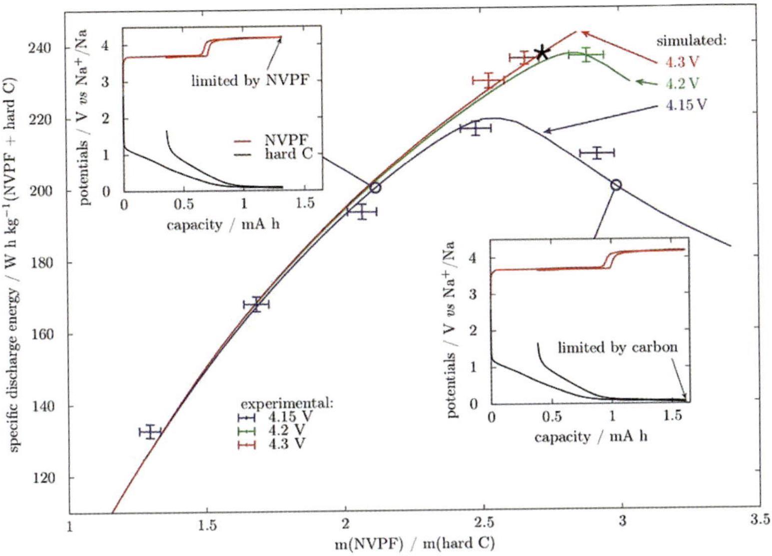

Figure 23. Simulated (lines) and experimental (points) specific energy densities of HC//NVPF cells for cut-off voltages of 4.15, 4.2, and 4.3 V. Insets show the working potential windows of anode and cathode for cells with m(NVPF)/m(hard C) = 2.1 and 3.0 and cell cut-off at 4.15 V. Taken from Ref. 223 with permission of ECS.

cost of sodium, the market prospects of SIBs will be related to both the overall cell cost, including active material synthesis and processing, and the practically attainable energy densities. While certain models provide very useful estimates, experimental validation is to be made.[224–226] Besides the openly disclosed academic development described above, industrial R&D has been disclosed to some extent and is summarized in Figure 24. In addition to press releases, from which the true performance always is difficult to evaluate, some reports on fabrication of $HC|NaFe_{0.4}Mn_{0.3}Ni_{0.3}O_2$ coin cells and laminated SIBs using 1 M $NaPF_6$ in PC as electrolyte, without any FEC additive included, were made available by Sumitomo.[227] These SIBs exhibit both good cycle life and rate capability, although metrics to compare with similar LIB cells are missing. Comparative heating and overcharging tests seem to indicate better results for SIBs, as 200% overcharge only resulted in swelling, without any bursts or ignitions. Some SIB cell performance was also recently disclosed by Faradion for their 3 Ah pouch cells using an HC negative electrode, a

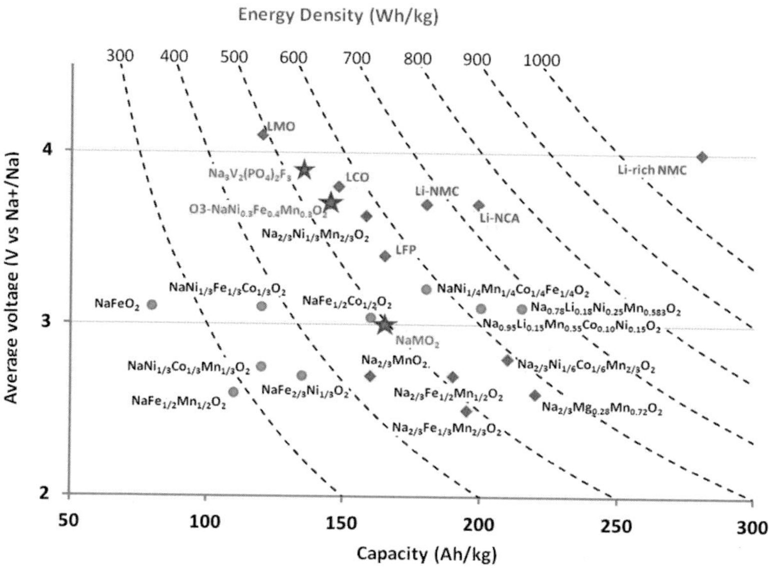

Figure 24. Graph showing the specific energies (based on discharge) of materials achieved in SIB cells. The stars and green labels correspond to prototypes reported in technical documents or press releases.

layered positive electrode material, and an electrolyte based on $NaPF_6$ dissolved in a carbonate solvent mixture — a cell that seems to be comparable to the LIB state of the art.[228]

VI. Conclusions and Perspectives

Growing sustainability concerns coupled to emerging interest in large-scale energy storage applications have boosted the interest in the SIB field. There are also suggestions made that SIBs can well compete with LIBs *both* in terms of energy density and sustainability.[5] Furthermore, progress in the SIB field is achieved at a quick pace, catalyzed by the chemical analogies between lithium and sodium and the wide accumulated know-how of LIBs. Yet, some points are crucial to put in perspective in order to assess the practical viability of the SIB technology — and here the focus is on the materials creating the basic opportunities.

The choice on the negative side is relatively restricted, today only HC exhibits any realistic application prospects. While the feasibility of SIBs using HC negative electrodes is beyond any doubt, remembering that the first generation of LIBs used HC, the ability to compete in performance toward alternative contenders relies in overcoming practical bottlenecks. The most critical issue is the irreversible capacity loss upon the first cycle, which severely penalizes the practical cell energy density achievable and indeed was also the cause of the gradual decrease in use of HC for LIBs, finally favoring graphite. Today, no real contender to HC can be foreseen, but given that many materials were suggested for LIBs before and LIBs might use, at least partly, Si-based anodes in a near future — but it took over 25 years, there ought to be other future alternatives also for SIBs.

With respect to the positive electrode, the versatility of the chemical composition of layered Na_xMO_2 allows, by taking benefit from synergetic effects of both active and inactive elements, to tune electrochemical performances reaching the one obtained with cathodes used in LIBs. Remaining challenges such as sensitivity of compounds to the atmosphere, low operating voltages, and limited reversible capacity will, with no doubt, be solved by adequate mastering of structure type and composition and even promoting the activation of anionic redox couple in addition to the conventional cationic ones. Most of these drawbacks are solved using polyanionic-based compounds that, despite undoubtedly suffering from lower gravimetric capacities than layered one, clearly benefit for long-term cycling stability and fast ion diffusion thanks to stable and open 3D frameworks and higher operating voltages due to strongly covalent polyanionic groups. The use of the know-how already acquired for LIBs enables fast research progress, which with no doubt will lead to further improved performance with the expectation that this improvement will be achieved in less time than for LIBs.

As for the choice of components and the composition for the SIB electrolyte, it can be seen as a mere "blueprint" from the LIB field. The dominant salt is $NaPF_6$ and the solvents are chosen from linear and cyclic carbonates — and even the optimal concentration is more or less the same as for LIBs: 1 M. The exception toward the LIB field is the continued use of PC as solvent, but this might change as binary or ternary solvent combinations seem to offer better performance. While there are optional concepts suggested, the SIB future will most likely be based on organic liquid electrolytes, possibly with some cosolvents being based on ILs to promote safety aspects. Given a general increased attention to safety aspects, perhaps a nonfluorinated salt will be promoted, but this should in such a case most likely also happen simultaneously for LIBs. The perspective on the future and any larger differences relate to the nature of the SEIs of SIBs as compared to LIBs. This affects how to choose the right (SEI-forming/enhancing/stabilizing) additives as well as solvents and salts, but will ultimately also affect the usage patterns for the SIB cells. A lesson learned from the LIB field is that any such optimization must be made using the real relevant SIB electrodes.

All the materials above must be put in the context of technological development to validate the attractive figures of merit. This can be and is made at the level of prototyping, to obtain real SIB cell energy densities at a larger preindustrial scale, and the preliminary information available prompts for some optimism. Together with the promises of improvements in sustainability and cost, validated by both sustainability indicator calculations and life-cycle assessments, certainly justify the expectations of a bright future for the SIB technology.[5,229]

VII. References

1. Larcher, D.; Tarascon, J.-M. *Nat. Chem.* **2015**, *7*, 19.
2. Jow, T. T.; Shacklette, L. W.; Maxfield, M.; Vernick, D. *J. Electrochem. Soc.* **1987**, *134*, 1730.
3. Ma, Y.; Doeff, M. M.; Visco, S. J.; De Jonghe, L. J.; *J. Electrochem. Soc.* **1993**, *140*, 2726.
4. Yabuuchi, N.; Kubota, K.; Dahbi, M.; Komaba, S. *Chem. Rev.* **2014**, *114*, 11636.
5. Ponrouch, A.; Monti, D.; Boschin, A.; Steen, B.; Johansson, P.; Palacin, M. R. *J. Mater. Chem. A.* **2015**, *3*, 2242.
6. Kundu, D.; Talaie, E.; Duffort, V.; Nazar, L. *Angew. Chem.* **2015**, *54*, 3431.
7. Adelhelm, P.; Hartmann, P.; Bender, C. L.; Busche, M.; Eufinger, C.; Beilstein, J. *J. Nanotechnol.* **2015**, *6*, 1016.
8. Hueso, K. B.; Rojo, T.; Armand, M. *Energy Environ. Sci.* **2013**, *6*, 734.
9. Kim, H.; Hong, J.; Park, K. Y.; Kim, H.; Kim, S. W.; Kang, K. *Chem. Rev.* **2014**, *114 (23)*, 11788.
10. Goldsztaub, S. *C. R. Acad. Sci.* **1933**, *196*, 280.
11. Mishra, S. K.; Ceder, G. *Phys. Rev. B* **1999**, *59*, 6120.
12. Parant, J.-P.; Olazcuaga, R.; Devalette, M.; Fouassier, C.; Hagenmuller, P. *J. Solid State Chem.* **1971**, *3*, 1.

13. Delmas, C.; Fouassier, C.; Hagenmuller, P. *Phys.* **1980**, *99(1–4)*, 81.
14. Zhang, B.; Dugas, R.; Rousse, G.; Rozier, P.; Abakumov, A.; Tarascon, J.-M. *Nat. Comm.* **2016**, *7*, 10308.
15. Blesa, M.-C.; Moran, E.; Leon, C.; Santamaria, J.; Tornero, J. D.; Menendez, N. *Solid State Ionics* **1999**, *126*, 81.
16. Zhao, J.; Zhao, L.; Dimov, N.; Okada, S.; Nishida, T. *J. Electrochem. Soc.* **2013**, *160*, A3077.
17. Delmas, C.; Maazaz, A.; Fouassier, C.; Réau, J.-M.; Hagenmuller, P. *Mater. Res. Bull.* **1979**, *14*, 329.
18. Yabuuchi, N.; Kajiyama, M.; Iwatate, J.; Nishikawa, H.; Hitomi, S.; Okuyama, R.; Usui, R.; Yamada, Y.; Komaba, S. *Nat. Mater.* **2012**, *11(6)*, 512.
19. Lei, Y. C.; Li, X.; Liu, L.; Ceder, G. *Chem. Mater.* **2014**, *26*, 5288.
20. Shibata, T.; Fukuzumi, Y.; Kobayashi, W.; Moritomo, Y. *Sci. Rep.* **2015**, *5*, 9006.
21. Carlier, D.; Blangero, M.; Menetrier, M.; Pollet, M.; Doumerc, J. P.; Delmas, C. *Inorg. Chem.* **2009**, *48*, 7018.
22. Xia, X.; Dahn, J. R.; *Electrochem. Solid-State Lett.* **2012**, *15*, A1.
23. Kubota, K.; Ikeuchi, I.; Nakayama, T.; Takei, C.; Yabuuchi, N.; Shiiba, H.; Nakayama, M.; Komaba, S. *J. Phys. Chem. C* **2015**, *119*,166.
24. Vassilaras, P.; Ma, X.; Li, X.; Ceder, G. *J. Electrochem. Soc.* **2012**, *160*, A207.
25. Hamani, D.; Ati, M.; Tarascon, J. M.; Rozier, P. *Electrochem. Comm.* **2011**, *13*, 938.
26. Guignard, M.; Didier, C.; Darriet, J.; Bordet, P.; Elkaïm, E.; Delmas, C. *Nat. Mater.* **2013**, *12*, 74.
27. Didier, C.; Guignard, M.; Darriet, J.; Delmas, C. *Inorg. Chem.***2012**, *51*, 11007.
28. Braconnier, J.-J.; Delmas, C.; Fouassier, C.; Hagenmuller, P. *Mater. Research Bull.* **1980**, *15*, 1797.
29. Berthelot, R.; Carlier, D.; Delmas, C. *Nat. Mater.* **2011**, *10*, 74.
30. Ding, J. J.; Zhou, Y. N.; Sun, Q.; Yu, X. Q.; Yang, X. Q.; Fu, Z. W. *Electrochim. Acta* **2013**, *87*, 388.
31. Rai, A. K.; Anh, L. T.; Gim, J.; Mathew, V.; Kim, J. *Ceram. Int.* **2014**, *40*, 2411.
32. Billaud, J.; Clément, R. J.; Armstrong, A. R.; Canales-Vázquez, J.; Rozier, P.; Grey, C. P.; Bruce, P. G. *J. Am. Chem. Soc.* **2014**, *136*, 17243.
33. Guo, S.; Yu, H.; Jian, Z.; Liu, P.; Zhu, Y.; Guo, X.; Chen, M.; Ishida, M.; Zhou, H. *ChemSusChem.* **2014**, *7*, 2115.
34. Ma, X. H.; Chen, H. L.; Ceder, G. *J. Electrochem. Soc.* **2011**, *158*, A1307.
35. Su, D.; Wang, C.; Ahn, H. J.; Wang, G. *Chem. Eur. J.* **2013**, *19*, 10884.
36. Pang, W. K.; Kalluri, S.; Peterson, V. K.; Sharma, N.; Kimpton, J.; Johannessen, B.; Liu, H. K.; Dou, S. X.; Guo, Z. P. *Chem. Mater.* **2015**, *27*, 3150.
37. Zhao, J.; Xu, J.; Lee, D. H.; Dimov, N.; Meng, Y. S.; Okada, S. *J. Power Sources* **2014**, *264*, 235.
38. Komaba, S.; Yabuuchi, N.; Nakayama, T.; Ogata, A.; Ishikawa, T.; Nakai, I. *Inorg. Chem.* **2012**, *51*, 6211.
39. Yuan, D. D.; Wang, Y. X.; Cao, Y. L.; Ai, X. P.; Yang, H. X. *ACS Appl. Mater. Interfaces* **2015**, *7*, 8585.

40. Oh, S. M.; Myung, S. T.; Yoon, C. S.; Lu, J.; Hassoun, J.; Scrosati, B.; Amine, K.; Sun, Y. K. *Nano Lett.* **2014**, *14*, 1620.
41. Yoshida, H.; Yabuuchi, N.; Komaba, S. *Electrochem. Comm.* **2013**, *34*, 60.
42. Su, J. C.; Pei, Y.; Yang, Z. H.; Wang, X. Y. *Comp. Mater. Sci.* **2015**, *98*, 304.
43. Sathiya, M.; Hemalatha, K.; Ramesha, K.; Tarascon, J.-M.; Prakash, A.-S. *Chem. Mater.* **2012**, *24*, 1846.
44. Li, X.; Wu, D.; Zhou, Y. N.; Liu, L.; Yang, X. Q.; Ceder, G. *Electrochem. Comm.* **2014**, *49*, 51.
45. Yuan, D.; Liang, X.; Wu, L.; Cao, Y.; Ai, X.; Feng, J.; Yang, H. *Adv. Mater.* **2014**, *26*, 6301.
46. Yoshida, H.; Yabuuchi, N.; Kubota, K.; Ikeuchi, I.; Garsuch, A.; Schulz-Dobrick, M.; Komaba, S. *Chem. Commun.* **2014**, *50*, 3677.
47. Wang, H.; Yang, B.; Liao, X. Z.; Xu, J.; Yang, D.; He, Y. S.; Ma, Z. F. *Electrochim. Acta* **2013**, *113*, 200.
48. Billaud, J.; Singh, G.; Armstrong, A. R.; Gonzalo, E.; Roddatis, V.; Armand, M.; Rojo, T.; Bruce, P. G. *Energy Environ. Sci.* **2014**, *7*, 1387.
49. Yabuuchi, N.; Hara, R.; Kubota, K.; Paulsen, J.; Kumakura, S.; Komaba, S. *J. Mater. Chem. A* **2014**, *2*, 16851.
50. Kataoka, R.; Mukai, T.; Yoshizawa, A.; Sakai, T. *J. Electrochem. Soc* .**2013**, *160*, A933.
51. Liu, H.; Xu, J.; Ma, C.; Meng, Y. S. *Chem. Commun.* **2015**, *51*, 4693.
52. Rozier, P.; Sathiya, M.; Paulraj, A.-R.; Foix, D.; Desaunay, T.; Taberna, P.-L.; Simon, P.; Tarascon, J.-M. *Electrochem. Com.* **2015**, *53*, 29.
53. Mortemard de Boisse, B.; Liu, G.; Ma, J.; Nishimura, S.; Chung, S.-C.; Kiuchi, H.; Harada, Y.; Kikkawa, J.; Kobayashi, Y.; Okubo, M.; Yamada, A. *Nat. Comm.* **2016**, *7*, 11397.
54. Manthiram, A.; Goodenough, J. B. *J. Solid State Chem.* **1987**, *71*, 349.
55. Manthiram, A.; Goodenough, J. B. *J. Power Sources* **1989**, *26*, 403.
56. Hong, H. Y. P. *Mater. Res. Bull.* **1976**, *11*, 173.
57. Goodenough, J. B.; Hong, H. Y. P.; Kafalas, J. A. *Mater. Res. Bull.* **1976**, *11*, 203.
58. Reiff, W. M.; Zhang, J. H.; Torardi, C. C. *J. Solid State Chem.* **1986**, *62*, 231.
59. Torardi, C. C.; Prince, E. *Mater. Res. Bull.* **1986**, *21*, 719.
60. Nadiri, A.; Delmas, C.; Salmon, R.; Hagenmuller, P. *Rev. Chim. Miner.* **1984**, *21*, 537.
61. Delmas, C.; Cherkaoui, F.; Nadiri, A.; Hagenmuller, P. *Mater. Res. Bull.* **1987**, *22*, 631.
62. Delmas, C.; Nadiri, A. *Mater. Res. Bull.* **1988**, *23*, 65.
63. Delmas, C.; Nadiri, A.; Soubeyroux, J. L. *Solid State Ionics* **1988**, *28*, 419.
64. Moreau, P.; Guyomard, D.; Gaubicher, J.; Boucher, F.; *Chem. Mater.* **2010**, *22*, 4126.
65. Lee, K. T.; Ramesh, T. N.; Nan, F.; Botton, G.; Nazar, L. F. *Chem. Mater.* **2011**, *23*, 3593.
66. Zaghib, K.; Trottier, J.; Hovington, P.; Brochu, F.; Guerfi, A.; Mauger, A.; Julien, C. M. *J. Power Sources* **2011**, *196*, 9612.
67. Le Page, Y.; Donnay, G. Can. *Mineral.* **1977**, *15*, 518.

68. Sun, A.; Beck, F. R.; Haynes, D.; Poston, J. A.; Narayanan, S. R.; Kumta, P. N.; Manivannan, A. *Mater. Sci. Eng. B* **2012**, *177*, 1729.
69. Saravanan, K.; Mason, C. W.; Rudola, A.; Wong, K. H.; Balaya, P. *Adv. Energy Mater.* **2013**, *3*, 444.
70. Zhu, C.; Song, K.; van Aken, P. A.; Maier, J.; Yu, Y. *Nano Lett.* **2014**, *14*, 2175.
71. Liu, J.; Tang, K.; Song, K.; van Aken, P. A.; Yu, Y.; Maier, J. *Nanoscale* **2014**, *6*, 5081.
72. Ellis, B. L.; Makhanouk, W. R. M.; Makimura, Y.; Toughilland, K.; Nazar, L. F. *Nat. Mater.* **2007**, *6*, 749.
73. Recham, N.; Chotard, J.-N.; Dupont, L.; Djellab, K.; Armand, M.; Tarascon, J.-M. *J. Electrochem. Soc.* **2009**, *156*, A993.
74. Ellis, B.; Ryan, D. H.; Nazar, L. F. *Chem. Mater.* **2010**, *22*, 1059.
75. Park, Y. U.; Seo, D. H.; Kwon, H. S.; Kim, B.; Kim, J.; Kim, H.; Kim, I.; Yoo, H. I.; Kang, K. *J. Am. Chem. Soc.* **2013**, *135*, 13870.
76. Bianchini, M.; Brisset, N.; Fauth, F.; Weill, F.; Elkaim, E.; Suard, E.; Masquelier, C.; Croguennec, L. *Chem. Mater.* **2014**, *26*, 4238.
77. Le Meins, J. M.; Crosnier-Lopez, M. P.; Hemon-Ribaud, A.; Courbion, G. *J. Solid State Chem.* **1999**, *148*, 260.
78. Barker, J.; Gover, R. K. B.; Burns, P.; Bryan, A. J. *J. Electrochem. Soc.* **2007**, *154*, A882.
79. Bianchini, M.; Fauth, F.; Brisset, N.; Weill, F.; Suard, E.; Masquelier, C.; Croguennec, L. *Chem. Mater.* **2015**, *27*, 3009.
80. Ponrouch, A.; Dedryvere, R.; Monti, D.; Demet, A. E.; Ateba Mba, J. M.; Croguennec, L.; Masquelier, C.; Johansson, P.; Palacin, M. R. *Energy Environ. Sci.* **2013**, *6*, 2361.
81. Xu, M.; Xiao, P.; Stauffer, S.; Song, J.; Henkelman, G.; Goodenough, J. B. *Chem. Mater.* **2014**, *26*, 3089.
82. Dacek, S. T.; Richards, W. D.; Kitchaev, D. A.; Ceder, G. *Chem. Mater.* **2016**, *28*, 5450.
83. Barpanda, P.; Ye, T.; Avdeev, M.; Chunga, S. C.; Yamada, A. *Mater. Chem. A* **2013**, *1*, 4194.
84. Barpanda, P.; Oyama, G.; Nishimura, S.; Chung, S. C.; Yamada, A. *Nat. Commun.* **2014**, *5*, 4358.
85. Pizarro-Sanz, J.; Dance, J.; Villeneuve, G.; Arriortua-Marcaida, M. *Mater. Lett.* **1994**, *18*, 327.
86. Tripathi, R.; Ramesh, T. N.; Ellis, B. L.; Nazar, L. F. *Angew. Chem. Int. Ed.* **2010**, *49*, 8738.
87. Barpanda, P.; Chotard, J.-N.; Recham, N.; Delacourt, C.; Ati, M.; Dupont, L.; Armand, M.; Tarascon, J.-M. *Inorg. Chem.* **2010**, *49*, 7401.
88. Tripathi, R.; Gardiner, G. R.; Islam, M. S.; Nazar, L. F. *Chem. Mater.* **2011**, *23*, 2278.
89. Sanz, F.; Parada, C.; Rojo, J. M.; Ruíz-Valero, C. *Chem. Mater.* **2001**, *13*, 1334.
90. Kim, H.; Park, I.; Seo, D.-H.; Lee, S.; Kim, S.-W.; Kwon, W. J.;Park, Y.-U.; Kim, C. S.; Jeon, S.; Kang, K. *J. Am. Chem. Soc.* **2012**, *134*, 10369.

91. Liu, J.; Chang, D.; Whitfield, P.; Janssen, Y.; Yu, X.; Zhou, Y.; Bai, J.; Ko, J.; Nam, K. W.; Wu, L.; Zhu, Y.; Feygenson, M.; Amatucci, G.; Ven, A. V.; Yang, X. Q.; Khalifah, P. *Chem. Mater.* **2014**, *26*, 3295.
92. Lu, Y.; Wang, L.; Cheng, J.; Goodenough, J. B. *Chem. Commun.* **2012**, *48*, 6544.
93. Takachi, M.; Matsuda, T.; Moritomo, Y. *Appl. Phys. Express* **2013**, *6*, 25802.
94. Lee, H. W.; Wang, R. Y.; Pasta, M.; Lee, S. W.; Liu, N.; Cui, Y. *Nat. Commun.* **2014**, *5*, 5280.
95. Shacklette, L. W.; Toth, J. E.; Murthy, N. S.; Baughman, R. H. *J. Electrochem. Soc.* **1985**, *132*, 1529.
96. Shacklette, L. W.; Maxfield, M.; Gould, S.; Wolf, J. F.; Jow, T. R.; Baughman, R. H. *Synth. Met.* **1987**, *18*, 611.
97. Liu, Q.; Tang, Z.; Wu, M.; Zhou, Z. *Polym. Int.* **2013**, *63*, 381.
98. Xu, F.; Chen, X.; Tang, Z.; Wu, D.; Fu, R.; Jiang, D. *Chem. Commun.* **2014**, *50*, 4788.
99. Sakaushi, K.; Hosono, E.; Nicker, G.; Gemming, T.; Zhou, H.; Kaskel, S.; Eckert, J. *Nat. Commun.* **2013**, *4*, 1485.
100. Roncali, J. *Macromol. Rapid Commun.* **2007**, *28*, 1761.
101. Zhao, R.; Zhu, L.; Cao, Y.; Ai, X.; Yang, H. X. *Electrochem. Commun.* **2012**, *21*, 36.
102. Guo, C.; Zhang, K.; Zhao, Q.; Pei, L.; Chen, J. *Chem. Commun.* **2015**, *51*, 10244.
103. Palacin, M. R. *Chem. Soc. Rev.* **2009**, *38*, 2565.
104. Pillot, C. Battery Market Development for Consumer Electronics, Automotive, and Industrial: Materials Requirements and Trends, Presentation at China International Battery Fair (18 to 20 June 2014, Shenzhen); http://doc.mbalib.com/view/43447d8528f6f7f9ec048d87d65754b9.html.
105. Di Vincenzo, D. P.; Mele, E. J. *Phys Rev. B.* **1985**, *32*, 2538.
106. Nobuhara, K.; Nakayama, H.; Nose, M.; Nakanishi, S.; Iba, H. *J. Power Sources* **2013**, *243*, 585.
107. Jache, B.; Adelhelm. P. *Angew. Chem. Int. Ed.* **2014**, *53*, 10169.
108. Kim, H.; Hong, J.; Yoon, G.; Kim, H.; Park, K. Y.; Park, M. S.; Yoon, W. S.; Kang, K. *Energy Environ. Sci.* **2015**, *8*, 2963.
109. Kim, H.; Hong, J.; Park, Y. U.; Kim, J.; Hwang, I., Kang, K. *Adv. Funct. Mater.* **2015**, *25*, 534.
110. Franklin, R. E. *Proc. R. Soc. London Series A.* **1951**, *209*, 196.
111. Kinoshita, K. In *Handbook of Battery Materials*, 2nd ed.; Daniel, C.; Besenhard, J. O., Eds.; Wiley, 2011.
112. Irisarri, E.; Ponrouch, A.; Palacin, M. R. *J. Electrochem. Soc.* **2015**, *162*, A2476.
113. Liu, Y.; Xue, J. S.; Zheng, T.; Dahn, J. R. *Carbon* **1996**, *34*, 193.
114. Dahn, J. R.; Xing, W.; Gao, Y. *Carbon* **1997**, *35*, 825.
115. Berger, D.; Carton, B.; Herold, A.; Metrot, A. *C. R. Acad. Sci. Paris Serie C.* **1970**, *271*, 837.

116. Dahn, J. R.; Sleigh, A. K.; Shi, H.; Reimers, J. N.; Zhong, Q.; Way, B. M. *Electrochim. Acta*. **1993**, 38, 1179.
117. Zheng, T.; Xing, W.; Dahn, J. R. *Carbon* **1996**, 34, 1501.
118. Stevens, D. A.; Dahn, J. R. *J. Electrochem. Soc*. **2001**, *148*, A803.
119. Alcantara, R.; Ortiz, G. F.; Lavela, P.; Tirado, J. L. *Chem. Mater*. **2006**, *18*, 2293.
120. Gotoh, K.; Ishikawa, T.; Shimadzu, S.; Yabuuchi, N.; Komaba, S.; Takeda, K.; Goto, A.; Deguchi, K.; Ohki, S.; Hashi, K.; Shimizu, T.; Ishida, H. *J. Power Sources*. **2013**, *225*, 137.
121. Komaba, S.; Murata, W.; Ishikawa, T.; Yabuuchi, N.; Ozeki, T.; Nakayama, T.; Ogata, A.; Gotoh, K.; Fujiwara, K. *Adv. Funct. Mater*. **2011**, *21*, 3859.
122. Stevens, D. A.; Dahn, J. R. *J. Electrochem. Soc*. **2000**,*147*, 4428.
123. Tsai, P.; Chung, S.; Lin, S.; Yamada, A. *J. Mater. Chem. A* **2015**, *3*, 9763.
124. Zheng, T.; Sue, J.; Dahn, J. R. *Chem. Mater*. **1996**, *8*, 389.
125. Ge, P.; Fouletier, M. *Solid State Ionics* **1988**, *28–30*, 1172.
126. Doeff, M. M.; Ma, Y.; Visco, S. J.; DeJonghe, L. C. *J. Electrochem. Soc*. **1993**, *140*, L169.
127. Stevens, D. A.; Dahn, J. R. *J. Electrochem. Soc*. **2000**, *147*, 1271.
128. Alcantara, R.; Mateos, J. M.; Tirado, J. L. *J. Electrochem. Soc*. **2002**, *149*, A201.
129. Ponrouch, A.; Marchante, E.; Courty, M.; Tarascon, J. M.; Palacin, M. R. *Energy Environ. Sci*. **2012**, *5*, 8572.
130. Ponrouch, A.; Palacin, M. R. *Electrochem. Commun*. **2015**, *54*, 51.
131. Xing, W.; Xue, J. S.; Dahn, J. R. *J. Electrochem. Soc*. **1996**, *143*, 3046.
132. Jin, J.; Shi, Z. Q.; Wang, C. Y. *Electrochim. Acta* **2014**, *141*, 302.
133. Jin, J.; Yu, B. J.; Shi, Z. Q.; Wang, C. Y.; Chong, C. B. *J. Power Sources* **2014**, *272*, 800.
134. Larcher, D.; Mudalige, C.; Gharghouri, M.; Dahn, J. R. *Electrochim. Acta* **1999**, *44*, 4069.
135. Bommier, C.; Luo, W.; Gao, W. Y.; Greaney, A.; Ma, S.; Ji, X. *Carbon* **2014**, *76*, 165.
136. Winter, M.; Besenhard, J. O. In *Handbook of Battery Materials*, 2nd ed.; Daniel, C. Besenhard, J. O., Eds.; Wiley, 2011.
137. Gibaud, A.; Xue, J. S.; Dahn, J. R. *Carbon* **1996**, *34*, 499.
138. Simone, V.; Boulineau, A.; de Geyer, A.; Rouchon, D.; Simonin, L.; Martinet, S. *J. Energy Chemistry* **2016**, *25*, 761–768.
139. Buiel, E.; George, A. E.; Dahn, J. R. *Carbon* **1999**, *37*, 1399.
140. Dahn, J. R.; Sleigh, A. K.; Shi, H.; Reimers, J. N.; Zhong, Q.; Way, B. M. *Electrochimi. Acta* **1993**, *38*, 1179.
141. Simone, V. Ph.D. Thesis, Communaute Universite Grenoble Alpes, Novembre 2016.
142. Buiel, E.; Dahn, J. R. *J. Electrochem. Soc*. **1998**, *145*, 1977.
143. Li, Y.; Xu, S.; Wu, X.; Yu, J.; Wang, Y.; Hu, Y. S.; Li, H.; Chen, L.; Huang, X. *J. Mater. Chem. A* **2015**, *3*, 71.

144. Zhao, J.; Zhao, L.; Chihara, K.; Okada, S.; Yamaki, J. I.; Matsumoto, S.; Kuze, S.; Nakane, K. *J. Power Sources* **2013**, *244*, 755.
145. *Electrochemical Series in CRC Handbook of Chemistry and Physics* [Online]; David R. L., Ed.; CRC Press: Boca Raton, FL, 2005. http://www.hbcpnetbase.com.
146. Iermakova, D.; Dugas, R.; Palacin, M. R.; Ponrouch, A. *J. Electrochem. Soc.* **2015**, *162*, A2476.
147. Chevrier, V. L.; Ceder, G. *J. Electrochem. Soc.* **2011**, *158*, A1011.
148. Ellis, L. D.; Hatchard, T. D.; Obrovac, M. N. *J. Electrochem. Soc.* **2012**, *159*, A1801.
149. Ellis, L. D.; Hatchard, T. D.; Obrovac, M. N. *J. Electrochem. Soc.* **2014**, *161*, A416.
150. Allan, P. K.; Griffin, J. M.; Darwiche, A.; Borkiewicz, O. J.; Wiaderek, K. M.; Chapman, K. W.; Morris, A. J.; Chupas, P. J.; Monconduit, L.; Grey, C. P. *J. Am. Chem. Soc.* **2016**, *138*, 2352.
151. Darwiche, A.; Marino, C.; Sougrati, M. T.; Fraisse, B.; Stievano, L.; Monconduit, L. *J. Am. Chem. Soc.* **2012**, *134*, 20805.
152. Xiong, H.; Slater, M. D.; Balasubramanian, M.; Johnson, C. S.; Rajh, T. *J. Phys. Chem. Lett.* **2011**, *2*, 2560.
153. Nava-Avendaño, J.; Morales-Garcia, A.; Ponrouch, A.; Rousse, G.; Frontera, C.; Senguttuvan, P.; Tarascon, J. M.; Arroyo-de-Dompablo, M. E.; Palacin, M. R. *J. Mater. Chem. A.* **2015**, *3*, 22280.
154. Alcántara, R.; Jaraba, M.; Lavela, P.; Tirado, J. L. *Chem. Mater.* **2002**, *14,* 2847.
155. Cabana, J.; Monconduit, L.; Larcher, D.; Palacin, M. R. *Adv. Mater.* **2010**, *22*, E170.
155. Song, Z.; Zhou, H. *Energy Environ. Sci.* **2013**, *6*, 2280.
157. Lopez-Herraiz, M.; Castillo-Martinez, E.; Carretero-Gonzalez, J.; Carrasco, J.; Rojo, T.; Armand, M. *Energy. Environ. Sci.* **2015**, *8*, 3233.
158. Zhao, L.; Zhao, J.; Hu, Y. S.; Li, H.; Zhou, Z.; Armand, M.; Chen, L. *Adv. Energy Mater.* **2012**, *2*, 962.
159. Castillo-Martinez, E.; Carretero-Gonzalez, J.; Armand, M. *Angew. Chem Int. Ed.* **2014**, *126*, 5445.
160. Xu, K. *Chem. Rev.* **2014**, *114*, 11503.
161. Suo, L.; Borodin, O.; Gao, T.; Olguin, M.; Ho, J.; Fan, X.; Luo, C.; Wang, C.; Xu, K. *Science* **2015**, *350*, 938.
162. Yamada, Y.; Yaegashi, M.; Abe, T.; Yamada, A. *Chem. Commun.* **2013**, *49*, 11194.
163. Jónsson, E.; Johansson, P. *Phys. Chem. Chem. Phys.* **2012**, *14*, 10774.
164. Krause, L. J.; Lamanna, W.; Summerfield, J.; Engle, M.; Korba, G.; Loch, R.; Atanasoski, R. *J. Power Sources* **1997**, *68*, 320.
165. Vevlin, D. J.; Herley, P. *J. React. Solids* **1987**, *3*, 75.
166. Evans, T.; Olson, J.; Bhat, V.; Lee, S.-H. *J. Power Sources* **2014**, *269*, 616.
167. Bhide, A.; Hofmann, J.; Dürr, A. K.; Janek, J.; Adelhelm, P. *Phys. Chem. Chem. Phys.* **2014**, *16*, 1987.
168. Izutsu, K. *Electrochemistry in Nonaqueous Solutions*; Wiley-VCH Verlag GmbH, 2002. ISBN: 3-527-30516-5.

169. Pearsons, R. G. *J. Am. Chem. Soc.* **1963**, *85*, 3533.
170. Shakourian-Fard, M.; Kamath, G.; Smith, K.; Xiong, H.; Sankaranarayanan, S. K. R. S. *J. Phys. Chem. C* **2015**, *119*, 22747.
171. Fong, R.; von Sacken, U.; Dahn, J.-R. *J. Electrochem. Soc.* **1990**, *137*, 2009.
172. Alcantara, R.; Lavela, P.; Ortiz, G. F.; Tirado, J. L. *Electrochem. Solid-State Lett.* **2005**, *8*, A222.
173. Komaba, S.; Murata, W.; Ishikawa, T.; Yabuuchi, N.; Ozeki, T.; Nakayama, T.; Ogata, A.; Gotoh, K.; Fujiwara, K. *Adv. Funct. Mater.* **2011**, *21*, 3859.
174. Vidal-Abarca, C.; Lavela, P.; Tirado, J. L.; Chadwick, A. V.; Alfredsson, M.; Kelder, E. *J. Power Sources* **2012**, *197*, 314.
175. Ponrouch, A.; Goñi, A. R.; Palacín, M. R. *Electrochem. Commun.* **2013**, *27*, 85.
176. Komaba, S.; Ishikawa, T.; Yabuuchi, N.; Murata, W.; Ito, A.; Ohsawa, Y. *Appl. Mater. Interfaces* **2011**, *3*, 4165.
177. Wang, Y. X.; Lim, Y. G.; Park, M. S.; Chou, S. L.; Kim, J. H.; Liu, H. K.; Dou, S. X.; Kim, Y. J. *J. Mater. Chem. A* **2014**, *2*, 529.
178. Qian, J.; Wu, X.; Cao, Y.; Ai, X.; Yang, H. *Angew. Chem. Int. Ed.* **2013**, *52*, 4633.
179. Bitner-Michalska, A.; Michalczewski, K.; Zdunek, J.; Ostrowski, A.; Zukowska, G.; Trzeciak, T.; Zero, E.; Syzdek, J.; Marcinek, M. *Electrochim. Acta* **2016**, *210*, 395.
180. Hosokawa, T.; Matsumoto, K.; Nohira, T.; Hagiwara, R.; Fukunaga, A.; Sakai, S.; Nitta, K. *J. Phys. Chem. C* **2016**, *120*, 9628.
181. Monti, D.; Jónsson, E.; Palacin, M. R.; Johansson, P. *J. Power Sources* **2014**, *245*, 630.
182. Ding, C. S.; Nohira, T.; Kuroda, K.; Hagiwara, R.; Fukunaga, A.; Sakai, S.; Nitta, K.; Inazawa, S. *J. Power Sources* **2013**, *238*, 296.
183. Ding, C. S.; Nohira, T.; Hagiwara, R.; Matsumoto, K.; Okamoto, Y.; Fukunaga, A.; Sakai, S.; Nitta, K.; Inazawa, S. *J. Power Sources* **2014**, *269*, 124.
184. Noor, S. A. M.; Howlett, P. C.; MacFarlane, D. R.; Forsyth, M. *Electrochim. Acta* **2013**, *114*, 766.
185. Yoon, H.; Zhu, H.; Hervault, A.; Armand, M.; MacFarlane, D. R.; Forsyth, M. *Phys. Chem. Chem. Phys* **2014**, *16*, 12350.
186. Wu, L.; Moretti, A.; Buchholz, D.; Passerini, S.; Bresser, D. *Electrochim. Acta* **2016**, *203*, 109.
187. Matsumoto, K.; Hosokawa, T.; Nohira, T.; Hagiwara, R.; Fukunaga, A.; Numata, K.; Itani, E.; Sakai, S.; Nitta, K.; Inazawa, S. *J. Power Sources* **2013**, *265*, 36.
188. Plashnitsa, L. S.; Kobayashi, E.; Noguchi, Y.; Okada, S.; Yamaki, J. I. *J. Electrochem. Soc.* **2010**, *157*, A536.
189. Chagas, L. G.; Buchholz, D.; Wu, L. M.; Vortmann, B.; Passerini, S. *J. Power Sources* **2014**, *247*, 377.
190. Wongittharom, N.; Lee, T. C.; Wang, C. H.; Wang, Y. C.; Chang, J. K. *J. Mater. Chem. A* **2014**, *2*, 5655.

191. Monti, D.; Ponrouch, A.; Palacín, M. R.; Johansson, P. *J. Power Sources* **2016**, *324*, 712.
192. West, K.; Zachau-Christiansen, B.; Jacobsen, T.; Atlung, S. *J. Electrochem. Soc.* **1985**, *132*, 3061.
193. Fergus, J. W. *J. Power Sources* **2010**, *195*, 4554.
194. Serra Moreno, J.; Armand, M.; Berman, M. B.; Greenbaum, S. G.; Scrosati, B.; Panero, S. *J. Power Sources* **2014**, *248*, 695.
195. Boschin, A.; Johansson, P. *Electrochim. Acta* **2015**, *175*, 124.
196. Boschin, A.; Johansson, P. *Electrochim. Acta* **2016**, *211*, 1006.
197. Villaluenga, I.; Bogle, X.; Greenbaum, S.; Gil de Muro, I.; Rojo, T.; Armand, M. *J. Mater. Chem. A* **2013**, *1*, 8348.
198. Kumar, D.; Hashmi, S. A. *Solid State Ionics* **2010**, *181*, 416.
199. Patel, M.; Chandrappa, K. G.; Bhattacharyya, A. J. *Solid State Ionics* **2010**, *181*, 844.
200. Egashira, M.; Asai, T.; Yoshimoto, N.; Morita, M. *Electrochim. Acta* **2011**, *58*, 95.
201. Song, S.; Duong, H. M.; Korsunsky, A. M.; Hu, N.; Lu, L. *Sci. Rep.* **2016**, *6*, 32330.
202. Ueno, K.; Yoshida, K.; Tsuchiya, M.; Tachikawa, N.; Dokko, K.; Watanabe, M. *J. Phys. Chem. B* **2012**, *116*, 11323.
203. Forsyth, M.; Chimdi, T.; Seeber, A.; Gunzelmann, D.; Howlett, P. C. *J. Mater. Chem. A* **2014**, *2*, 3993.
204. Kubota, K.; Nohira, T.; Hagiwara, R. *Electrochim. Acta* **2012**, *66*, 320.
205. Nohira, T.; Ishibashi, T.; Hagiwara, R. *J. Power Sources* **2012**, *205*, 506.
206. Kubota, K.; Nohira, T.; Goto, T.; Hagiwara, R. *Electrochem. Comm.* **2008**, *10*, 1886.
207. Fukunaga, A.; Nohira, T.; Kozawa, Y.; Hagiwara, R.; Sakai, S.; Nitta, K.; Inazawa, S. *J. Power Sources* **2012**, *209*, 52.
208. Yamamoto, T.; Nohira, T.; Hagiwara, R.; Fukunaga, A.; Sakai, S.; Nitta, K.; Inazawa, S. *Electrochim. Acta* **2014**, 135, 60.
209. Chen, C. Y.; Matsumoto, K.; Nohira, T.; Hagiwara, R.; Fukunaga, A.; Sakai, S.; Nitta, K.; Inazawa, S. *J. Power Sources* **2013**, *237*, 52.
210. Kubota, K.; Nohira, T.; Hagiwara, R. *Electrochim. Acta* **2012**, *66*, 320.
211. Philippe, B.; Valvo, M.; Lindgren, F.; Rensmo, H.; Edstrom, K. *Chem. Mater.* **2014**, *26*, 5028.
212. Baggetto, L,; Allcorn, E.; Unocic, R.; Manthiram, A.; Veith, G. M. *J. Mater. Chem. A* **2013**, *1*, 1116.
213. Baggetto, L,; Ganesh, P.; Sun, C. N.; Meisner, R. A.; Zawodzinski, T. A.; Veith, G. M. *J. Mater. Chem. A* **2013**, *1*, 7985.
214. Darwiche, A.; Bodenes, L.; Madec, L.; Monconduit, L.; Martinez, H. *Electrochim. Acta* **2016**, *207*, 284.
215. Shacklette, L. W.; Jow, T. R.; Townsend, L. *J. Electrochem. Soc.* **1985**, *135*, 2669.
216. Barker, J.; Saidi, Y.; Swoyer, J. L. *Electrochem. Solid-State Lett.* **2003**, *6*, A1.
217. Kim, D.; Lee, E.; Slater, M.; Lu, W.; Rood, S.; Johnson, C. S. *Electrochem. Commun.* **2012**, *18*, 66.

218. Zhu, Z.; Cheng, F.; Hu, Z.; Niu, Z.; Chen, J. *J. Power Sources* **2015**, *293*, 626.
219. Oh, S. M.; Myung, S. T.; Yoon, C. S.; Lu, J.; Hassoun, J.; Scrosati, B.; Amine, K.; Sun, Y. K. *Nano Letters,* **2014**, *14*, 1620.
220. Darwiche, A.; Dugas, R.; Fraisse, B.; Monconduit, L. *J. Power Sources* **2016**, *304*, 1.
221. Abouimrane, A.; Weng, W.; Eltayeb, H.; Cui, Y.; Niklas, J.; Poluektov, O.; Amine, K. *Energy Environ. Sci.* **2012**, *5*, 9632.
222. Fukunaga, A.; Nohira, T.; Hagiwara, R.; Numata, K.; Itani, E.; Sakai, S.; Nitta. K. *J. Appl. Electrochem.* **2016**, *46*, 487.
223. Dugas, R.; Zhang, B.; Rozier, P.; Tarascon, J. M. *J. Electrochem. Soc.* **2016**, *163*, A867.
224. Nelson, P. A.; Gallagher, K. G.; Bloom, I. BatPaC (Battery Performance and Cost) Software, Argonne National Lab, htt5://www.cse.anl.gov/BatPaC/
225. Eroglu, D.; Ha, S.; Gallagher, K. G. *J. Power Sources* **2014**, *267*, 14.
226. Berg, E. J.; Vilevieille, C.; Streich, D.; Trabesinger, S.; Novak, P. *J. Electrochem. Soc.* **2015**, *162*, A2468.
227. Kuze, S.; Kageura, J. I.; Matsumoto, S.; Nakayama,T.; Makidera, M.; Saka, M.; Yamaguchi, T.; Yamamoto, T.; Nakane, K. *Sumitomo Kagaku,* **2013**, 1.
228. Barker, J.; Heap, R.; Roche, N.; Tan, C.; Sayers, R.; Liu, Y. 17th IMLB, Como, Italy, June 10–14, 2014, Abstr. 266.
229. Peters, J.; Buchholz, D.; Passerini, S.; Weil, M. *Energy Environ. Sci.* **2016**, *9*, 1744.

5 Battery Systems Based on Multivalent Metals and Metal Ions

Doron Aurbach,* Romain Berthelot,† Alexandre Ponrouch,‡ Michael Salama* and Ivgeni Shterenberg*,§

*Department of Chemistry, Bar-Ilan University, Ramat-Gan 5290002, Israel
†Institut Charles Gerhardt de Montpellier (UMR 5253 CNRS Université de Montpellier), 2, Place Eugène Bataillon — CC1502 — 34095 Montpellier, France
‡Institut de Ciència de Materials de Barcelona (ICMAB-CSIC), Campus UAB, E-08193 Bellaterra, Catalonia, Spain
§Ivgeni.Shterenberg@gmail.com

Table of Contents

List of Abbreviations

AN	acetonitrile
APC	all phenyl complex
BPC	N-(1-butyl)pyridinium chloride
CTAB	cetyltrimethylammonium bromide
CTAC	cetyltrimethylammonium chloride
CuHCF	copper hexacyanoferrate
DCC	dichloro complex
DEC	diethyl carbonate
DMC	dimethyl carbonate
DME	dimethoxyethane
DMF	N,N-dimethylformamide
$DMSO_2$	dimethylsulfone
EC	ethylene carbonate
EQCM	electrochemical quartz crystal microbalance
EMC	ethyl methyl carbonate
EMIC	1-ethyl-3-methylimidazolium chloride
EPB	ethylpyridinium bromide
GBL	γ-butyrolactone
HOMO	highest occupied molecular orbital

HRTEM	high-resolution transmission electron microscopy
LUMO	lowest unoccupied molecular orbital
MACC	magnesium aluminum chloride complex
MCMB	mesocarbon microbeads
$MgTFSI_2$	magnesium bis(trifluoromethanesulfonyl)imide
NBS	national bureau of standards
ORR	oxygen reduction reaction
PB	prussian blue
PC	propylene carbonate
PG	pyrolytic graphite
PTFE	polytetrafluoroethylene
PP14TFSI	N-methyl-N-butylpiperidinium bis(trifluoromethanesulfinyl)imide
PTMA	poly (2,2,6,6-tetramethyl-piperidinyloxy-4-ylmethacrylate)
RT	room temperature
RTIL	room-temperature ionic liquid
SEI	solid electrolyte interphase
SEM	scanning electron microscopy
STM	scanning tunnelling microscopy
THF	tetrahydrofuran
TMPAC	trimethylphenylammonnium chloride
XPS	X-ray photoelectron spectroscopy

I. Introduction

The success in the development and commercialization of lithium batteries (mostly primary systems) and then rechargeable Li-ion batteries and the increasing demand for more types of power sources for versatile uses stimulate research and development efforts related to battery systems based on other active metals. Highly interesting is the possibility to employ light, multivalent active metals as anodes for batteries. Magnesium, calcium, and aluminum are light, active, and highly abundant metallic elements in earth crust. Indeed, during the last decades, efforts have been devoted to research that aimed at considering the possibility to use these metals in batteries. The most interesting and intriguing are attempts to develop rechargeable batteries based on these metals and ionic compounds, in operational modes with resemble that of Li-ion batteries. This chapter contains 11 main sections that describe systematically and extensively the studies related to Mg, Ca, and Al as anodes in batteries. Both fundamental and practical/technical aspects are included.

II. Passivation Phenomena Related to Calcium, Magnesium, and Aluminum

The utilization of Li-based energy storage devices is possible mainly due to the unique properties of the solid electrolyte interphase (SEI) layers formed on the various cell components.[1] The SEI in Li-based systems conducts Li^+ ions but does not conduct electrons, and therefore the passivation in Li-based systems is not "true" passivation.[2] This SEI layer is imperative in Li-based systems as it prohibits further degradation of the electrolyte solutions and all other cell's components during the cell's high voltage operation.

The same cannot be applicable to systems where the carrier ion is of a multivalent nature, such as magnesium, calcium, and aluminum.[3,4] In these multivalent systems, the passivation layers are truly passivating and conduct neither ions nor electrons. This in turn dictates that in multivalent-based systems, passivation-free conditions must be upheld. Generally, this means that the electrolyte solution, that is, solvent and salt, should be stable with all the relevant cell components, including the active metal anode. These mandatory limitations severely limit the selection of appropriate solvents, salts, anodes, cathodes, and current collectors.

Aluminum is naturally passivated by a highly stable oxide layer. This oxide layer formation is one of aluminum's most desired properties, along with its low weight and high thermal and electrical conductance. The stable oxide passivation layers make aluminum highly resistant to corrosion.[5] When considering aluminum as an anode material for a rechargeable battery however, the passivation layers are a major hindrance to the cell performance. More specifically, passivation significantly increases the aluminum electrode's reduction potential causing an overall lower operational voltage. The oxide passivation layer can be removed by addition of alkali solutions[6] or completely avoided by using an aluminum amalgam.[6] Since aluminum is highly reactive, both of these approaches ultimately result in the rapid corrosion of the aluminum electrode.[7,8] Therefore, in passivation controlled environments, the use of an aluminum anode is fairly limited in primary batteries, whereas completely impossible in secondary batteries. When considering the use of aluminum in secondary batteries, nonaqueous electrolytes that are stable with aluminum must be employed. Nonaqueous electrolyte-based aluminum halides, aluminum hydrides, room temperature (RT) molten salts, and organoaluminum compounds show reversible aluminum deposition dissolution processes that in some cases reach 99.9999% Coulombic efficiency.[9,10] However, no groundbreaking full rechargeable aluminum-based battery prototype has yet been demonstrated.

Much like with magnesium and aluminum, the passivating layers that are formed on calcium, inhibit any calcium ions or electrons transfer. Unlike magnesium and aluminum, no reversible calcium deposition dissolution processes have ever been demonstrated at RT. This is mainly due to the inability to develop an electrolyte that is stable enough with calcium and its very low reduction potential of –2.87 V. Calcium was initially tested to replace Li in primary Li/thionyl chloride ($SOCl_2$) cells.[11] However, the rapid corrosion of calcium made the feasibility of such cells unpractical. Calcium anodes were also screened with conventional organic electrolytes: salts and solvents that are normally used in Li batteries. Perchlorate (ClO_4^-)- and tetrafluoroborate (BF_4^-)-based salts were tested in acetonitrile (AN), tetrahydrofuran (THF), γ-butyrolactone (GBL), and propylene carbonate (PC). TBA(BF_4^-) in THF was the only combination that did not result in surface film formation on the calcium electrode.[12] Calcium dissolution was indeed observed in a TBA(BF_4^-)/THF solution, indicating that a reversible calcium deposition/dissolution processes could be observed with a $Ca(BF_4^-)_2$/THF, though the electrochemical performance of such solution has not been demonstrated in that research. In 2016, Palacin *et al.* demonstrated some reversible Ca electrochemistry with $Ca(BF_4^-)_2$ in EC:PC (ethylene carbonate/propylene carbonate) at elevated temperatures (100°C). It is possible that at high enough temperatures, surface films formed on calcium do not fully passivate the active metal and hence allow some Ca ions transport through them.[13]

Contrary to aluminum and calcium, a fully feasible prototype of a magnesium-based battery has been demonstrated.[14] Reversible Mg^{2+} ion migration between the anode and the solution was only realized by incorporating electrolyte solutions that do not react with magnesium metal. Some ether solvents, more specifically, THF and glymes [dimethoxyethane (DME)], were found to be completely stable with Mg metal.[15] Most conventional magnesium salts are either incompatible with magnesium metal or do not dissolve in ethers and therefore are unusable. Highly reductive magnesium compounds such as organomagnesium halides[14,16] and magnesium borohydrides[17] exhibit great stability with magnesium, and magnesium can be easily deposited and dissolved reversibly from electrolyte solutions comprising such compounds in THF. Interestingly, $MgTFSI_2$, magnesium bis(trifluoromethanesulfonyl)Imide, can be used as an electrolyte in glyme solvents, enabling both bulk and interfacial magnesium ions' transport. Thermodynamically, the bis(trifluoromethanesulfonyl) imide (TFSI) anion is expected to react and passivate magnesium; however, a recent research showed[18] that magnesium can be reversibly deposited and dissolved, to some extent, in $MgTFSI_2$/DME solutions. Very large overpotentials, for both deposition and dissolution, are observed, suggesting the formation of passivating layers. This passivation, however, is not as formidable and does allow limited migration of Mg^{2+} ions. Even more fascinating is the fact that addition of chloride ions[19] to $MgTFSI_2$/DME solutions (e.g., my dissolving therein $MgCl_2$) forms solutions with good electrochemical properties comparable to the Grignard-based solutions (Grignard-based solutions are regarded as passivation free for Mg electrodes). It may be possible that chlorides are adsorbed strongly to magnesium metal surface and inhibit the $TFSI^-$ anions from approaching magnesium electrodes, thus avoiding their reduction and subsequent Mg surface passivation.

III. First Calcium and Magnesium Batteries — $SOCl_2$-Based Electrolytes

Several unsuccessful attempts to electrodeposit divalent cations have demonstrated the difficulty of these systems.[20] Indeed, unlike Li, which can be electroplated in a large variety of aprotic solvents, the first electrolyte that has been identified to allow for Mg plating was based on Grignard reagents with the addition of boranes in ether solutions.[21] A few years later, Auborn *et al.* and Behl *et al.* studied solutions of lithium inorganic salts ($LiBCl_4$, $LiAlCl_4$, etc.) in phosphorous oxychloride ($POCl_3$) and $SOCl_2$ as electrolytes for Li–chlorine cells.[22,23] Following these results on Li metal electrode, Peled *et al.*[24] and Staniewicz[25] first reported on the electrochemical behavior of M-$SOCl_2$ (with M = Mg or Ca) systems. The overall cell reaction has been suggested in a study of the cell discharge products and appears to be quite reasonable considering the analogy with the Li-$SOCl_2$ technology[25]:

$$2\ M + 2\ SOCl_2 \rightarrow 2\ MCl_2 + SO_2 + S \qquad (1)$$

In terms of specific energy density Mg and Ca-$SOCl_2$ (ca. 300 Wh/kg) compare favorably with their Li analog and present improved safety due to the absence of dendrite formation.[26] In addition, Ca and Mg anodes have the prospect for higher temperature operation given the higher melting point of Ca (839°C) and Mg (650°C) when compared with Li (180°C), preventing from anode melting under abused temperature conditions.

Although this technology was considered for the development of primary cells, it is worth noting that the effort to electroplate Ca in $SOCl_2$-based electrolytes also failed. Indeed, although high current density can be recorded upon anodic polarization of a Ca electrode with relatively low overpotential and efficient oxidation of the metal electrode, very high overpotential was recorded upon cathodic polarization (Figure 1) and

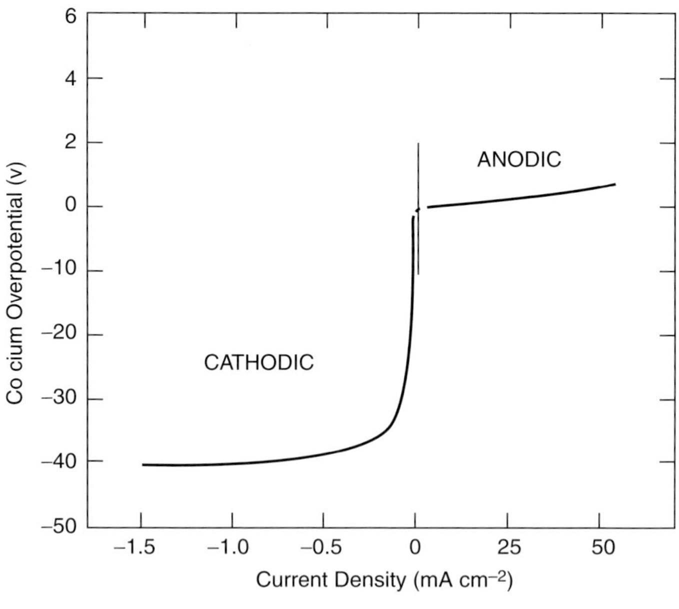

Figure 1. I–V characteristic of a Ca/Ca^{2+} SEI in $SOCl_2$ "diode-like" behavior). Taken from Ref. 27 with permission of Elsevier.

strong solvent reduction occurs leading to SO_2 formation, the Coulombic efficiency for the Ca plating reaction being ca. 5%.[26,27] The low reversibility of the Ca plating/stripping reactions has been mainly attributed to the poor diffusion of divalent cations through the passivation layer as well as to the mixed ionic conductor nature of the latter that precludes Ca plating.[25,27] It is worth mentioning that Mg electrodeposition was found to be possible and the obtained deposit to be less prone to corrosion than Ca in $SOCl_2$-based electrolytes. Indeed, a Ca deposit is completely consumed in less than 7 min after plating,[26] whereas Mg electroplated on Ni appears to be stable for at least 2 h.[24]

Only MCl_2 (with M = Mg or Ca) was identified as a possible component of the metal electrode passivation layer, in agreement with the low standard free energy of formation of MCl_2 from MO and $SOCl_2$.[24,25] Cation transport number in alkaline earth halides is low[28] and $CaCl_2$ is more covalent than LiCl, thus further limiting the divalent cation mobility. These parameters led Staniewicz to propose that the anion transport number is close to unity in $CaCl_2$.[25] Ca deposition should thus proceed through the destruction of the $CaCl_2$ at the current collector–passivation layer interface and growth of $CaCl_2$ at the passivation layer — electrolyte interface (Figure 2), this process being highly inefficient. This model was further developed and extended to Mg by Peled and coworkers suggesting that upon cathodic polarization of a Ca or Mg electrode vacancies accumulate at the current collector–passivation layer interface and Cl^- at the passivation layer — electrolyte interface.[27] Both processes eventually lead to ionic current blocking through the passivation layer and result in the diode-like behavior of Ca and Mg-$SOCl_2$ system that can deliver high current density under anodic polarization of the metal anode while behaving as a strong resistor in the cathodic direction. It is worth mentioning that this blocking phenomenon did not appear in Li cells[29] or in Mg or Ca cells[26] in $LiAlCl_4$-$SOCl_2$ solution most probably due to the good mobility of Li^+ through the LiCl, $MgCl_2$, or $CaCl_2$ passivation layers.

The rate-determining step for Ca and Mg plating/stripping in $SOCl_2$-based electrolyte is the ionic migration through the passivation layer formed on the metal electrode. The formation and properties of such layers at open-circuit voltage (OCV) and under anodic or cathodic polarization were extensively studied, mainly by the group of Peled.

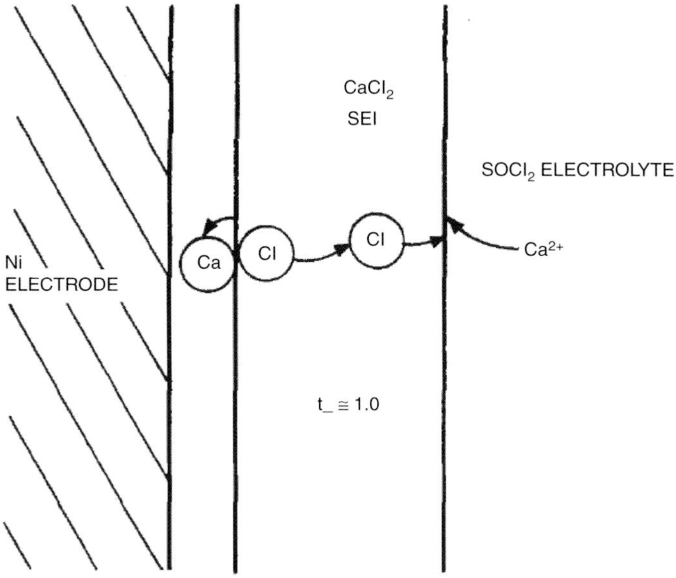

Figure 2. Model for calcium deposition through a SEI layer of $CaCl_2$. Ionic current is carried by the anion as a result of the low transport number of the calcium cation. Taken from Ref. 25 with permission of the Journal of the Electrochemical Society.

The apparent passivation layer thickness was estimated through the evaluation of the differential capacitance of the metal electrode assuming an average and fixed value of the dielectric constant of the SEI components and the equation from parallel-plate capacitor.[30] For Li, the passivation layer resistivity (assumed to be proportional to its thickness) was found to increase with time until it reaches a steady value after about 20 days (Figure 3a).[31] By contrast, Mg in $SOCl_2$ solutions reached steady state after about 50 h at OCV (Figure 3b),[30] whereas Ca passivation layer does not appear to fully stabilize (Figure 3c).[27] Under steady-state conditions, the passivation layer should exhibit the highest integrity and compactness leading to the lowest ionic and electronic conductivities. It is worth mentioning that at OCV much higher passivation layer resistivity was recorded for Ca and Mg (ca. 10^8–10^{10} Ω/cm) when compared with Li (ca. 10^7–10^8 Ω/cm). However, upon anodic polarization, the apparent resistivity of Mg and Ca passivation layers decreases by more than three

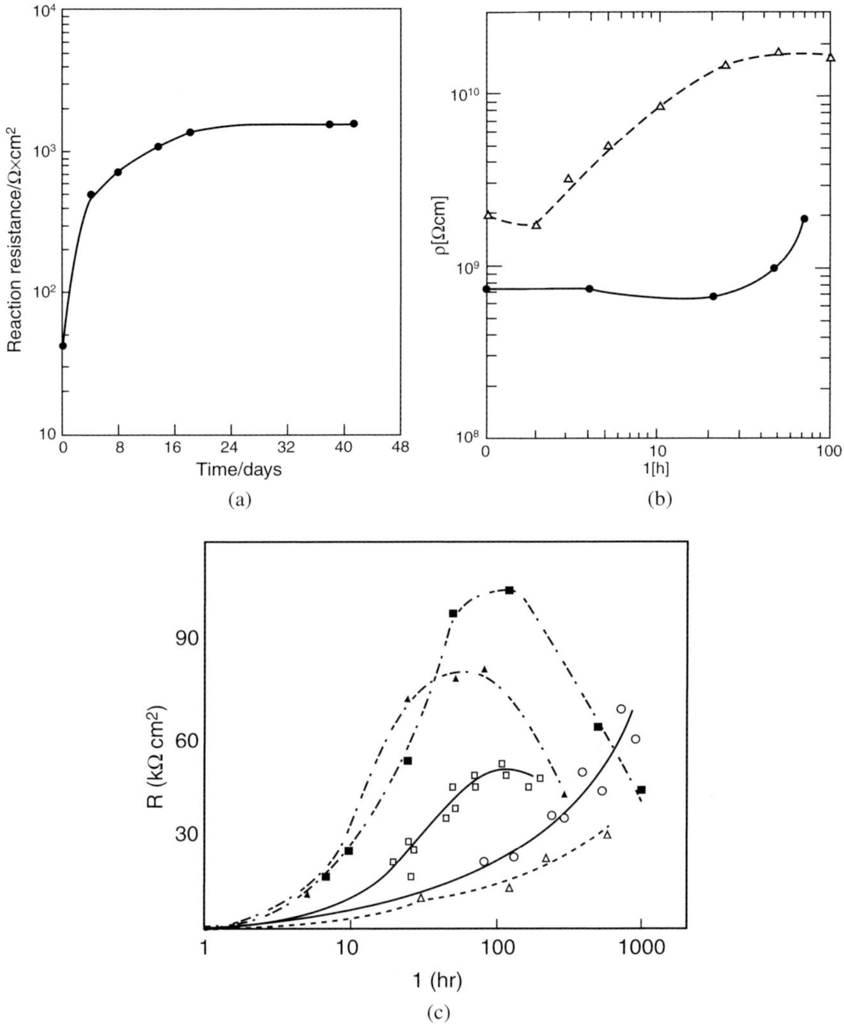

Figure 3. (a) A typical variation of the reaction resistance (R_r) with time of Li-$LiAlCl_4$ system stored at RT. Taken from Ref. 31 with permission of the Journal of the Electrochemical Society. (b) The resistivity of the SEI of magnesium in saturated and unsaturated (acidic) solutions vs the immersion time. Δ, saturated solution, 0.6 M Mg $(Al < Cl_4)_2$ in $SOCl_2$; •, unsaturated solution, 1.2 M $AlCl_3$ and 0.3 M $MgCl_2$–$SOCl_2$. Taken from Ref. 30 with permission of the Journal of the Electrochemical Society. (c) Reaction resistance *versus* immersion time of calcium electrode in TC solutions containing various electrolytes including (O) 1.9 M $Ca(AlCl_4)_2$. Taken from Ref. 27 with permission of the Journal of the Electrochemical Society.

orders of magnitude.[27] As it can be seen from Figure 3b, the passivation layer resistivity of Mg is significantly decreased in unsaturated (acidic) solution such as 1.2 M $AlCl_3$ and 0.3 M $MgCl_2$–$SOCl_2$ when compared with saturated (neutral) solution such as 0.6 M $Mg(AlCl_4)_2$ in $SOCl_2$, indicating an important impact of the $AlCl_3$ Lewis acid on the passivation layer resistivity (thickness) and leading to significantly enhanced metal electrode corrosion rate.[30]

Corrosion of the metal electrode is an important issue in $SOCl_2$-based electrolytes. Indeed, the rate of the metal anode corrosion is determinant for the cycle life and self-discharge during storage of the cell. As it can be seen from Figure 4, rather flat discharge curves are obtained for Ca-$SOCl_2$-based cells with a plateau at ca. 3 V (reliant on the electrolyte composition and temperature)[26] and the cell voltage decline more or less abruptly depending on the cell end-of-life mechanism (constant and homogeneous oxidation of the anode and cathode limitation, see Figure 4a or concomitant corrosion and mechanical failure of the anode, see Figure 4b). Nonnegligible self-discharge of Ca-$SOCl_2$-based cells has also been reported as they lost about 10% of their capacity after 2 weeks of storage at 70°C.[32] In addition, possible corrosive character of the electrolyte could lead to the dissolution of freshly deposited metal (if any) formed upon charge further limiting the rechargeability of $SOCl_2$-based system.[33] In addition, possible corrosion of the Ca-$SOCl_2$ cell can material [stainless steel (SS)], has been highlighted by Peled *et al.*[34] and was attributed to the loss of the protective nature of the SS native oxide layer through partial reduction at the calcium potential. As expected, SS corrosion is very harmful to Li- and Ca- $SOCl_2$ cells due to the anode SEI doping by iron impurities that enhanced the corrosion rate.

As mentioned above, the addition of Lewis acids, such as $AlCl_3$, to the electrolyte leads to dramatic enhancement of the Mg and Ca corrosion, a Mg electrode being completely dissolved after ca. 70-h immersion at RT in 1 M $AlCl_3$-$SOCl_2$.[24] This is either related to the reactivity of the Lewis acid with the MCl_2-based passivation layer, allowing for further reaction of the metal with the electrolyte solvent, or to the direct reaction of $AlCl_3$ with Ca leading to the formation of Al and $CaCl_2$.[25] Interestingly, Mg practically does not corrode in solution saturated with $MgCl_2$. Similar strategy was used for Ca system by ensuring stoichiometric (1:2) ratio of

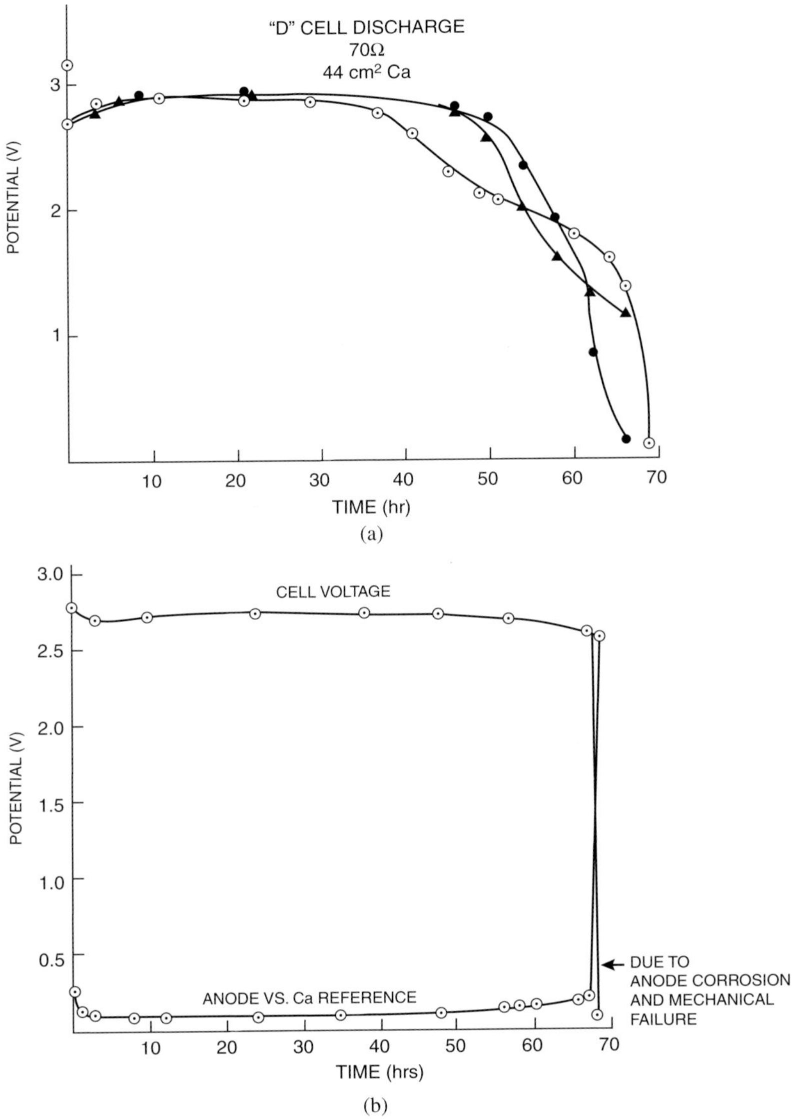

Figure 4. (a) Typical discharge profile for a $Ca/Ca(AlCl_4)_2$-$SOCl_2$ cell and (b) cell failure resulting from the anode corrosion. Taken from Ref. 25 with permission of the *Journal of the Electrochemical Society.*

Ca:Al when mixing the halide ($CaCl_2$) with the Lewis acid ($AlCl_3$) within the $SOCl_2$ solution leading to the complete in situ transmetalation of the Lewis acid.[25]

Elster *et al.* evaluated the use of $Sr(AlCl_4)_2$ or $Ba(AlCl_4)_2$ instead of $Ca(AlCl_4)_2$ in $SOCl_2$ solution in order to modify the passivation layer composition and stability aiming at improving corrosion resistance. They stated that the use of $Sr(AlCl_4)_2$ or $Ba(AlCl_4)_2$ leads to the formation of passivation layers composed of 80% $SrCl_2$/20% $CaCl_2$ or 1%–4% $BaCl_2$/96%–99% $CaCl_2$, respectively. Lower corrosion rate and self-discharge were obtained from a system with $SrCl_2$-containing passivation layer and was attributed to the lower solubility and higher resistivity of the latter when compared with $CaCl_2$.

Overall, even though Mg and Ca-$SOCl_2$ appear as a very interesting primary cell technologies with similar capacity and improved safety when compared with their Li analog (commercialized by SAFT and TADIRAN Batteries Ltd.), these technologies did not reached the market, most probably due to the fact that Li-$SOCl_2$ batteries are used in niche applications such as military and low temperature environments, for which Mg and Ca $SOCl_2$ battery systems would not provide significant advantage.

IV. Nonaqueous Electrochemistry of Magnesium — Grignard-Type Solutions

As mentioned in the previous section, formation of surface films must be avoided in secondary battery systems based on Mg anodes. Therefore, the electrolytic solutions should be stable with Mg metal. Ethers were found to be sufficiently stable with Mg but most Mg salts are not. The first electrochemical, reversible Mg deposition was demonstrated in 1927.[35] The solutions that were used in that experiment were various types of Grignard reagent dissolved in THF. Grignard reagents are very strong reductants and therefore are stable with Mg metal. They also act as scavengers for various ubiquitous contaminants (such as water and oxygen) that can passivate the Mg electrodes. Such solutions have, however, major flaws: very low ionic conductivity, in the range of micro siemens, and very low anodic stability that make them irrelevant for batteries applications. The solution's low conductivity[36] is due to the nonionic nature of Grignard reagents. In addition to these electrochemical drawbacks, Grignard solutions are highly pyrophoric and are unsafe to handle under ambient atmosphere. In addition, ethers and especially THF are extremely volatile.

In order to increase conductivity to acceptable levels, the Grignard reagents must be ionized. This is accomplished by adding to these strong Lewis bases and Lewis acids such as organoaluminum chloride compounds (e.g., R_yAlX_{3-y}, where R is an organic group and X is an halide ion).[21] The aluminum moiety reacts with the Grignard reagent (a Lewis base) in an acid/base transmetallation reaction. During the transmetallation reaction, the organic groups, R, migrate to the Al atoms to form complex anions $[R_yAlX_{3-y}]^-$, whereas the halides, X, interact with the Mg cations to form complex ions such as $[MgX]^+$, $[Mg_2X_3]^+$ coordinated (and stabilized) by THF ligands.[37] These solution structures are termed magnesium organo-halo-aluminate complexes. The conductivity of these solutions is in the 1-mS range,[38] owed to the charge separation in the newly formed complexes.

The electrochemical performance of the organo-halo-aluminate-based solutions is highly dependents on the nature of the halide and organic group, and the Mg:Al:R:X ratios and concentrations.[38] The number of metal–organic bonds, in this case Al–R, directly affects the anodic stability of the solutions as those are the weakest bonds. Also, fewer organic ligands result in solutions with a wider anodic stability.[39] In addition, as the aluminum ions are bound to more halides, their anodic stability is higher. As the halogen, X, are more electropositive, they should endow the AlX_3 with higher acidity thus making aluminate anions harder to be oxidized. The most electropositive halogen atom is fluorine. However, it is unsuitable because it forms the highly insoluble MgF_2 that instantly passivates Mg electrodes. Chlorine, the next "in line" in electropositive values, produces electrolytic solutions with the best electrochemical characteristics[39] out of all the halides that were tested. The organic, R, group can also affect the electrochemical window depending on its resistance toward oxidation. The most direct path to oxidation is through β-hydrogen elimination reaction. Using organic ligands that do not contain hydrogen on the β position produce electrolyte solutions with higher electrochemical windows. For example, solutions that contain butyl and ethyl ligands (that have hydrogen on β position) can reach an electrochemical window up to 2.5 V,[14] whereas solutions that contain phenyl ligands (that do not have hydrogens on β positions) exhibit electrochemical window up to 3.2 V.[16]

The organo-halo-aluminate-based electrolyte solutions are still the most commonly used so far, in research related to nonaqueous electrolyte solutions. These electrolytic solutions, nevertheless, are unsuitable for practical batteries due to their being corrosive for metallic current collectors, their toxicity, volatility, and relative high price.

V. First Attempt to Explore Rechargeable Magnesium Batteries

Magnesium possesses many of the desired properties expected from a next-generation anode materials for secondary batteries: low reduction potential (−2.37 V), high volumetric specific capacity (3833 mAh/cm^3), high stability in ambient atmosphere, abundance, competitive price, and low toxicity.[40] Despite its potential applicability as a secondary battery anode, no real breakthrough in the field was demonstrated before the research of Gregory *et al.* in 1990.[41] This was a promising attempt to utilize magnesium as an anode for secondary batteries. Magnesium is known to form native surface films that block Mg ions transport from Mg electrodes to the solution and vice versa. These surface films are formed by the reaction of the Mg metal with any ubiquitous reducible species such as trace water and oxygen. Mg also react with most organic and inorganic solvents, along with most conventional Mg salt anions. Only ethereal solvents were found to be sufficiently stable with Mg. The selection of Mg salts was initially limited to highly reductive Mg compounds such as Grignard reagents (organo Mg compounds) and Mg borohydride, as these compounds are reductive enough to be stable with Mg and are soluble in ethers.[39]

The first attempts to explore the feasibility of secondary magnesium batteries were conducted during the 1980s by Gregory *et al.*[6] They explored various cathode materials, magnesium metal anodes, and various electrolytes. It is important to note that until that studies, the knowledge about Mg electrochemistry was fairly limited.

For the screening of cathode materials, many known intercalation materials were tested both chemically and electrochemically. Chemically, the cathode materials were submerged in a 0.7 M solution of dibutylmagnesium in THF for over 72 h after which they were rinsed and measured for structural changes and Mg/cathode material molecular ratios via x-ray diffraction and atomic absorption. Electrochemically, the cathode materials were cycled galvanostatically with current densities up to 15 mA/cm^2. The results regarding the cathode materials provided in this research are mostly irrelevant since the researchers did not provide sufficient evidence that the suggested processes they studied were indeed related to Mg intercalation rather than, for example, Mg deposition. A limited reversibility was demonstrated by Gregory *et al.* with Co_3O_4; however, the potentials reported were too low.[26] The early work on V_2O_5 cathodes is mentioned. Novak *et al.* found that Mg ions can intercalate reversibly to some extent, with partially hydrated V_2O_5, in conventional solutions (e.g., AN/$Mg(ClO_4)_2$).[42] Passerini *et al.* demonstrated some reversible Mg intercalation into aerogel V_2O_5.[43] These studies, while showing some directions were impractical, because they were not conducted in electrolyte solutions in which magnesium anodes can work reversibly.

In parallel, in addition to the $Mg(BR_4)_2$/ether complex solutions studied by Gregory *et al.*,[26] several types of modified Grignard reagent/ether solutions were examined. The most notable of these solutions is the hexamethyldisilazane magnesium chloride (HMDS-Mg-Cl) in THF.[44] The HMDS-Mg-Cl is a unique Grignard reagent as the Mg atom is bonded to the nitrogen rather than the carbon, which is the reason why the reagent is less nucleophilic. The less nucleophilic nature of this reagent allows for the implementation of previously unsuitable cathode materials, such as sulfur.[45]

Grignard/ether-based electrolyte solutions, without any additives, show poor electrochemical performance, which can be extensively improved by the addition of $AlCl_3$. Grignard solutions with added $AlCl_3$ exhibited much higher conductivities along with a reasonable Coulombic efficiency of Mg deposition/dissolution processes. Electrochemical deposition from these solutions could produce highly crystalline Mg-deposited layers. The general formula of such electrolyte solutions is $RMgCl/AlCl_3$ in THF as the best ether solvent. Various R ligands and $AlCl_3$:RMgCl ratios and concentrations were examined and are summarized in Table 1. Another important finding was the Magnesium

Table 1. Electrochemical performance of various Grignard reagents with different amount of $AlCl_3$ additive. Adopted from Ref 41.

R-Group	RMgCl (M)	$AlCl_3$ (M)	Cell Voltage	Anodic Current Efficiency	Cathodic Current Efficiency	Mg Deposit Quality
Methyl	1.5	0.1	0.5	100	101	Dendritic
Ethyl	0.8	0.1	1.5	98.9	100	Excellent, small grains
Ethyl	1.5	**0.1**	0.6	99.5	100	Excellent, small grains
Ethyl	2	0.2	0.6	90.1	–	Very dendritic
Butyl	1	0.1	1	104	98.4	Excellent, small grains

halide-organoboron complexes in mixed DME/THF. They were found to support reversible Mg deposition/dissolution cycles at 98.8% efficiency. The high toxicity of organoboron complexes seems to exclude them from practical use.

The basis for composing suitable electrolyte solutions for magnesium secondary batteries was established by a systematic work of Aurbach *et al.* starting 1999. The first important published work of this group included a demonstration of working magnesium battery prototypes in 2000.[14]

VI. First Prototypes of Rechargeable Mg Batteries

Aurbach *et al.*[14] demonstrated the first prototypes of secondary magnesium batteries by two major discoveries.

The first one was the development of new electrolyte solutions that have the necessary adequate properties. The basic requirements are inertness with Mg metal, high conductivity, high Coulombic efficiency of the Mg deposition/dissolution process, and a wide enough electrochemical window. The development of the desired electrolytic solution was related the work done by Gregory *et al.*[41] and focused on complex electrolytes formed by the Lewis acid/base reactions in ethers (mostly THF) between $R_{2-n}MgX_n$ (Lewis base) and R_nAlX_{3-n} (Lewis acid), where R is an organic group and X is a halide. Combinations of different Lewis acids/bases were thoroughly studied and screened along with their total concentrations and ratios in solutions. The best solution derived initially from Aurbach's studies was termed the dichloro complex (DCC) solution. It comprised the reaction products between Bu_2Mg and $EtAlCl_2$ at 1:2 ratio in THF[14] with the general formula of $Mg(AlCl_2EtBu)_2$. These electrolytes have relatively high ionic conductivity of 1–1.4 mS/cm, a cycling efficiency of Mg deposition/dissolution = 100%, and an anodic stability window of 2.5 V.

The second discovery was the implementation of a suitable cathode material that could (1) facilitate reversible insertion/deinsertion of Mg^{2+} ions and (2) is fully stable in the highly reductive and highly corrosive DCC solution. The only suitable cathode material was of the Chevrel phase family, Mo_6T_8 (where T = S, Se).[14,46,47] The Mo_6T_8 Chevrel phase structure can be described as stacking of Mo_6S_8 blocks with every block having an octahedral cluster of Mo atoms inside a cube of sulfur atoms. This unique structure allows for 12 sites between each two Mo_6S_8 units into which magnesium ions could be intercalated (up to two Mg^{2+} ions per Mo_6S_8 unit). The Cheverel phase cathode structure allows the intercalation of multivalent ions in relatively fast kinetics. This is owed to the unique capability of Mo_6T_8 to redistribute charge without inflicting major structural changes.[46,47] This Chevrel phase material exhibits a theoretical specific capacity of 122 mAh/g capacity and an average voltage of 1.2 V *versus* magnesium. Further work of Aurbach *et al.* concentrated on understanding all aspects of multivalent ions' intercalation mechanisms into Chevrel phases.[47–54] Also, Aurbach *et al.* demonstrated improved cathode materials in which one or two sulfur atoms were substituted by Se atoms (i.e., $Mg_2Mo_6S_6Se_2$). Such substitution enabled to realize the entire specific capacity and obtain very good rate capability.[47,48,54] Various practical synthetic modes for these cathode materials were developed as well. It was also demonstrated that $CuMo_6S_8$ insert highly reversibly and fast Mg ions, via a complicated displacement mechanism.[50] To date, the Chevrel phases are still the only viable cathode materials for Mg batteries, which can work properly at RT. None of the other materials that were studied so far and reported as potential cathode materials for Mg batteries could demonstrate reasonable Mg ions' insertion rates at RT and true stability during cycling. An operation scheme of full cells, comprising Mg metal anodes, Chevrel phase cathodes and DCC electrolyte solution showing the

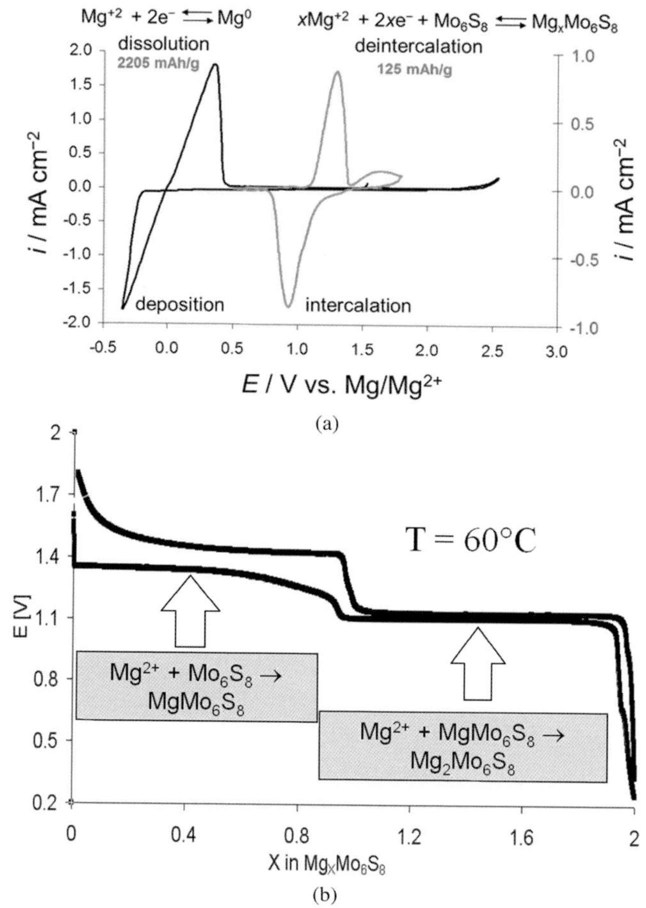

Figure 5. (a) An operation scheme of secondary Mg battery comprising cheverel phase (Mo_6S_8) cathode, Mg metal anode, and DCC electrolyte solution. The orange line represents the Mo_6S_8 cathode response during intercalation/deintercalation of Mg ions. The black line represents the magnesium deposition dissolution process at the anode side. (b) Galvanostatic charge–discharge curve for a secondary Mg battery at 60°C, showing 100% reversibility. Taken from Ref. 40 with permission of the authors.

voltammograms for both the anodic and cathodic processes is presented in Figure 5.

Such cells could demonstrate thousands of fully reversible charge–discharge cycles at 0.1–1 mA/cm^2 rate and within a temperature range of −10°C to 60°C.

VII. Development of New Families of Electrolyte Solutions for Rechargeable Mg Batteries During Recent Years

Electrolyte solutions based on the organo-halo-aluminate complexes[14] are endowed with severe disadvantages: they are highly corrosive, extremely toxic, have high susceptibility toward oxidation, and relatively expansive. Such downsides make these electrolyte solutions to be unpractical for secondary magnesium batteries. During the last few years, new concepts for electrolyte solutions that stray from the organo-halo-aluminate format have been developed. The most important and interesting of which are MACC[55] (magnesium aluminum chloro complex), magnesium monocarboranes,[56] and $MgTFSI_2$-based electrolytes.[18,19]

The magnesium aluminum chloro complex (MACC) electrolyte solutions comprise entirely of inorganic salts and show reversible magnesium deposition. The most practical MACC electrolyte solution comprises 0.075 M $AlCl_3$ and 0.25 M $MgCl_2$ in THF. At first, the resulting electrolyte solution exhibits poor electrochemical performance, most notably characterized by a very high overpotential for deposition and poor reversibility.[55] The MACC solution needs to be activated by cycling it at 1 mV/s between −1.2 and 2.8 V (*versus* Mg), this process was termed "conditioning." The conditioning process highly improves the electrochemical properties of the solution, putting it quite comparable to the organo-haloaluminate solutions in terms of performance. The postconditioned electrolyte solution exhibits a Coulombic efficiency of >99%, low overpotential for Mg deposition, and an electrochemical window of 3.1 V. Also, reversible Mg intercalation/deintercalation into Mo_6S_8 Chevrel phase cathodes was demonstrated with the MACC electrolytic solution.[55] Voltammograms comparing the conditioned and unconditioned MAAC solutions are shown in Figure 6. As of now, the conditioning process effect on the solution is not fully understood. One theory[40] claims that the conditioning process eliminates various impurities from the solution (such as trace water and oxygen) that their reduction processes would otherwise passivate the electrodes. Minor impurities are generally not an issue in organo-haloaluminate-based solutions, as these solutions have highly reductive compounds that eliminate most impurities. Other theories claim

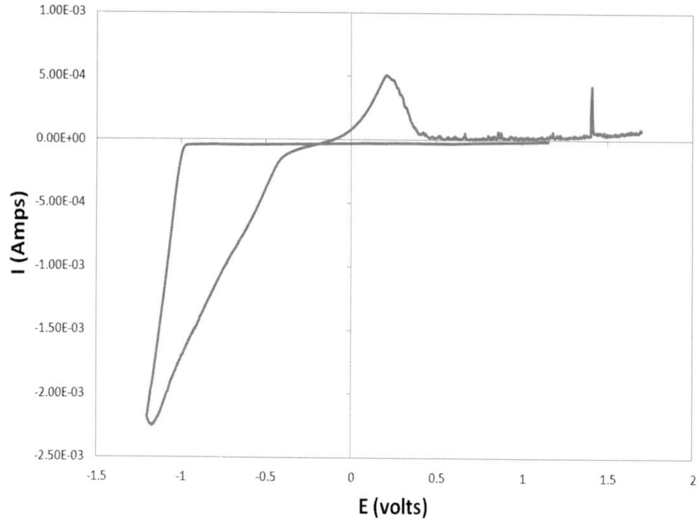

(a)

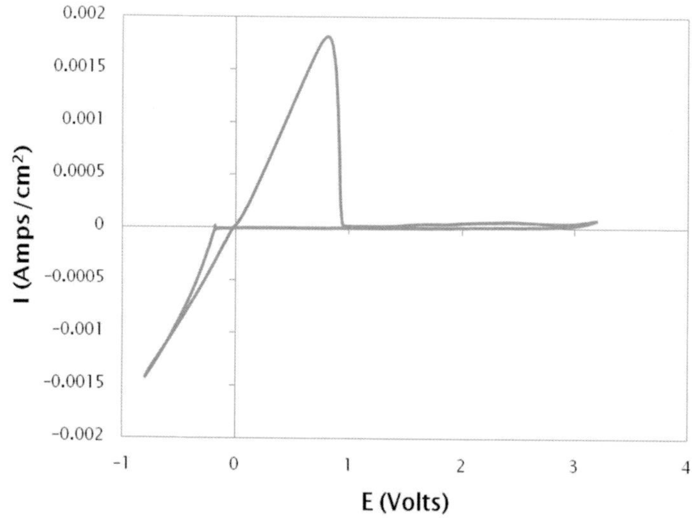

(b)

Figure 6. Steady-state cyclic voltammograms of $MgCl_2$–$AlCl_3$/THF (MACC) solution with Pt electrodes: (a) unconditioned electrolyte solution (b) after a conditioning process. Pt working electrodes. Taken from Ref. 40 with permission of the authors.

that the conditioning process changes the stoichiometry[57] of the solution toward a more favorable Al:Mg ratio. It is believed that the active species in the MACC solution are the products of a Lewis acid/base reaction, where $AlCl_3$ acts as the acid and $MgCl_2$ as the base. In essence it is, as with the organo-haloaluminate solutions, a transmetallation reaction where the chlorides migrate from the less electronegative Mg atom to the more electronegative Al atom, to form $MgCl^+$ and $AlCl_4^-$ ions. The $MgCl^+$ ion can react further with $MgCl_2$ to form $Mg_2Cl_3^+$.[16]

$MgTFSI_2$ is the only known "simple" magnesium salt that readily dissolves in ethereal solvents.[5] This salt has high thermal stability and high resistivity toward oxidation/reduction.[19] It also forms highly conductive solutions when dissolved in glymes. The electrochemical performance of a $MgTFSI_2$/DME solution is rather poor, plagued by high overpotentials for deposition/dissolution and low reversibility (Figure 7). Even so, the fact that even some electrochemical activity is observed means that the electrodes are not fully passivated, even though a reaction between the $TFSI^-$ ion and the Mg metal is highly plausible.

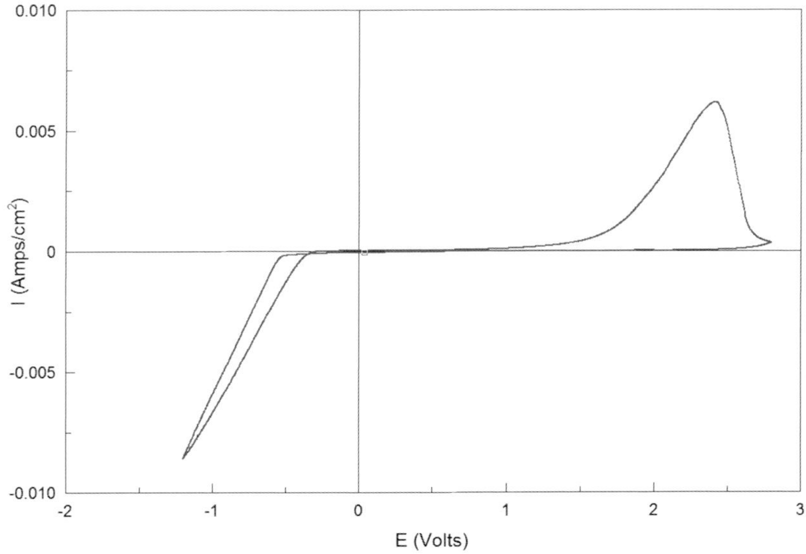

Figure 7. A typical steady-state cyclic voltammogram of $MgTFSI_2$ 0.5 M solution in DME with a Pt working electrode. Taken from Ref. 19 with permission of the Journal of the Electrochemical Society.

The $MgTFSI_2$/DME electrolyte solution can be augmented by the addition of $MgCl_2$.[19] Although $MgCl_2$ in practically insoluble in glymes, it can be dissolved in $MgTFSI_2$/DME solutions. The addition of $MgCl_2$ improves the electrochemical properties of these solutions. Most noticeable is the minimization of the overpotential for Mg dissolution. As with the MACC solutions, a conditioning process is usually required to produce a solution with adequate electrochemical performance. A post conditioned $MgTFSI_2/MgCl_2$/DME solution exhibits high anodic stability, good Coulombic efficiency, and a high conductivity (Figure 8).

Another important family of electrolyte solutions that has shown great progress lately is the magnesium borohydride (in glymes)-based electrolyte solutions, proposed by Mohatadi *et al.*[17] Earlier studies demonstrated reversible Mg deposition from solutions based on $Mg(BH_4)_2$ in DME/THF. The big disadvantage of such solutions is their narrow electrochemical window of <2 V invoked by the low oxidative resistance of the $Mg(BH_4)_2$ salt. Further studies have introduced promising solutions comprising magnesium monocarborane,[56] $Mg(CB_{11}H_{12})_2$, in tetraglyme. These solutions exhibit reversible Mg deposition/dissolution with a Coulombic efficiency >99% and an exceptionally wide electrochemical window of 3.8 V. Initial examinations have also suggested that the Mg monocarborane solutions are not corrosive toward current collectors such as SS and aluminum.

VIII. Interfacial Aspects Related to Nonaqueous Mg Batteries — The Challenge

Mg metal anodes in conventional organic electrolytes are surface-film controlled, as is the case for Li. However, Mg plating is virtually impossible in the presence of a passivation layer, which was attributed to the lack of divalent cation transport through the latter.[58] Therefore, all electrolytes developed for Mg cells should not build such interphases and bare electrode should be maintained. This specific feature of the Mg system brings along a major difference with the Li system. Upon plating, the desolvation of Li^+ takes place at the interface between the electrolyte and

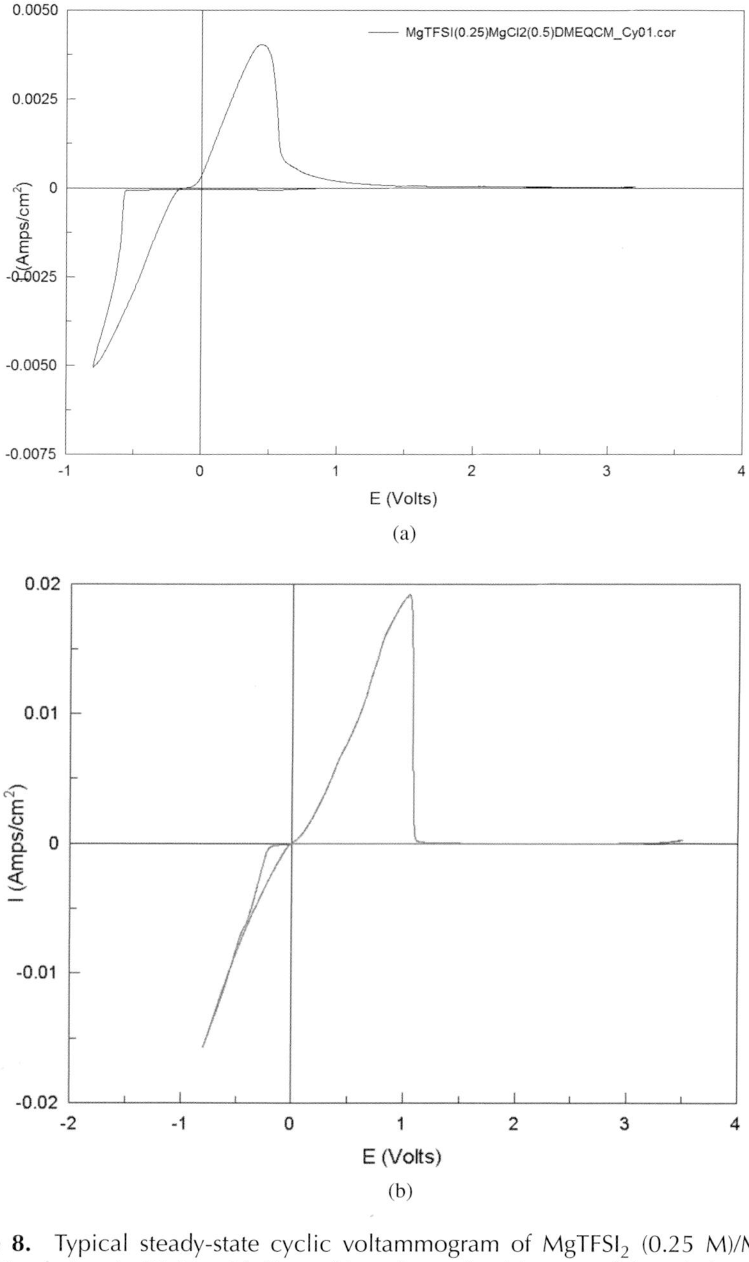

Figure 8. Typical steady-state cyclic voltammogram of $MgTFSI_2$ (0.25 M)/$MgCl_2$ (0.5 M) solution in DME, with Pt working electrodes (a) unconditioned electrolyte solution (b) $MgTFSI_2$ (0.25 M)/$MgCl_2$ (0.5 M) electrolyte solution after conditioning. Taken from Ref. 19 with permission of the Journal of the Electrochemical Society.

the SEI, whereas in the case of Mg, the cation complex would migrate all the way to the electrode interface where adsorption, desolvation, and charge transfer would occur. In addition, the different polarizabilities of the Li and Mg cations in conjunction with the nature of the electrolyte being used in Li and Mg cells lead to two distinct cation complex chemistries, which is of crucial importance for the plating/stripping process. In this section, the current state of knowledge on the nature of the Mg-based complexes migrating toward the electrode is reviewed. Furthermore, Mg nucleation and growth mechanism in such interphase free system are discussed together with its influence on dendrite formation.

A. Electro-Active Species in Nonaqueous Mg Cells

Over the last century, several electrolyte solution families have been developed allowing for the absence of passivation layer on Mg electrode and reversible plating/stripping behavior.[39,59] As for all other metals, the Mg plating/stripping properties (Coulombic efficiency, kinetics, etc.) are strongly dependent on the nature of the electrolyte (Figure 9). Indeed, the latter dictates the formation of the electro-active species: the metal cation complexes (polarizability, solvation shell, etc.). As discussed in Section B "Mg Nucleation and Growth Models," the metal cation complex plays a key role in the Mg nucleation and growth mechanism. Therefore, in this section, the electrolyte families that have been developed for Mg-based batteries and the resulting cation complexes they tend to form are briefly reviewed.

Some early success (~1920) was achieved in depositing magnesium from solutions based on Grignard reagents (R-MgX, X = Br, Cl).[35] However, this came with only a very narrow useful potential window, limited to <1.5 V *versus* Mg^{2+}/Mg, and a low ionic conductivity due to the very low dissociation level of RMgX species in ether solvents. In such electrolyte formulations, RMg^+ or XMg^+ were found to be the migrating cations. Later on, significant improvement of the electrochemical stability window (<2 V *versus* Mg^{2+}/Mg) was achieved by researchers at Dow Chemical Company[41] using solutions containing Mg organoborate compounds such as $Mn(BBu_2Ph_2)_2$ in THF, in which solvated MgR^+ and $Mg(BR_4)^+$ are formed with R = alkyl or aryl.

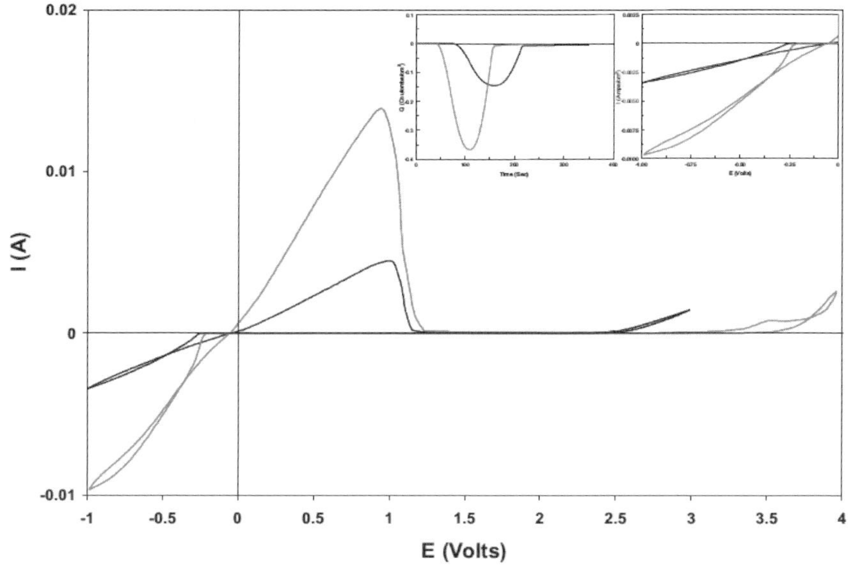

Figure 9. Comparison of the voltammetric behavior of THF solutions containing 0.25 M of the reaction product between 1:2 $MgBu_2$ and $AlCl_2Et$ (ethyl-butyl complex, standard solutions) and 0.4 M of the reaction product between 1:2 $AlCl_3$ and PhMgCl [all phenyl complex (APC) solution]; 25 mV/s, Pt wire working electrode, 25°C. Right insert: Enlargement of the cyclic voltammograms near 0 V *versus* Mg R.E., comparing the overpotential for Mg deposition in the two solutions. Left insert: The charge balance upon typical Mg deposition–dissolution cycles in these solutions (100% cycling efficiency of Mg deposition). Taken from Ref. 60 with permission of the Journal of the Electrochemical Society.

The group of Prof. Aurbach developed complex organo-haloaluminate electrolytes based on $AlCl_{3-n}R'_n$ Lewis acids reacting with R_2Mg Lewis bases in etheric solvents (THF or glymes). This allowed for an expansion of the potential window up to ca. 2.2 V *versus* Mg^{2+}/Mg. The individual species coexisting in equilibrium depend on many factors, such as the ratio between the components, the solvent(s) used, the concentration, the temperature, and the identity of both R and R'. The coordination number for aluminum and magnesium is always found to be four and six, respectively. The electroactive species in such electrolyte have been identified as solvated monomeric $MgCl^+$ and/or dimeric $Mg_2Cl_3^+$ species (Figure 10).[45,61] Interestingly, for Al:Mg ratios lower than 2, pure magnesium deposits are achieved on the

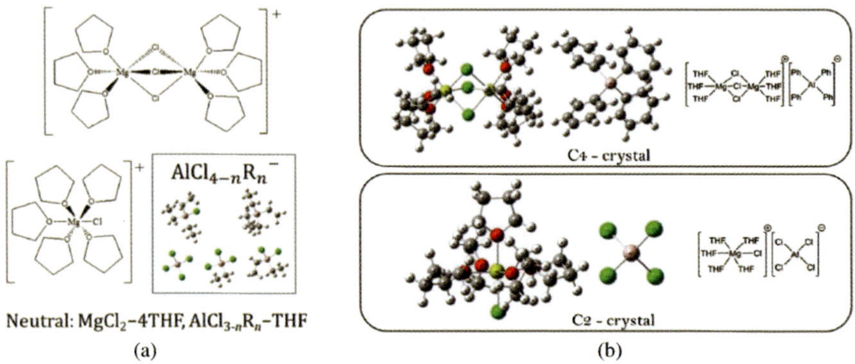

Figure 10. Structure of the major magnesium and aluminum species in (a) organo-haloaluminate or (b) phenyl complex electrolyte solutions. Taken from Ref. 39 with permission of the Royal Society of Chemistry.

metal anode, which is not a trivial observation since the standard aluminum deposition potential (−1.66 V *versus* NHE) is appreciably higher than that of magnesium (−2.37 V *versus* NHE). The anionic nature of the Al complexes is crucial for the achievement of pure Mg electrodeposition since Al complexes are migrating toward the cathode during plating.

Another interesting approach consists of the use of inorganic ionic salt, such as $Mg(BH_4)_2$ or $Mg(CB_{11}H_{12})_2$, which is expected to lead to the formation of $MgBH_4^+$ and possibly to magnesium ions solely coordinated to the solvent molecules.[17,62]

Overall, two distinct species enabling for reversible magnesium deposition/stripping are known: 1) the $Mg_2Cl_3^+$ and/or $MgCl^+$ in organo/non-organo haloaluminate and organoborate electrolytes and 2) the $MgBH_4^+$ in borohydride electrolytes. For most of the electrolytes reported thus far, there is no evidence supporting the existence of Mg^{2+} solely coordinated to the solvent molecules.[63]

It is surprising that significant differences in terms of Mg plating/stripping performances are observed in electrolytes such as organo-haloaluminates (Figure 9), in which very similar electro-active species appear to be formed (mainly $MgCl^+$·5THF and $Mg_2Cl_3^+$·6THF). However, the equilibrium between the magnesium cations and other ionic or neutral species present in the solution could affect the Mg electrochemistry and, thus, the assumption that the $MgCl^+$·5THF and $Mg_2Cl_3^+$·6THF cations are solely responsible for magnesium deposition cannot be confidently made.[64] The exact structure of the electrochemically active species thus remains unknown.

Aside from the nature of the Mg-based cations (electro-active species), several parameters can affect the plating/stripping processes. Among them, the cation mobility is obviously a crucial one. Indeed, anions are usually more weakly solvated than cations, by virtue of having smaller charge/radius ratios, and are more mobile. Therefore, cation transference numbers are always inferior.

The diffusion coefficient of the magnesium ionic species (e.g., $Mg_2Cl_3^+ \cdot 6THF$) was found to be very low (2.26×10^{-7} cm^2/s in 0.2 M solution, which is 10 times lower than that observed in 1 M $LiPF_6$-based electrolyte). Interestingly, the cation transference number t_+, which plays an important role on the rate at which reversible magnesium deposition/stripping takes place, ranges between 0.018 and 0.19 at 0.40–0.15 M, respectively.[65] The reduction in t_+ values when increasing the electrolyte concentration was attributed to lowered mobility of the dimeric magnesium ions and an increased number of counter and nonmagnesium ions at high Lewis acid concentrations. For the sake of comparison, most of the Li- and Na-based electrolytes present cation transference number of ca. 0.2–0.4.[66,67] To the best of our knowledge, a systematic evaluation of the cation transference number for such complicated solution chemistries as those used in Mg batteries is still lacking and would be necessary to better understand the electrochemical behavior, particularly the Mg plating/stripping kinetics.

B. Mg Nucleation and Growth Models

Electrodeposition is commonly assumed to occur via the so called nucleation and growth mechanism where the electrocrystallization takes place after succession of steps: diffusion/migration of the complex cation to the electrode, adsorption with at least partial desolvation, charge transfer, surface diffusion of an adatom, and crystal growth. Each step is associated with an activation energy barrier. Nucleation and growth mechanisms can be mainly divided in two types: 1) interfacial (or charge) controlled, where the nucleus growth is limited by the charge transfer and 2) diffusion controlled, with the rate-determining step being the diffusion/migration of the complex cation toward the electrode.

Mg plating nucleation and growth mechanism were first systematically investigated by the group of Prof. Aurbach by means of electrochemical quartz crystal microbalance (EQCM) and Pt microelectrodes.[61,68] These works were initially motivated by the observation of a seemingly complicated adsorption during Mg deposition. Indeed, very high impedances of the Mg electrode (thousands of Ω/cm^2) were recorded when the electrode was left at open-circuit potential which then collapse to very small values during dynamic Mg plating/stripping. In addition, in situ scanning tunnelling microscopy (STM) measurements pointed at interesting adsorption processes taking place before charge transfer can occur at 0 V *versus* Mg^{2+}/Mg (Figure 11).[68] The following mechanism for Mg plating was suggested (Figure 12). First, $MgCl^+$ migrates in the vicinity of

Figure 11. In situ STM image of (a) pristine gold electrode in a tetraglyme/BuMgCl 1 M solution under OCV conditions, (b) same electrode and solution, the electrode being polarized to 0 V *versus* Mg^{2+}/Mg during imaging, and (c) after Mg deposition in which 0.15 C/cm^2 was involved. Scales for (a) and (b) are 0.5 µm × 0.5 µm × 200 Å and for c) 5 µm ´ 5 µm ´ 0.2 µm. Taken from Ref. 68 with permission of the Journal of the Electrochemical Society.

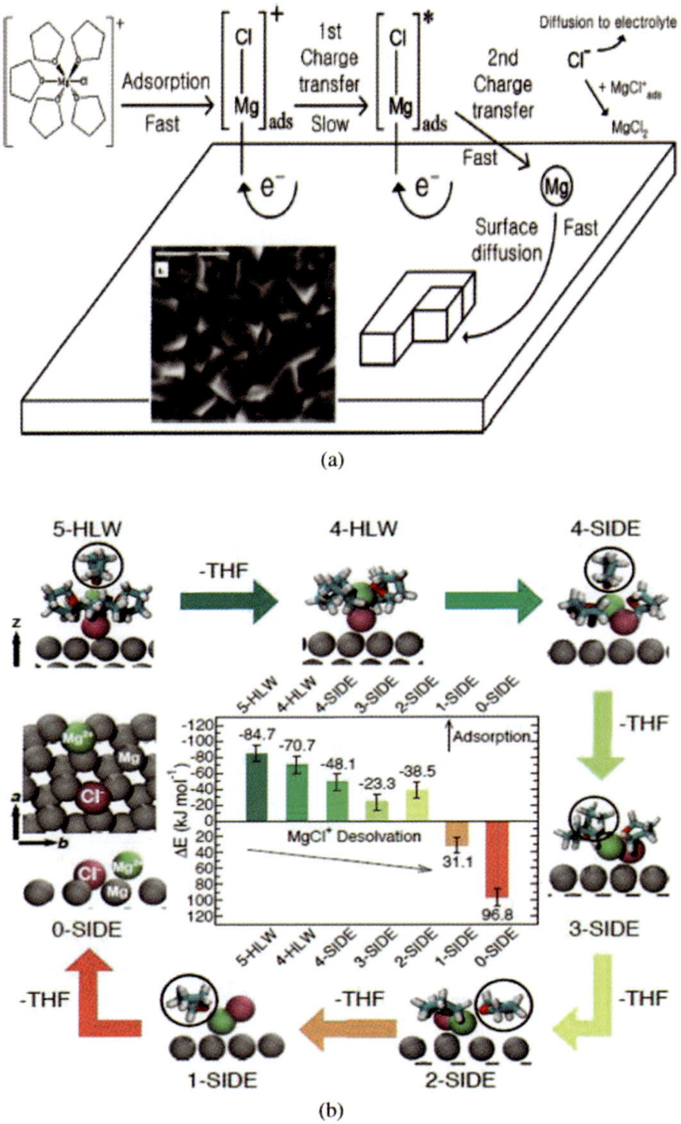

Figure 12. Possible Mg-electrodeposition mechanism from complex ethereal electrolyte solutions. (a) Taken from Ref. 39 and (b) from Ref. 70 with permission of the Royal Society of Chemistry and the American Chemical Society.

the electrode. This step is considered to be fast (at least during microelectrode test) and the chloride ligand remains in the cation complex. It is followed by the adsorption of the cation complex and two successive charge transfer steps, the last one leading to the liberation of the Cl^- ligand. The Mg adatom is then eventually diffuse on the surface of the electrode to reach a kink site and nucleate. Facile diffusion of the Mg adatom has been suggested and is in agreement with the well-defined crystal structure of the Mg deposit obtained (see scanning electron microscopy image in Figure 12a) and with recent theoretical insights into the surface diffusion of Mg adatoms.[69] At low and high overpotentials, the rate-determining steps are considered to be the adsorption of $MgCl^+$ on the electrode and the first charge transfer, respectively.

Since then, a few experimental and theoretical investigations have broaden our understanding of the Mg plating/stripping processes. For instance, Benmayza *et al.* introduced the possible participation of dimeric Mg complex ($Mg_2Cl_3^+ \cdot 6THF$) and Ceder *et al.* reported on a more complete description of the adsorption mechanism pointing at the desolvation of the last solvent molecules and of the Cl^- ligand of $MgCl^+$ as the most energy consuming steps during plating (Figure 12b).[65,70]

Up to now, a complete understanding of the impact of the ligand desolvation is still lacking. Several plausible recombinations of the Cl ligand after its desolvation from the cation complex were suggested by Aurbach for organohaloaluminate solutions.[68] However, how exactly the nature of the electrolyte could impact the desolvation energy of various ligands (solvent or ionic species), and could these desolvation steps become the rate-determining ones for Mg plating remain open questions? Also, could the presence of ligand acceptors in the vicinity of the electrode decrease the activation energy for the "desolvation" and thus improve the Mg plating kinetics?

Recently, a few theoretical studies on the desolvation energy in multivalent cation-based systems have been reported. Nakai *et al.* compared the calculated desolvation energy for Li, Na, and Mg cations in similar electrolyte solutions (Figure 13a).[71] They concluded that Na ion desolvation energy is generally smaller than for Li by ca. 40–70 kJ/mol, mainly due to the weaker Lewis acidity of the former cation. On the other hand, desolvation of Mg ion is commonly more energy consuming than for Li

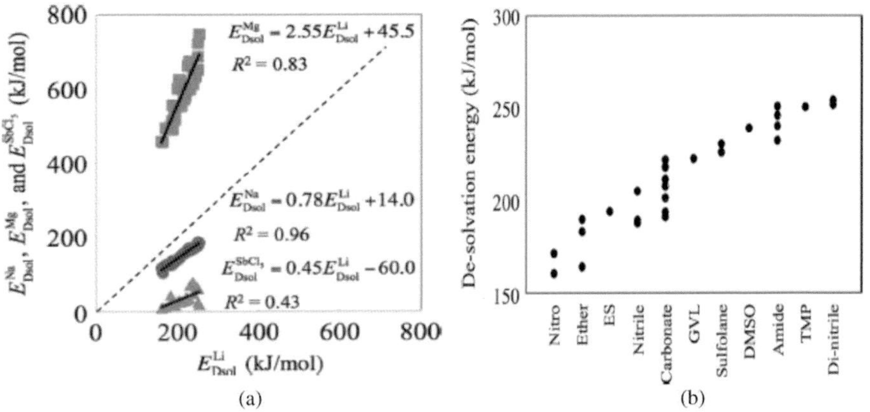

Figure 13. Relationship between (a) desolvation energies of Li ion and those of Na (blue circle) and Mg (red square) ions, as well as $SbCl_5$ (green triangle), the dashed line corresponds to the energies of Li ion and (b) desolvation energies of Li ion in typical solvents. Taken from Ref. 71 with permission of Elsevier.

ion by ca. 340–560 kJ/mol, as expected from the higher polarizability of the divalent cation. The authors also evaluated the influence of various solvents present in the solvation shell. Interestingly, the most common solvent family used in Mg batteries, namely ethers, exhibit one of the lowest desolvation energies (Figure 13b).

Similar investigations based on conventional carbonate solvents were reported by Sankaranarayanan *et al.*, confirming the fairly high energy cost for Mg^{2+} desolvation.[72] In addition, they stated the highest occupied molecular orbital (HOMO)–lowest unoccupied molecular orbital (LUMO) energy gap of the solvents is considerably affected by the interaction with the Mg cation. Thus, Mg solvated by carbonate molecules appears to be easier to oxidize while more resistant to reduction.

Unfortunately, realistic electro-active species, such as those present in Mg-based batteries (i.e., $MgCl^+ \cdot 5THF$ and $Mg_2Cl_3^+ \cdot 6THF$), were not considered in the previously described theoretical works. Nonetheless, the report from Ceder *et al.*[70] on the adsorption of $MgCl^+ \cdot 5THF$ onto Mg electrode already reveals that desolvation of ionic ligands (e.g., Cl^-) could imply a significant energy penalty and affect the plating kinetics. For the sake of comparison, it is interesting to note that in Li system, desolvation energy is identified as the main resistance contribution for the Li transport at the graphite/electrolyte junction.[73]

C. Origin of the Absence of Mg Dendrites

While the rechargeability of elemental electrodes may seem a simple matter at first sight, this is by no means the case.[74] Indeed, smooth plating of a metal is well known to be a complex issue from classical electrochemistry. Beside the influence of the presence of a possible surface layer (as the SEI) on the electrodes, the location of the electrodeposit may not be at the same specific place where stripping takes place due to issues related to current distribution, fluid dynamics, and crystal growth.

Metal dendrite formation upon cycling was, however, early identified as an impeding factor for the development of batteries based on metallic negative electrodes. Already observed in in situ optical microscopy studies in 1979,[75] dendrites have been thoroughly studied ever since in order to better understand their growth mechanisms and prevent their formation and the side effects associated with it.[76,77] Among the more important side effects, the low Coulombic efficiency and capacity loss can be associated to electrical contact loss of metal dendrites (also called dead Li in Li cells). However, the most challenging side effect is related to safety issues due to the so-called "fuse effect" of the dendrites.[78] Indeed, when dendrites penetrate the separator resulting in short circuit, severe local heat generation may occur through a thermo-fusible effect and could constitute the first step for thermal runaway of the battery.

Dendrite formation is not specific to the negative electrodes of batteries and the thermodynamic explanation of their growth can be traced back to the notion of surface free energies developed by Gibbs in 1873,[79] who suggested that "(. . .) an anisotropic surface free energy implied a non-spherical equilibrium shape (. . .) which minimizes the total surface free energy." The specific case of electroplating has later on been intensely studied and several nucleation and growth models have been developed.[80,81] The dendrite electrodeposition results from a diffusion-limited process, the deposition potential being significantly shifted negative from the equilibrium potential of the M^{x+}/M couple. Under such conditions, an anisotropic growth is highly favored by the very limited surface diffusion of the adsorbed cations before charge transfer occurs, and the dendrite formation is therefore facilitated by an improved hemispherical diffusion at the tip of the dendrite.

For Li-battery negative electrodes, the general principles of dendrite formation are unchanged, although the system is significantly more complicated due to the presence of an SEI layer. Current distribution associated with the cell geometry, cycling conditions, and M^{x+} diffusion properties are key parameters controlling the formation of dendrites.[76] Also, the composition, the homogeneity, and the mechanical properties of the SEI layer are important factors.[82] Indeed, an SEI layer presenting uniform and high ionic conductivity favors the formation of a safer, mossy, or granular-like M deposit, by inducing a more uniform cation saturation level at the M/SEI interface. Good mechanical properties of the SEI are advantageous in order to accommodate volume changes upon cycling, while limiting the dendrite formation due to surface-tension forces.[83] This, along with an improved M^{x+} transport number, explains the improved performance of polymer-based or solid-state electrolytes in terms of dendrite formation. For liquid electrolytes, the composition [salt(s), solvent(s), and additive(s)] and the purity play a major role in the formation of dendrites. Although most of the studies on metallic plating and dendrite formations deal with Li cells, the Mg case is focused here.

Unlike Li, Mg does not appear to be plagued by dendrite formation.[84,85] Indeed, even for plating achieved at relatively high current density (2 mA/cm^2) no evidence of irregular, dendritic growth can be observed (Figure 14b). Several parameters could explain this fundamental difference between Li and Mg systems. Obviously, the absence of an SEI at the surface of Mg electrode is a positive feature improving the mass transport homogeneity toward the electrode. Theoretical investigations of the surface mobility of Li and Mg adatoms were performed in order to grasp more information on the absence of Mg dendrite formation.[69,86] However, they present rather contradictory conclusions and the calculations did not consider the entire metal nucleation and growth mechanism since only ligand-free adatoms were taken into account.

The complicated Mg nucleation and growth mechanism is most probably playing an important role in the absence of dendrites. As mentioned above, dendrite formation results from diffusion-limited process and their

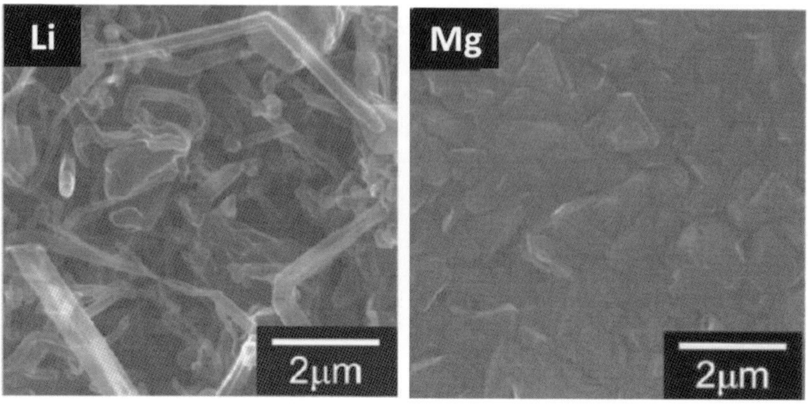

Figure 14. SEM images of electrodeposited lithium and magnesium (current density: 2 mA/cm^2). Taken from Ref. 84 with permission of Elsevier.

absence is in agreement with the nucleation and growth mechanism proposed by Aurbach *et al.*, the charge transfer (or the cation adsorption at high overpotential) being the rate-determining step.

Matsui systematically studied and compared Li and Mg plating/stripping processes, paying particular attention at two critical parameters for dendrite formation, namely the current density and the electrolyte salt concentration.[84] Although Mg dendrites were never observed, interesting trends were reported. An increase of the current density from 0.5 up to 2 mA/cm^2 was found to result in an evolution of the deposit morphology changing from granular-like into well-defined faceted crystals. Concomitant enhancement of the (001) preferential orientation of the deposit was observed, demonstrating that even under high overpotentials, the adatoms still have time to diffuse at the surface of the electrode before nucleation occurs. However, minimum energy surface diffusion pathways are favored, which leads to preferential orientation of the deposit; the latter finding being in agreement with density functional theory calculations.[86]

Overall, in the experimental conditions evaluated so far, Mg dendrite seems not to form, owing to the charge transfer limitation associated with Mg nucleation and growth mechanism. However, an open question remains: could the kinetics of Mg plating be further improved by improving the charge transfer process? And if it was so, would dendrite formation also become an issue for Mg system?

IX. R&D for New Cathode Materials for Rechargeable Mg Batteries

A. Introduction

The development of high-energy density rechargeable Mg batteries necessarily depends on the search for high-voltage positive electrode materials. Such quest is, however, strongly dependent from designing advanced electrolytes with extended electrochemical stability window (see Section VII). The up-to-date reference material Chevrel-type phases Mo_6S_8 exhibits limited specific capacity and a low working voltage around 1 V. Therefore, it has to be substituted by alternative cathode materials in order to move toward more performant rechargeable devices. This section aims on presenting the current researches devoted on various classes of positive electrode materials, as well as providing guidelines for future research based on very recent significant results.

B. V_2O_5 and MnO_2 Oxides and the Special Role of Crystal Water

As it is reminded earlier in this chapter, simple binary oxides were ones of the first materials to be evaluated as electrode in magnesium battery, especially with the seminal work of Gregory *et al.*[41] The V_2O_5 family of materials with different morphologies, phases, and structures has been extensively investigated for energy storage applications, and also for Mg intercalation.[87–89] However, the magnesiation mechanism was not really characterized. These studied have particularly concluded that nanostructuration is one of the key points to boost the cation diffusion. Cheng and coworkers followed this strategy by developing composite electrode materials of highly dispersed nanoclusters of V_2O_5 supported on porous carbon.[90] Excellent capacities exceeding 300 mAh/g at 40 mA/g were reported. Interestingly, microscopic, spectroscopic, and electrochemical characterizations suggested a surface-controlled mechanism that opens the gate to new approaches based on molecular energy storage.

In parallel, manganese dioxide was first investigated for Mg electrochemical intercalation in early 2000s. Like V_2O_5, several polymorphs of MnO_2 were also examined as Mg electrode material owing to the richness of crystalline forms, such as hollandite α-MnO_2 (2×2 tunnel),[91–93] spinel λ-MnO_2,[93] or birnessite δ-MnO_2 (layered).[94] The potassium-stabilized hollandite-type MnO_2 showed an initial discharge capacity of 282 mAh/g that, however, rapidly fades upon subsequent cycles. Unlike following typical cation insertion similar to Li^+, recent studies based on various experimental characterization techniques explain that a conversion reaction occurs, generating amorphous magnesium and manganese dioxide and finally participating to the decrease of the performance.

Interestingly, it has been found that traces of water in the electrolyte can strongly influence in a positive manner the electrochemical performance of the given material. Such observation might be unexpected as at a first thought water containing in the electrolyte may increase the passivation film of Mg electrode. The positive effect of water for magnesium intercalation was first identified for V_2O_5-based electrodes. Indeed, water molecules preferentially solvate Mg^{2+} cations and thus facilitate their intercalation within the oxide lattice owing to low electrostatic repulsion.[88,95] This unusual approach was put back on stage for the special case of birnessite MnO_2-based electrode for Mg battery[94,96] (Figure 15).

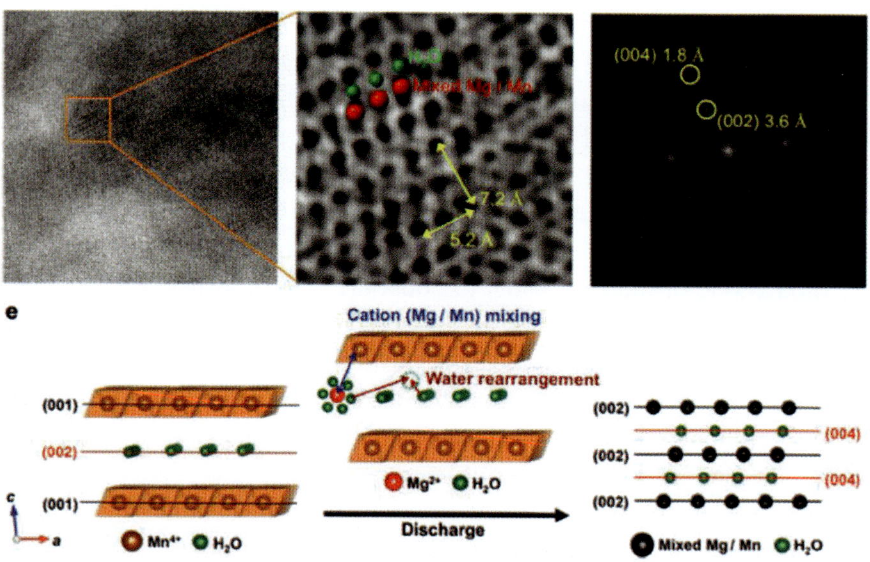

Figure 15. ABF-STEM images showing Mg intercalation in crystal water containing birnessite MnO_2 and proposed mechanism of the Mg intercalation. Taken from Ref.94 with permission of the American Chemical Society.

Playing with water molecule in the electrolyte should be definitely considered in the future to help achieving simultaneously high voltage and efficient kinetics. It might also open the gate to Mg-ion systems operating with friendly aqueous electrolyte.

C. Polyanionic Compounds

Polyanionic frameworks gather a wide family of mineral compounds characterized by a three-dimensional (3-D) anionic framework, such as phosphates and derivated, silicates, and sulfates. These compounds have been intensively studied as positive electrode materials for lithium and sodium batteries. The chemical richness, with especially various transition metals and anionic lattices, offers a huge playground for solid-state chemists looking for new synthesis procedures and interesting structure-properties relationships. Today, lithium iron phosphate, $LiFePO_4$, is established as one of the most performance cathode materials for lithium-ion systems. Polyanionic compounds containing magnesium cations also represent an impressive number of compounds. However, there were only few reports on the electrochemical activity *versus* magnesium of some polyanionic compounds. Such lack is first obviously explained by the absence of suitable electrolyte to carry on the systematic investigation on positive electrode materials.

Li_2MSiO_4 (where M is a transition metal) has been reported as interesting cathode for lithium-ion battery. Lithium reversible deintercalation occurs without significant structural change, and the strong inductive effect caused by the polyanion can moderate the transition metal redox couple to higher operating voltage. Analogically, the group of NuLi investigated a series of magnesium orthosilicates $MgMSiO_4$ and evaluated them as positive electrode of Mg battery.[97–102] The $MgMSiO_4$ structure reveals that both Mg and M occupy same crystallographic octahedral sites sharing edges or corner, and surrounded by SiO_4 tetrahedra. By tuning the synthesis protocol, high-temperature solid-state to sol-gel routes, as well as silica template-assisted process in order to create mesoporous materials, different morphologies were obtained. The best electrochemical performance was observed for mesoporous materials, with for example 300 mAh/g in a first discharge with $Mg_{1.03}Mn_{0.97}SiO_4$, and reversible capacity along cycling between 150 and 225 mAh/g. Attempts in creating Mg/M cationic ordering to boost the Mg diffusion were unsuccessful so far. This work was extended with unmagnesiated silicate compound by Orikasa and coworkers when they used $FeSiO_4$ obtained beforehand by lithium electrochemical deintercalation from Li_2FeSiO_4. In a three-electrode cell with $MgTFSI_2$/AN electrolyte, a reversible high capacity above 300 mAh/g was reported during cycling at 55°C. This work needs, however, to be confirmed using electrolytes that are compatible with Mg anode.[103]

Focusing on the phosphates family, the potential application of tavorite-$FeSO_4F$ and $MgVPO_4F$ as cathode material in Mg battery was evaluated by first principles calculation. A discharge voltage of 2.52 V was predicted for the electrochemical formation of $Mg_{0.5}FeSO_4F$, whereas a two-step mechanism with plateaus at 2.6 and 1.5 V is expected for $MgVPO_4F$[104,105] (Figure 16). In parallel to this simulation approach, the electrochemical behavior of $MgFePO_4F$ prepared by solid-state carbothermal method was experimentally evidenced. This fluorophosphates present an operating voltage of 2.6 V, and a good cyclability. The capacity is, however, rather low (60 mAh/g) but might be improve if the Mg^{2+}/Fe^{2+} antisite mixing is decrease by optimized synthesis routes.[106]

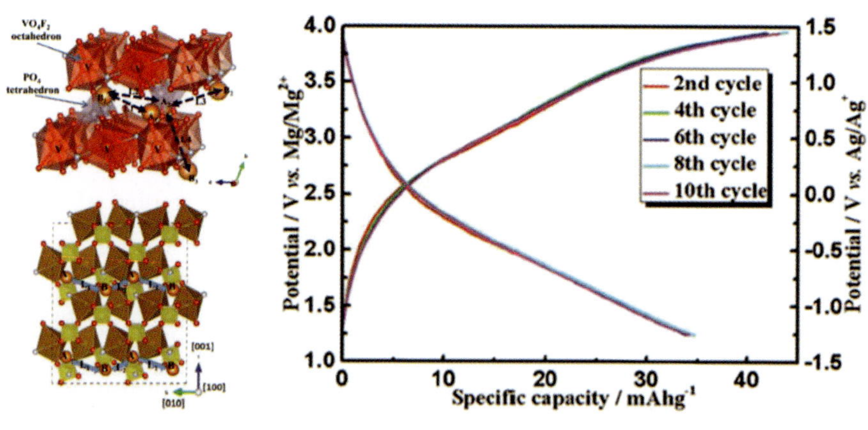

Figure 16. Left, diffusion pathways of Mg cations into $MgVPO_4F$ and $FeSO_4F$ lattices, and right galvanostatic curves obtained with a $Mg/MgFePO_4F$ cell at C/20. Adopted from Refs. 104–106.

D. Sulfur-Based Cathodes

Last paragraphs showed that achieving high-capacity and reversible magnesium electrochemical insertion into an oxide-based lattice (either in oxides or in polyanionic frameworks) is very challenging. Besides the crystallographic structure that helps the magnesium cations diffusion, the ability of Chevrel phases also lies on the chalcogenides framework (S, Se, or Te) which is less electronegative than an oxygen-based one. Therefore, it is also interesting to focus on sulfur itself as the cathode material. Moreover, sulfur is abundant on earth and thus a low-cost cathode material. The concept of a metal–sulfur battery has been investigated for a long time now, and Li-S system has been listed among the most promising concepts to explore beyond the lithium-ion technology.[107] There are relatively few works on Mg–S systems. Theoretically, sulfur can react with magnesium to form MgS with corresponding high capacities of 1671 mAh/g^{-1} and 3459 mAh/cm^3. Mg electrolytes based on Grignard-related chemistry are nucleophilic and therefore cannot be used with electrophilic cathode material such as sulfur. In 2011, the Toyota Research Institute of North America proposed a new electrolyte that fulfills the requirements to be used with magnesium metal and sulfur cathode. In this pioneering work, the association of hexamethyldisiliazide magnesium chloride and aluminum trichloride in THF enables the synthesis of a nonnucleophilic electrolyte that allows the reversible electrochemical magnesium stripping/deposition process. From crystallization, the electroactive species was identified to be $[Mg_2(\mu\text{-}Cl)_3 \cdot THF]^+$. With elemental sulfur dispersed in black carbon as simple positive electrode, an excellent capacity of 1200 mAh/g was observed for the first discharge, with an average plateau at 0.89 V. The rapid capacity fading upon cycling was explained by the dissolution of unreduced sulfur and polysulfides, which are common phenomena largely observed in Li–S battery. This proof of concept opens the gate to performant electrochemical storage devices, even if important improvements are still needed. Zhao-Karger and coworkers followed this seminal work by reporting a modified nonnucleophilic electrolyte directly prepared from commercial bis-hexamethyldisilazide magnesium with aluminum trichloride, and with a glyme/tetraglyme mixture as solvent.[108] The addition of ionic liquid N-methyl-N-butylpiperidinium bis(trifluoromethanesulfinyl) imide (PP14TFSI), as well as using highly ordered mesoporous carbon CMK-3 for the encapsulation of sulfur, enabled to enhance the electrochemical performance. A discharge potential of about 1.65 V, which is close to the theoretical value, has been achieved with this new Mg/S battery. The same group also investigated graphene–sulfur nanocomposite electrode, and reported a higher reversible capacity of 448 mAh/g, a longer cyclabilty (236 mAh/g after 50 cycles), and a better rate capability than prior Mg/S batteries.[109] XPS was the most direct technique used to check for a sulfur oxidation state change during battery cycling. After the first discharge, the presence of MgS as the final product was confirmed as well as some MgS_x sulfides that have still to be identified. Cycling improvements of S-based electrode might also come from combining sulfur with its neighbor element and more conducting selenium. Indeed, Se was found to be electrochemically active in Mg-batter, and it exhibits promising performance such as reversible capacity of 480 mAh/cm^3. Preliminary results showed that electrodes made SnS_2 confined into mesoporous carbon exhibit capacity around 600 mAh/g.[110]

The identification of Chevrel phase Mo_6S_8 has electrode material for magnesium battery by the group of Aurbach in 2000 initiated the idea that employing soft anions lattice might enable to obtain better mobility of divalent cation.[14] Chemical reduction by the action of organometallic reagents highlighted that titanium disulfide (TiS_2) is a good candidate to magnesium intercalation.[111,112] Following seminal works carried out on nanotubes with a nonoptimized AN-based electrolyte,[113] in 2016, Nazar *et al.* reinvestigated the electrochemical behavior of layered TiS_2 and obtained 115 mAh/g of stabilized capacity in a Mg full cell.[114] In the same spirit, they also focused on empty cubic thiospinel Ti_2S_4 — beforehand obtained by copper extraction from cubic $CuTi_2S_4$ — as electrode material.[115] At C/10 current rate, high capacities of 200 mAh/g (corresponding to $Mg_{0.84}Ti_2S_4$) were evidenced on the first cycles of a full Ti_2S_4–Mg battery. The 150 mAh/g stabilized capacity on subsequent cycles corresponds to a specific energy of 180 Wh/kg, almost twice that of Chevrel phase (Figure 17). In addition, good capacity retention was also noticed during rate capability tests. However, the relatively low operating voltage (1.2 V) is still insufficient for high-energy positive electrode material.

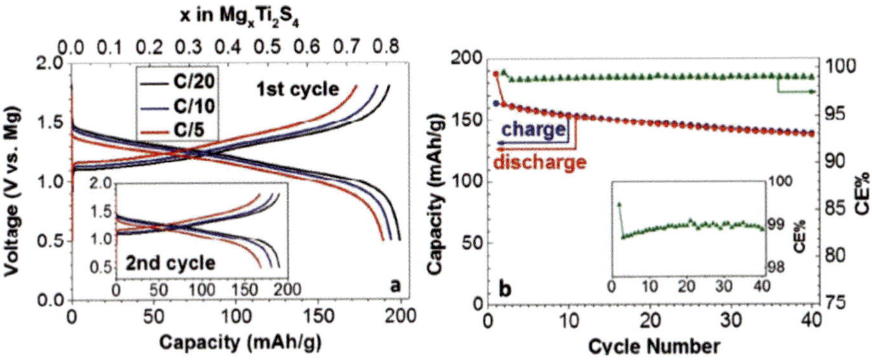

Figure 17. Electrochemistry of Mg/Ti_2S_4 coin cells with APC electrolyte at 60°C. Adopted from Ref. 115.

E. Organic Cathodes

Organic compounds might bring new possibilities for cost-effective and environmental friendly high energy/power density. Hence, they are now more and more investigated to substitute classical electrode materials in lithium-ion batteries.[116] Only few reports focused on organic-based cathode for Mg battery. However, these works should be followed in light of the difficulties in identifying performant inorganic material for positive electrodes. Ha and coworkers first proposed the use of a radical polymer, poly(2,2,6,6-tetramethyl-piperidinyloxy-4-ylmethacrylate) (PTMA), as a cathode material for evaluation new Mg electrolyte.[18] Organosulfur compounds were also pointed out; however, the use of organohaloaluminate-based electrolytes close to their stability limits questions the obtained results.[117,118] Following previous studies on Li-ion battery cathode material,[119] several recent groups reported that quinone-based polymers can be seen as reliable cathode material candidates.[120–123] As an example, Bitenc *et al.* investigated poly(antraquinoyl) sulfide and poly(hydroquinonyl-benzoquinonyl) sulfide. Using nonnucleophilic electrolytes with widened operating window, they obtained interesting capacities ranging from 150 to 200 mAh/g.[121,123] Such results should be followed by a better understanding of charge transfer and storage mechanisms and quinone-based organic materials. Improving the capacity retention and the cycling stability are the main challenges.

F. Hybrid Systems as an Alternative?

Despite a growing interest and significant worldwide efforts, the search for positive electrode materials that can reversibly accommodate magnesium cation at high voltage is still a major obstacle to the development of performant Mg battery. An interesting alternative is the hybrid Mg/Li battery that still uses magnesium metal as a superior anode material but bypasses the need for Mg cathodes with well-known lithium intercalation positive electrode materials. This concept arises in 2014 when different groups reported hybrid cells by associating Mg anode with various Li-cathodes such as Mo_6S_8,[124,125] TiS_2,[126,127] $LiFePO_4$.[128] Obviously, in such configuration, the electrolyte must contains both Li^+ and Mg^+ salts. Keeping in mind the compatibility with magnesium metal still, the authors proposed an APC-based solution (PhMgCl and $AlCl_3$ in THF) with lithium salt (LiCl or $LiBF_4$) for ensuring lithium conduction. During the electrochemical discharge only magnesium is reduced as the negative electrode owing to its less electropositive potential, whereas the positive electrode material is delithiated as in a classical Li-ion battery. On the contrary during, magnesium dissolution occurs at the negative side, and lithium cations accommodate the polyanionic framework as the positive side as their diffusion is way faster than magnesium cations.

The interest of such hybrid systems can be only evaluated by comparing their energy density, and some authors expected high values that are almost comparable with up-to-date Li-ion standards. In addition, given the current context surrounding the lithium resources and its intensive exploitation, such hybrid systems can give birth to lithium-free batteries, as it has been done by Walter and coworkers when associating Mg anode with FeS_2 nanocrystals and using a Na/Mg dual-salt electrolytes. The authors reported an encouraging energy density up to 210 Wh/kg which overcome the state-of-the-art Mg battery.[129] The development of such systems highly depends on designing optimized dual-salt electrolytes.

G. Conclusion

The systematic screening of positive electrode materials for Mg battery is strongly influenced by the research related to Li-ion technology. For these reasons, a lot of materials that exhibits interesting features with lithium were expected to present an analog behavior with magnesium. Unfortunately, the difficult challenge of designing performant electrolytes compatible with magnesium anode really slows down the research in this direction. Moreover, it is important to mention that intercalating double-charged species such as M^{2+} is still more difficult than playing with Li^{+} or Na^{+} cations. Although a growing interest is now observed on the search for high-voltage electrode materials, the route for future performant Mg batteries might be still a long one.

Concerning oxide-based cathode, the sluggish diffusion of magnesium cation might decrease by effort on the nanostructuration of the electroactive elements or during the electrode formulation. For such compound, a better understanding of the electrochemical mechanism is also important in order to design optimized electrode and to achieve the theoretical performance of the active materials. Besides the electrode nanostructuration, and as it is shown with Chevrel phase Mo_6S_8 reference positive electrode, moving toward soft anions frameworks such as sulfides seems also to be a promising direction. However for both cases, the design of new electrolytes that are compatible with magnesium anode and stable at high voltage is still the crucial point that needs to be solved in order to search for performant positive electrode materials.

X. Calcium Secondary Batteries: First Initiatives

Calcium is an attractive alternative divalent system to the more extensively studied magnesium system. Indeed, Ca possess high theoretical electrochemical capacities (both gravimetric and volumetric) and Ca^{2+}/Ca have a 0.5 V lower standard potential than Mg^{2+}/Mg. Ca is more abundant than Mg (respectively, the fifth and eighth most abundant elements on the Earth's crust) and Ca^{2+} would hold promise for faster reaction kinetics than Mg^{2+} (and thus better power performance) due to its lower polarizing character. This last point is crucial for the development of new cathode materials allowing for insertion deinsertion of divalent cation since slow solid-state diffusion is an important bottleneck to overcome for this technology.[39,130]

However, most of the attempts to electroplate Ca have failed. Deposition was found to be poorly efficient in $SOCl_2$-based electrolytes and to the best of our knowledge no report has been made on Ca-based Grignard reagent electrolytes that have been successful in obtaining efficient Mg plating/stripping. In contrast to magnesium, where PhMgCl is commercially available and is a precursor for numerous magnesium electrolytes, it has been reported that calcium powder shows no reactivity with chlorobenzene, so the corresponding heavy Grignard PhCaCl cannot be synthesized.[64] But heavy Grignards such as PhCaI can be prepared and have been suggested as a possible alternative to PhMgCl.[64] In conventional organic electrolytes, pioneering research work by Aurbach *et al.* allowed to conclude that the electrochemical behavior of Ca metal anodes is surface-film controlled, as is the case for Li, but Ca plating is virtually impossible (Figure 18) due to the lack of divalent cation transport through the SEI.[12,58]

Therefore, Ca-based systems rapidly fell into oblivion and only a few groups in the world have been pursuing research on Ca. These works have been focusing on the electrochemical behavior of active materials that operate within the electrochemical stability window of the electrolyte being used in order to prevent passivation layer formation. An additional issue to deal with in such system is the absence of reliable counter electrode that support the current generated at the working electrode (tested material). Indeed, metal electrode (i.e., Li, Na, or Mg) are widely used as counter electrodes in the so termed half-cell configuration, performed mostly at laboratory scale within the battery research community. Recently, a new approach allowing for Ca plating and stripping has been reported with encouraging results and constitutes the first steps toward the development of reliable counter electrode in half-cell configuration as well as for the viability of a battery relying on Ca metal anode.[13]

This section aims at reviewing all the computational and experimental work that has been done in the last 15 years, related to the quest for the development of Ca ion and Ca metal anode-based batteries.

A. Theoretical Insights into the Reactivity of Anode and Cathode Materials with Ca^{2+}

Neither a standard electrolyte nor a reliable counter electrode exists for Ca-based batteries. Therefore, theoretical studies currently appear as the most suitable strategy to acquire valuable knowledge in this field and

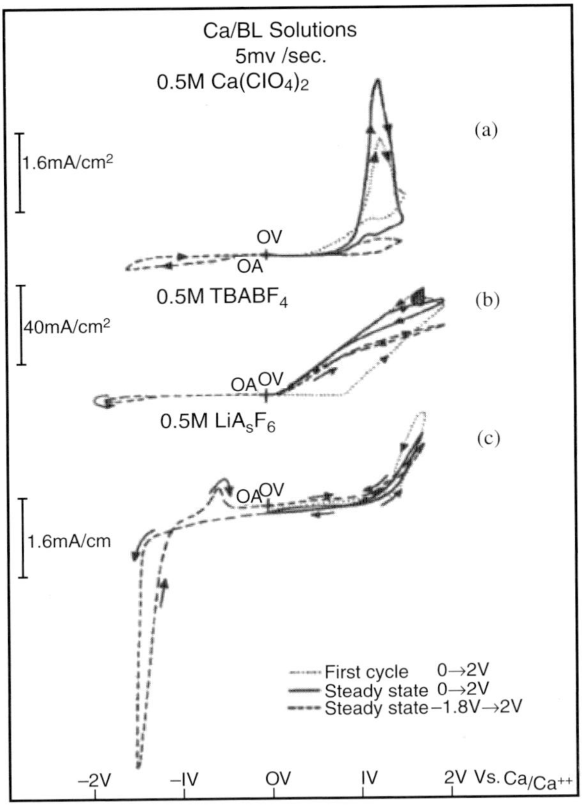

Figure 18. Comparison of typical voltammograms obtained with calcium electrodes and BL solutions of three different salts: (a) $Ca(ClO_4)_2$, (b) $TBABF_4$, and (c) $LiAsF_6$. Sweep rate: 5 mV/s. Taken with permission from Ref. 12 with permission of the Electrochemical Society.

achieve fast screening of possible new cathode and anode materials for Ca-based batteries.

Ceder *et al.* studied multivalent ions intercalation in the orthorhombic α- and δ-V_2O_5 polymorphs by means of first principles study. They determined the structural change, voltage, thermodynamic stability, and intercalant mobility for Li^+, Mg^{2+}, Zn^{2+}, Ca^{2+}, and Al^{3+} during the insertion process. Ca^{2+} intercalation was found to increase the layer spacing by more than 10% in both polymorphs. Although the mobility of Ca^{2+} in the α phase was found to be poor, intercalation of these ions in the δ phase seems promising, with sufficiently high voltage (3.02 V *versus* Ca^{2+}/Ca) and cation mobility, with migration barrier (E_m) value of ca. 200 meV for Ca^{2+}. It is interesting to note that an easier Ca^{2+} diffusion than Mg^{2+} (E_m ~ 600–800 meV) is expected (Figure 19).[131]

Similar investigation was carried out in order to grasp more information about possible Spinel compounds as multivalent battery cathodes and materials with the general formula AB_2O_4 (A = Mg, Ca, Zn, Al, or Y and B = Ti, V, Cr, Mn, Fe, Co, Ni) were considered.[132] The authors suggested that among all material tested, Mn_2O_4 was the most encouraging candidate in terms of insertion voltage (ca. 3 V *versus* Ca^{2+}/Ca), capacity (250 mAh/g), thermodynamic stability of the charged and discharged states, as well as the intercalating ion mobility (Ca^{2+} diffusion energy barriers being ca. 500 meV, similar to that of Li^+).

Chevrel phases (Mo_6X_8 with X = S, Se) have been studied by Zaghib *et al.* and present low energy barriers (ca. 500 meV for X = Se and

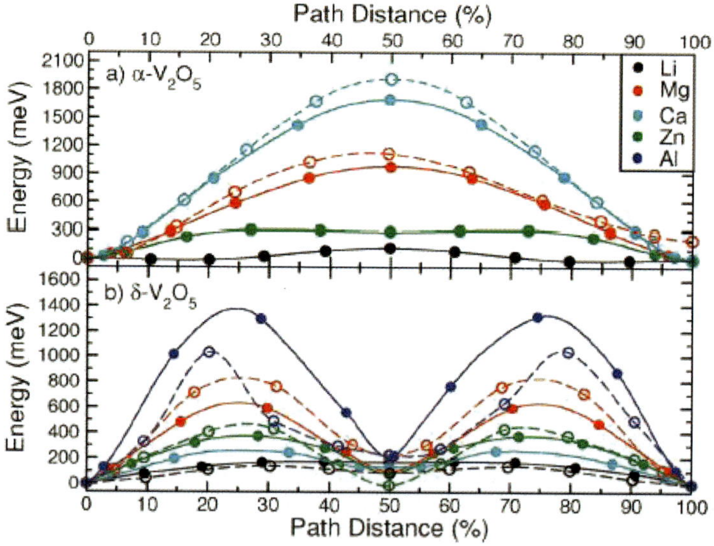

Figure 19. The activation barriers for the diffusion of the different intercalating ions in the α- and δ-V_2O_5 polymorphs are plotted in (a) and (b), respectively. The solid lines correspond to the empty lattice limit (charged state), whereas the hollow lines correspond to the full lattice limit (discharged state). Reprinted with permission from Ref. 131 with permission of the Royal Society of Chemistry.

800 meV for X = S).[133] Unfortunately, as expected, they present low operation voltage (<1.5 V *versus* Ca^{2+}/Ca). Arroyo-de Dompablo *et al.* investigated $CaMO_3$ perovskites (M = Mo, Cr, Mn, Fe, Co, Ni) by means of density functional theory (DFT) and identified $CaMoO_3$ as the most promising candidate. Insertion of Ca^{2+} was calculated to occur in the 2–3 V *versus* Ca^{2+}/Ca potential window. However, they also highlighted a prohibitively high Ca^{2+} diffusion energy barrier of 2 eV.[134]

Alternative anode materials to calcium metal have also been evaluated through theoretical considerations and the group of Prof. Shenoy investigated the adsorption of Na and Ca on graphene with divacancy and Stone–Wales defects.[135] They concluded that adsorption of Na and Ca is not possible on pristine graphene but it is enhanced by increasing the density of defects. Maximum calculated capacities were 1450 and 2900 mAh/g for Na^+ and Ca^{2+} adsorption, respectively. The same group considered other 2-D materials such as transition metal carbides or MXenes and focused on the adsorption of Li, as well as Na, K, and Ca on Ti_3C_2.[136] They claimed that a larger effective ionic radius increases the interaction between alkali atoms, thus penalizing the capacity and 319.8 mAh/g was calculated for Ca^{2+} adsorption. However, it is worth mentioning that, as it is the case for Li- and Na-ion system, the use of such high surface area anode materials is of very limited practical interest since they present very low first-cycle Coulombic efficiency resulting in unacceptable energy density loss at the cell level. Similar limitations are expected with Ca system, unless highly stable electrolytes are developed at the operation potential of those materials.

B. Experimental Evaluation of Cathode and Anode Materials for Secondary Ca-Ion Cells

As mentioned above, Ca metal electrode has not proved reversibility in most of the experimental conditions used in Ca system. Therefore, one of the ways for the assessment of the electrochemical properties of cathode or anode materials for Ca-based battery has been to use either supercapacitive (e.g., activated carbon) or platinum counter electrodes. For both types of counter electrodes and depending on the required current density, reduction/oxidation of the solvent/salt composing the electrolyte could occur and result in electrolyte contamination, passivation of the counter electrode, and so on. Even though these parasitic reactions occurring at the counter electrode could affect the cyclability of the tested cell, this parameter was not considered in most of the works described below.

V_2O_5 is the most studied intercalation host material for divalent cations and is among the handful of cathode materials for which reversible divalent cations insertion (mostly Mg^{2+}) has been claimed. Amatucci *et al.* were the first to report on an electrochemically reversible Ca^{2+} insertion into V_2O_5.[89] Even though the main focus of this study was the intercalation of trivalent yttrium, comparison with the electrochemical behavior of V_2O_5 in Li-, Mg-, Ca- and Y-based electrolytes was made. The authors used two interesting strategies in order to achieve reversible Ca^{2+} insertion. First they used activated carbon counter electrodes and carefully balanced the working and counter electrode capacity in order to prevent parasitic reactions with the electrolyte.[137] The second important strategy used in this work was to overcome the low diffusion coefficient of multivalent cations by synthesizing nanocrystalline V_2O_5 (primary particles of ca. 25 nm diameter), thus decreasing the diffusion pathway of the cations. Slightly higher reversible capacity was measured for Ca^{2+} (>200 mAh/g) when compared with Li^+ (<150 mAh/g) intercalation, but multivalent intercalation exhibited higher charge/discharge voltage hysteresis. The influence of the presence of water contamination was carefully considered and although the improved multivalent insertion kinetics in presence of water could not been ruled out, the contribution from proton insertion was stated to be negligible. Few years later, Amatucci *et al.* extended the study on vanadium oxide xerogel achieving capacity as high as 465 mAh/g and concluded that the presence of PC in the nanocomposite is critical to its performances.[138]

Hayashi *et al.* studied three type of V_2O_5 cathode (crystalline, amorphous, and amorphous P_2O_5 doped as crystallization inhibitor) evaluated in 1 M $Ca(ClO_4)_2$ dissolved in PC, dimethyl carbonate (DMC), diethyl carbonate (DEC), ethyl methyl carbonate (EMC), GBL, N,N-dimethylformamide (DMF), or AN.[139] For all electrolyte formulation tested while Ca stripping was found to be relatively easy, Ca plating was never observed in agreement with reference.[12] Therefore,

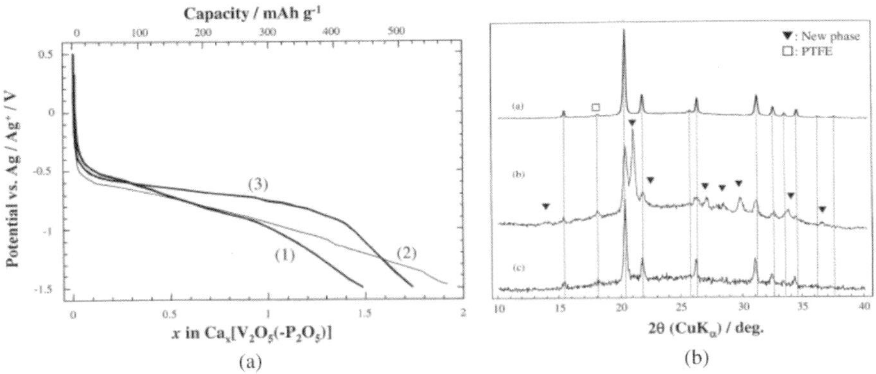

Figure 20. (a) Discharge curves of vanadium oxides [c-V_2O_5 (1), a-V_2O_5 (2), and a-V_2O_5-P_2O_5 (3)] under a galvanostatic condition of 50 mA/cm^2 in 1 M $Ca(ClO_4)_2$ dissolved in AN; (b) XRD patterns of c-V_2O_5 (a) untreated, (b) after discharging, and (c) after charging. Reprinted with permission from Ref. 139 with permission of Elsevier.

only Ca insertion was tested and the highest capacity was obtained for amorphous V_2O_5 compounds. Nonetheless, X-ray diffraction was performed on crystalline V_2O_5 and the appearance of a new phase was observed and ascribed to the apparently reversible insertion of Ca^{2+} (Figure 20). Regrettably, the electrochemical curves for the sample that seem to experience reversible insertion of Ca^{2+} was not shown and the water content of the electrolyte was not indicated.

More recently, Prussian blue (PB) open framework nanomaterials were investigated for calcium ion insertion. The general formula for these materials is $A_xPR(CN)_6 \cdot yH_2O$. They contain zeolitic water and their crystal structure is analogous to that of the ABX_3 perovskites. Cui *et al.* studied the insertion of Mg^{2+}, Ca^{2+}, Sr^{2+}, and Ba^{2+} ions in $KNiFe(CN)_6$ using aqueous-based electrolyte (pH = 2).[140] Fairly similar capacities (ca. 50 mAh/g) and voltage hysteresis (<100 mV) were recorded at C/5 rate in Mg-, Ca-, Sr-, and Ba-based electrolytes. In order to evaluate the possible contribution from proton insertion, the authors performed cyclic voltammetry in dilute nitric acid solution and inductively coupled plasma optical emission spectroscopy (ICP-OES) of dilute electrolytes but could not rule out the possible contribution from protons' insertion. Ingram *et al.* investigated the use of $Na_xMnFe(CN)_6$ that was electrochemically desodiated.[141] From energy dispersive X-ray spectroscopy performed on the discharge and charged sample in Ca-based electrolyte, it was found that the initial desodiation step was not complete or salt impurities were still present and generate Na contamination. From X-ray absorption near edge, it was concluded that only the Mn^{3+}/Mn^{2+} was active during cycling in Ca-based electrolytes. A Ca-ion cell was assembled using a precalciated tin anode (up to 40 mAh/g), but the full cell capacity droped to 50% of its initial value after 35 cycles. Shiga *et al.* prepared $K_{0.1}Mn_1Fe_{1.1}(CN)_6$, $4.2H_2O$ PB analog and tested Ca^{2+} insertion in ionic liquid-based electrolyte (N,N-diethyl-N-methyl-N-(2-methoxyethyl) ammonium bis(trifluoro-methanesulfonyl) amide) containing $Ca(CF_3SO_3)_2$ at 60°C.[142] For the sake of comparison, cells were also assembled using an analog Na-based electrolyte and similar voltage and capacities were recorded in Ca or Na cells (ca. 60 mAh/g) although Ca cell presented higher voltage hysteresis than the Na based one (ca. 1 and 0.5 V, respectively). Tojo *et al.* and Padigi *et al.* reported on the possible insertion of Ca^{2+} in $K_xNiFe(CN)_6$ and $K_2BaFe(CN)_6$ and about 50 mAh/g was recorded in both cases with fairly low Coulombic efficiencies (ca. 92%–93%).[143,144]

Overall, PB compound family appeared as an interesting candidate cathode material to demonstrate calcium insertion (at least to some extent) but, unlike V_2O_5, they failed to demonstrate any improved reversible capacity when compared with monovalent cation insertion. In addition, they present important drawbacks such as solubility issues, low capacity, and Coulombic efficiency.[140,142] The influence of water in the electrochemical properties of this family of compound also remain to be ascertained.

Interesting HRTEM experiments have been designed in order to observe in situ the reactivity of Li, Na, and Ca with WO_3.[145] An intercalation step before conversion was observed, regardless of the nature of the intercalant. However, in the case of Ca, the conversion only proceeded for ca. 2 nm into the film, which has been attributed to the poor diffusivity of big Ca^{2+} ions in the dense W metal layers.

Experimental work on new anodes is quasi inexistent and is limited to new approach to use Ca metal and a study on $CaSi_2$ decalciation.[146]

They were few attempts to assemble full Ca-based batteries. Seshadri *et al.* evaluated the electrochemical behavior of primary cell based on: Ca/ $Ca(ClO_4)_2$ in AN/sulfur infiltrated mesoporous carbon.[147] This cell exhibit ca. 600 mAh/g (calculated on the S mass basis) but no evidence of sulfur reactivity or the presence of discharge products was obtained. In addition, explosive hazards associated with handling dried discharged electrodes were mentioned. In the work from Ingram *et al.*, a Ca-ion cell was assembled using $Na_xMnFe(CN)_6$ cathode and a precalciated tin anode (see Section B "Experimental Evaluation of Cathode and Anode Materials for Secondary Ca-Ion Cells"). Unfortunately, precalciation of the tin electrode was achieved only up to 40 mAh/g and the Coulombic efficiency for the alloy formation was not mentioned.[141] A Ca-ion cells was also reported by Tirado *et al.* using V_2O_5 as an anode and $CaCo_2O_4$ as the cathode.[148] Even though, limited cycling stability was obtained, up to $x = 0.35$ Ca were extracted from $CaCo_2O_4$.

C. First Demonstration of Ca Plating, a First Step toward Ca Metal-Based Batteries

To develop viable calcium metal anodes, the electrolyte must contain Ca^{2+} ions and allow for reversible calcium metal plating/stripping (on reduction/oxidation). Considering an electrode covered with a surface passivation layer, the electrodeposition of Ca metal is enabled only if all the following requisites are fulfilled (Figure 21): solvated Ca^{2+} ions can diffuse/migrate within the electrolyte; the desolvation energy barrier at the

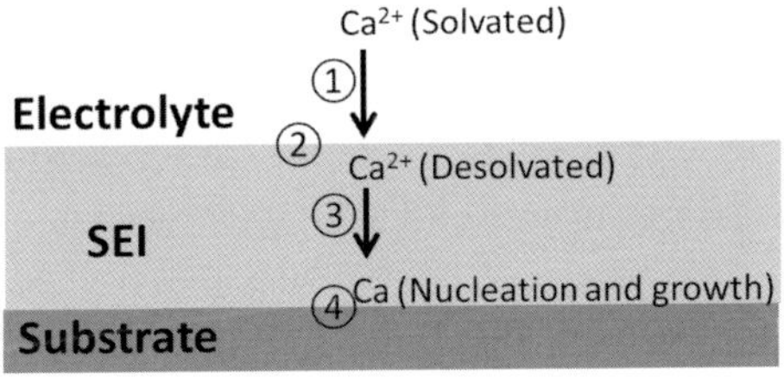

Figure 21. Schematic representation of the four main factors affecting the electrodeposition Ca at the surface of an electrode covered with a passivation layer. Reprinted with permission from Ref. 13 with permission of Nature Publishing Group.

electrolyte/passivation layer interface is low; the desolvated Ca^{2+} ions can migrate through the passivation layer; and the energy barrier for nucleation and growth of Ca at the electrode substrate interface is low. A number of factors can influence the feasibility of one, or most commonly more, of the above processes, including the composition of the electrolyte solvent and salt, its concentration, and the temperature and the nature of the substrate. These determine the tendency to ion pairing (in turn influencing the diffusion of Ca^{2+} ions within the electrolyte), the desolvation energy, the composition of the passivation layer (and its ionic conductivity) and the nucleation energy barrier.

Recent work from Ponrouch *et al.* demonstrated for the first time that calcium can be reversibly plated and stripped through a stable SEI layer using conventional organic electrolytes (Figure 22).[13] This work set out the basis for the development of new electrolyte for divalent cation based batteries with high resilience upon oxidation and a crucial parameter has been highlighted: the formation of ion pairs within the electrolyte, hindering Ca^{2+} migration. Indeed, while most of the studies on multivalent cations have been focussing on the cation solid state diffusion, the impact of the migration process within the electrolyte (significantly affected by ion pair formation) and the energy barrier associated with the cation desolvation at the SEI/electrolyte interface are seldom considered. In monovalent cation systems (Li- or Na-based electrolytes), ion pair formation is negligible but they could become the dominant limiting factors in Ca^{2+}-based electrolytes where the high charge/radius ratio of the cation induces much stronger Coulombic interactions with the solvent and the anion.

Finally, even though Ca-based batteries attracted much less attention than their Mg analog, interesting strategies for their development were already highlighted. Two parallel approaches are being pursued: (1) development of Ca-ion cells for which combinations of anode and cathode materials with electrolytes are selected in order to avoid parasitic reaction and passivation layer formation and (2) development of high voltage Ca metal-based battery owing to the formation of passivation layers and optimization of the divalent cation mobility in the electrolyte. As for today, V_2O_5 appears as the only reliable standard testing material for Ca-based cells although, Ca based battery technology is still in its infancy and important work remains to be done on the development of new cathode

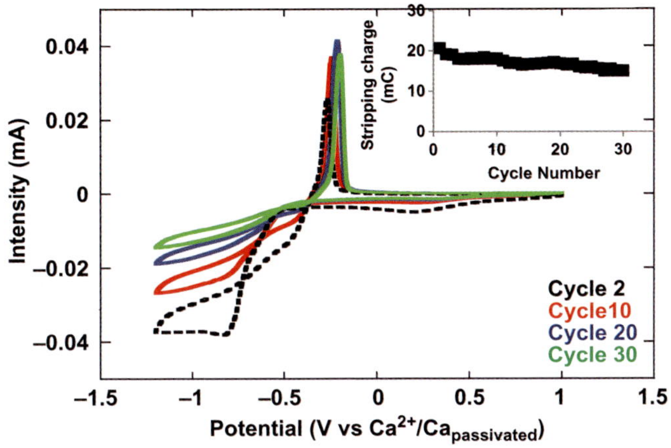

(a)

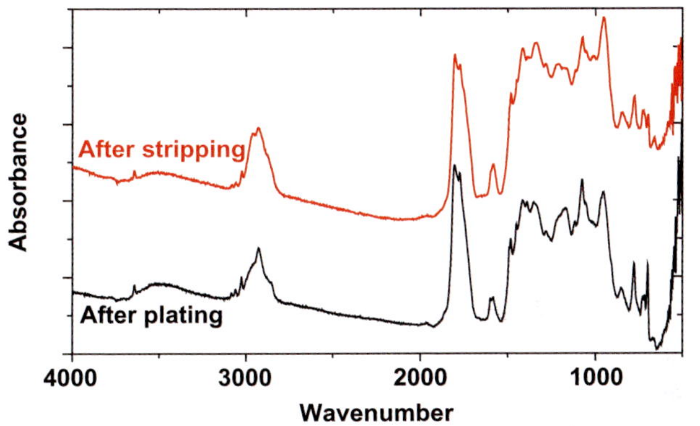

(b)

Figure 22. (a) Cyclic voltammograms (100°C, 0.2 mV/s) of a calcium deposit (grown by potentiostatic electrodeposition at −1.2 V *versus* Ca^{2+}/Ca passivated, 5 h, 100°C) in 0.45 M $Ca(BF_4)_2$ EC/PC electrolytes and (b) corresponding infrared spectra on calcium deposits after plating and stripping (−0.1 V *versus* Ca^{2+}/Ca passivated, 5 h, 100°C). The inset displays the evolution of the charge associated with calcium metal stripping upon cycling. Reprinted with permission from Ref. 13 with permission of Nature Publishing Group.

materials. In addition, a deep understanding of the cation transport properties in the solid state (active material and SEI), within the electrolyte and at their interfaces, is mandatory for further development of this technology (as it was also the case for Li).

XI. Nonaqueous Al Electrochemistry and Attempts to Develop Rechargeable Al Batteries

A. Introduction

After exploring magnesium- or calcium-based electrochemical devices, the hunt for high-energy density nonaqueous supervalent rechargeable batteries logically moves to Al. Indeed, it is a very attractive electrode material for energy storage and conversion. Owing to its low molar weight (26.98 g/mol) and its high density (2.7 g/cm), there is more matter packed per volume for aluminum compare to magnesium or calcium (1.00 *versus* 0.72 and 0.38 mol/cm^3, respectively). Such feature combined with the trivalent nature of the aluminum cation results in a superior theoretical volumetric capacity almost four times higher than lithium (8040 *versus* 2016 mAh/cm^3). Moreover, aluminum is the most abundant metal in the earth's crust, the third element after oxygen and silicon. Aluminum is also already intensively used in the industry especially due to its high strength, low density, and corrosion resistance (due to formation of oxides surface film preventing further reactions). Easily shapeable and nontoxic, it is mostly used for packing in agronomic industries (cans, foils, etc.) Its lightness makes it a material of choice for transportation vehicles or in construction industries. As an example, aluminum pieces represent around 80% by weight in modern airplanes. Highly electrical conductor, aluminum metal is already widely used by the batteries companies as a current collector foil.

In 1857, Buff discovered the growth of an oxide film on the surface of a polarized aluminum piece immersed in aqueous solution. It is the first example of the use of aluminum as anode material. Then aluminum was used in Leclanché-type primary cells with MnO_2 cathode and aqueous electrolytes. However, the passivation surface film altered the voltage and the cell efficiency. The use of concentration alkali solution enabled to dissolve the oxide film and restore the anode potential but it also entails enhanced corrosion and shelf life. Amalgamation to form alloy at the anode might also reduce or even cancel the passivation layer formation. Such anodes have been extensively studied for Al–air batteries and are detailed in Section XII. Consequently, nonaqueous electrolytes are mandatory to develop rechargeable Al battery. This first part of this section summarizes the different types of electrolyte reported in the literature, whereas the second part logically focuses on positive electrode materials.

B. Electrolyte Solutions for Al Batteries

Aluminum coating has been widely used in industry especially for corrosion protection in electronics. Therefore, the electrolytes that are currently used in the attempts to develop rechargeable Al batteries logically derive from industrial electroplating baths. In this section, the electrochemical aluminum deposition is detailed using either aluminum compounds dissolved in nonaqueous solvents or room-temperature ionic liquids (RTILs) as the electrolyte.

Nonaqueous electrolytic systems.

Similar to classical electrolytes used in lithium-ion batteries, these electrolytes consist on an aluminum salt dissolved in an organic solvent. Attempts to use aprotic solvents such as AN, pyridine, or PC were not successful. Indeed, small- and high-charge aluminum cation can be heavily solvated and it requires less energy to decompose solvent rather than cation desolvatation and metal electrodeposition. So far, three classes of organic solvents have been reported to successfully enable the electrochemical deposition of aluminum: aromatic hydrocarbons, ethers, and dialkylsulfones. For each type of solvent, the main works are briefly summarized in the following part. Exhaustive reviews deeply detail the characteristics of each electrodeposition baths.[149]

In 1899, Plotnikov was the first trying to deposit aluminum from organic solvents.[150] Aluminum deposition was found possible from solution of aromatic hydrocarbons such as benzene, or toluene, in which aluminum bromide was dissolved.[151,152] As those first aluminum electrodeposits were quality poor and plagued by relatively low efficiencies, the following works up to now aimed on increasing the deposit quality as well as improving the yield. Other halides used as additives (HBr, KBr) or complexing agent (R_4NBr) were investigated in order to either boost the conductivity or improve the deposition uniformity.[10,153–156] Organoaluminum compounds were also tested as electrolyte salt dissolved in aromatic solvents.[157,158]

In 1952, Brenner and Couch proposed an alternative bath with $AlCl_3$ and LiH dissolved in diethyl ether, subsequently adopted by the National Bureau of Standards (NBS) despite the solvent high flammability, limited lifetime, and possible hydrogen evolution.[159] Following improvements clearly focused on enhancing the coating quality ad reducing the fire hazard by substitution with safer solvents. Ishibashi and Yoshio suggested to use THF alone or combined with benzene, and reported better anode dissolution and improved bath life. In these works, $AlCl_3$ was combined with $LiAlH_3$, in the molar ratio 3:1 for the best deposits.[160,161]

Dialkylsulfones (RR'SO_2) exhibit interesting features for their use as electrolyte solvents. They are stable in temperature, and easily dissolved a wide range of metallic salts. Messina and coworkers evidenced that aluminum can be reversibly plated and stripped in a mixture of $AlCl_3$ and LiCl (2:1 mole ratio) with dimethylsulfone ($DMSO_2$). Continuous aluminum deposits were revealed by SEM observations. In 2015, Nakayama and coworkers investigated various dialkylsulfone-based solvents mixing and evidenced that lowering the $AlCl_3$ concentration with dilutants (toluene) enables to reduce the operating heating to RT.[162]

C. From Molten Salts to Room-Temperature Ionic Liquids

Similarly to nonaqueous electrolytes, molten salts used in aluminum electrodeposition uses aluminum halides ($AlCl_3$ or $AlBr_3$) that are here combined with another alkali or organic halide. Here, the molar ratio R = MX/AlX_3 (M = Na, K) defines the nature of the major anion species participating to the reversible Al stripping/deposition: either $Al_2X_7^-$ in case of acidic melts or AlX_4^- when the basicity increase.[163] Inorganic melts can be used above 100°C, whereas organic-based melts usually operates below 100°C and are ionic liquids. They exhibit enhanced chemical and thermal stabilities, high electrical conductivity, and low vapor pressure. Haloaluminates-based ionic liquids have been extensively investigated for aluminum electrodeposition and the following part briefly summarizes their characteristics.

Alkylpyridinium aluminates were the first electrolytes practically used for aluminum electroplating. In the early 1950s, Hurley and Wier reported that mixing $AlCl_3$ with ethylpyridinium bromide (EPB) leads to a RT melt that can be subsequently used for aluminum electroplating. The addition of aromatic hydrocarbons and/or ethers that formed supernatant layer helps reducing the operating temperature and protects the bath from moisture and oxidation. Despite decent efficiencies generally above 90%, these baths suffer from a narrow electrochemical window due to the easy oxidation of bromide. In addition, the melting point significantly increases when the $AlCl_3$ volume content differs from 2/3. Substituting EPC by N-(1-butyl)pyridinium chloride (BPC) widens the electrochemical window up to 2 V. The high viscosity can be reduced by addition or organic solvent such as benzene or toluene.[164–168] Chronoamperometry analysis evidenced that the aluminum deposition involved fast nucleation then followed by diffusion-controlled film growth.

Alkylimidazolium cations were found to be reduced less easily than alkylpyridium. As an example, coupling $AlCl_3$ with 1-ethyl-3-methylimidazolium chloride (EMIC) gives a melt with larger electrochemical window. It is considered today as the most promising electrolyte for investigating electrode material.[169–171] However, the use of EMI^+ cation is problematic due to its tendency to decompose during preparation. In addition, the oxidation stability is still not enough, around 2.7 V.[172] Chlorine gas might evolve if the potential exceeds this value.

Besides EMIC, trimethylphenylammonnium chloride (TMPAC) was found easy to prepare and TMPAC-$AlCl_3$ melts exhibit good electrochemical activity and high conductivities.[173]

D. Cathode Materials

Despites its interesting physical properties which making it a superior anode in terms of volumetric capacity, the development of rechargeable Al battery is hindered by the standard reduction potential of −1.66 V *versus* SHE. It is almost 1.4 V more positive than lithium and still 0.7 V higher than magnesium. Therefore, searching for high-voltage cathode materials is difficult. Indeed with a similar positive electrode materials, working with aluminum anode obviously involves a 0.7 V decrease of the working voltage in comparison with a system using a magnesium anode. In addition, it is known that moving from alkali to alkali-earth electrochemical intercalation is made more difficult because of the stronger Coulombic effect caused by the doubled-charge cation. It is even worse for aluminum intercalation, for which the trivalent state combined with a small radius involves a low polarizibility. Despites this drawback, there have been interesting studies on the electrochemical intercalation of Al into positive electrode hosts, and they are summarized in this section. Lot of the investigated materials have been also tested *versus* magnesium, which makes sense as Mg and Al electrochemical insertion gathers lot of similarities.

1. *Chevrel Phase Mo_6S_8*

The particular open structure of Chevrel phase molybdenum sulfide Mo_6S_8 combined with its sulfide anionic framework is the key element to explain that it can accommodate various type of cation (Li^+, Cu^+, Mg^{2+}, Sn^{2+}, Cd^{2+}, etc.). The group of Sadoway was the first to report an experimental attempt to electrochemically insert aluminum within Chevrel phases Mo_6S_8 or Mo_6Se_8.[174] Unfortunately, only one galvanostatic cycle and limited experimental details were provided. This encouraging work was revisited by Geng and coworkers when investigating the electrochemical behavior of Mo_6S_8 using Al foil as both counter and reference electrode and ionic liquid (1.5:1 mixture of $AlCl_3$ and [BMIm]Cl) as electrolyte.[175] The authors rapidly noticed that the electrochemical mechanisms were enhanced at 50°C. Similarly to Mg intercalation, a reversible two-set electrochemical is observed in cyclovoltammetry or during galvanostatic cycling. A first discharge capacity of 148 mAh/g is reported; however, a strong irreversible capacity certainly due to Al cations trapping lowers the subsequent charge to only 85 mAh/g (Figure 23). After that, Mo_6S_8 electrode exhibits interesting performance along cycling — with a stabilized capacity around 70 mAh/g after 50 cycles — as well as decent rate capability. SEM analysis of the active material after cycling shows cracks on particle surface, suggesting a significant mechanical stress upon electrochemical intercalation. The assumption of a full electrochemical intercalation leads to the formula of $Al_2Mo_6S_8$, with a corresponding specific capacity of 193 mAh/g.

2. *Titanium and Vanadium Binary Oxides*

Since the work of Perreira-Ramos *et al.* and Amatucci *et al.*, V_2O_5 is known to electrochemically accommodate various cations other than Li^+,

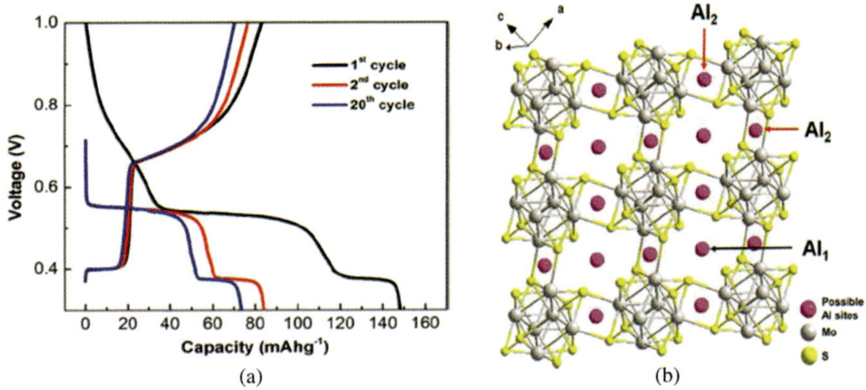

Figure 23. Galvanostatic cycles of Al-Mo_6S_8 battery at 50°C (a) and crystal structure of Al intercalated Mo_6S_8 (b). Adopted from Ref. 175 with permission of the American Chemical Society.

such as Mg^{2+}, Ca^{2+}, or even trivalent Y^{3+}. In 2011, Jayaprakash and coworkers evaluated the electrochemical behavior of V_2O_5 nanowires as cathode *versus* aluminum, with an ionic liquid-based electrolyte ($AlCl_3$ dissolved in [EMIm]Cl).[176] During the first discharge a capacity of 305 mAh/g is observed at the current rate of 125 mA/g, corresponding to the insertion of 0.7 Al per V_2O_5 formula unit. A stabilized reversible capacity around 270 mAh/g is recorded for 20 cycles, with insertion/deinsertion plateaus at 0.55 and 0.8 V, respectively. However, Reed and Menke questioned this work and suggested that corrosion of Fe and Cr within the SS mainly occurs.[177] Nevertheless, V_2O_5-based electrodes are still focusing the interest. Composite electrodes made from amorphous V_2O_5, and binder-free V_2O_5-cathodes prepared by hydrothermal process, were reported to deliver an initial discharge capacity of 200 and 239 mAh/g, respectively. In these works, experimental characterizations such as XPS evidences a redox activity of vanadium; however, a rapid capacity fading was observed upon cycling.[178,179]

Staying in vanadium oxides, Wang *et al.* evaluated the electrochemical intercalation of aluminum into VO_2 (B) nanorods prepared through low temperature hydrothermal method.[180] Computational simulations enabled to claim that inserting up to 2 Al per VO_2 does not affect too much the tunnel structure of monoclinic VO_2 (B). However, experimental electrochemical tests only showed an initial discharge capacity of 165 mAh/g (corresponding to a theoretical insertion of 0.255 Al per VO_2 unit). Moreover, a slight but regular capacity fading is observed along cycling, with almost 30% after 100 cycles, and the overall working voltage was (0.5 V and lower). It is interesting to note the addition benzyl sulfoxide (0.5 wt%) as corrosion inhibitor in the ionic liquid electrolyte (mixture $AlCl_3$ and 1-butyl-3-methylimidazolium chloride).

The electrochemical behavior of various nanostructured TiO_2 was also investigated. However, it is worth noting that aqueous electrolytes and three-electrode cells were used in these studies. In 2011, Liu *et al.* anodized metallic titanium foil to get TiO_2 nanotubes array and evidenced for the first time that trivalent Al ions can be reversibly inserted into TiO_2 anatase. However, the observed capacities were rather low (below 80 mAh/g, i.e., an insertion of 0.07 Al per TiO_2).[181] After that, He and coworkers obtained better result when investigating black mesoporous anatase nanoleaves prepared by solution plasma process.[182] The high electrical conductivity due to the presence of Ti^{3+} species facilitates the aluminum diffusion. In comparison with commercial anatase, better performance was observed, especially a stabilized reversible capacity above 250 mAh/g after 300 cycles. Such results need now to be adapted using non-aqueous electrolyte to validate the interest of nanostructured as electrode material.

3. *Graphite-Based Cathodes*

Covalent graphite fluorides have been extensively studied as cathode materials in lithium batteries. When tested *versus* lithium anode, they exhibit high discharge capacity of 900 mAh/g, with a decent working voltage above 2 V. When moving to aluminum batteries, Levitin *et al.* reported a discharge capacity is around 500 mAh/g, with however a logical decrease of the potential to only 1 V.[183] Using fluorinated natural graphite prepared electrochemically, Vatsala Rani and coworkers confirmed the interest of CF_x material by showing very stable electrochemical behavior with a discharge capacity above 200 mAh/g for 40 cycles at the constant current rate of 60 mA/g.[184] In 2015, Lin and coworkers constructed Al/pyrolytic graphite (PG) cells with ionic liquid electrolyte[185] (Figure 24).

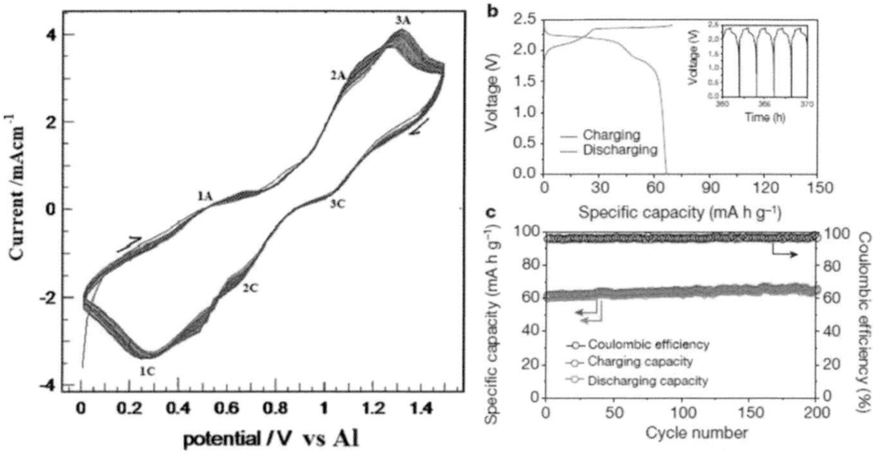

Figure 24. Left, cycling voltammogram of Al/CF_x battery at 10 mV/s. Right, galvanostatic curves and long-term stability of Al/PG cell at 66 mA/g. Taken from Refs. 184 and 185 with permission of the Electrochemical Society and Nature Publishing Group, respectively.

The 3-D graphitic-foam cathode was prepared by action of methane at high temperature on Ni foam that is then removed by acidic treatment. A specific capacity approaching 70 mAh/g, as well as 98% of Coulombic efficiency is observed for more than 7500 cycles. However, this outstanding long life of 7500 cycles has to be balanced by the ultrafast charging time. Nevertheless, owing to the well-defined 2-V plateau observed during discharge, tis promising work might provide new guidelines for the search of cathode materials. In the same time, a rechargeable Al-ion prototype with carbon paper as cathode material was shown to exhibit a high voltage plateau of 1.8 V and high reversible capacity of 70 mAh/g.[186]

4. *PB Analogs*

After the announcement of the use of PB analog as fast, highly reversible and stable cathode materials for aqueous batteries, there has been a growing interest around this kind of compounds. Their widely open crystal structure and their chemical versatility offer room for electrochemical intercalation. PBs are now widely studied for lithium, sodium, and even divalent batteries.[140,187,188] The reversible aluminum ions reversible intercalation was first demonstrated in 2015 on copper hexacyanoferrate (CuHCF) nanoparticles with an aqueous electrolyte.[189] Using CuHCF and an organic-based electrolyte (aluminum triflate dissolved in diglyme), Reed and coworkers reported an initial discharge capacity of 60 mAh/g that unfortunately rapidly faded down to 14 mAh/g with cycling. As the organic electrolyte based on calcium triflate salt might not be the most suitable electrolyte, this pioneer work needs confirmation.

5. *Conductive Polymers*

All the abovementioned examples showed that identifying suitable cathode materials that can reversibly insert trivalent aluminum ions is a real challenge. The solution might arise from conductive polymers. Indeed, they can be electrochemically oxidized or reduced in solutions, and consequently are more and more investigated for electrode materials in lithium-ion batteries. In the middle 1980s, the group of Osteryoung was the first to evidence an electrochemical activity of polypyrrole-based electrode in Al ionic liquids.[190] This work was then extended to other conducting polymers, such as polythiophene, polyaniline, or polyphenylene[191–197] (Figure 25).

E. Conclusion

In the quest for postlithium electrochemical energy storage, it is important to also focus on rechargeable Al batteries. Indeed, Al anode offers a three-time-higher volumetric capacity than lithium. In addition, aluminum is a very abundant element. The research for performant rechargeable battery had already and will continue to benefit from the intensive use of aluminum in industry, especially in for the design of electrolyte. There are existing electrolytic solutions — mainly ionic liquids — that are now extensively known and that can be used to investigate cathode materials. However, the redox potential of aluminum is 1.4 V higher than lithium, and therefore it appears very challenging to discover high-voltage electrode material with moreover the ability to reversibly transfer three electrons.

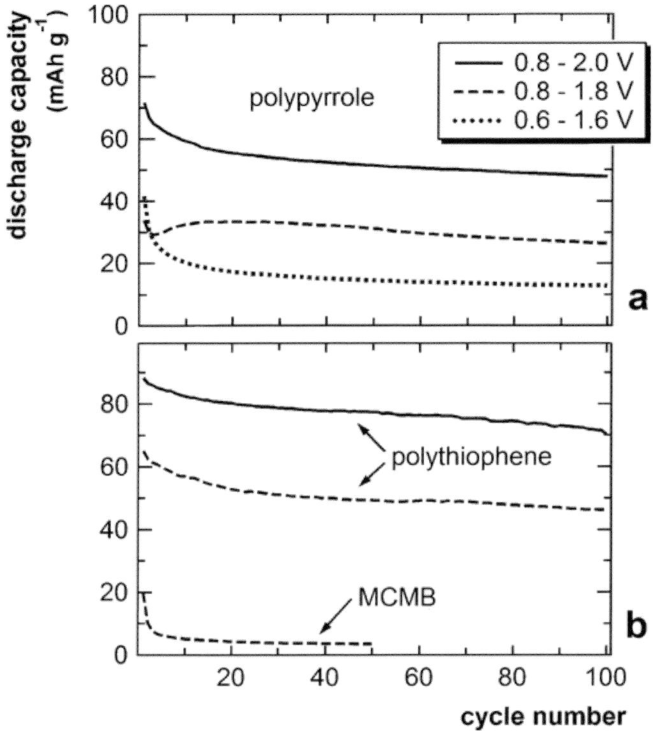

Figure 25. Discharge capacities obtained with conductive polymer-based electrodes. Specific currents were 30 mA/g for polypyrrole and MCMB, 16 mA/g for polythiophene. Taken from Ref. 197 with permission of American Chemical Society.

Moreover, the relatively low capacities reported so far still limit the development of such electrochemical storage systems.

XII. Potential Importance of Aluminum–air Batteries

A. Introduction

Metal–air batteries gather a metal as the negative electrode and oxygen from the air at the positive side. They have received a growing interest in the recent years due to their high theoretical specific densities, from 1 kWh/g for Zn/air batteries to more than 5 kWh/g for Li–air batteries.[198]

Indeed, endless oxygen involves that electrochemical performances theoretically directly derived from the negative metal electrode characteristics. Among them, Al–air cells are ranked second in terms of electrochemical performance, with a theoretical specific energy exceeding 4 kWh/g. Owing to the three-charge redox mechanism of aluminum and its lightness, Al–air systems also offer very promising performance with more than 8 kWh per kg of aluminum. A practical specific energy density of 400 W/kg is usually reported, which is still significantly higher than Li-ion cells.[6,199] Lastly, aluminum is abundant and not expensive, and the thin oxide film that passivates its surface in contact with air makes it easy to work with, to store or to transport.

The history of the development of Al–air batteries starts in the early 1960s from the work of Zaromb and coworkers.[200,201] Since this work, most of the research was carry out for military purposes for examples to power reserve unit of communications devices, or to equipped unmanned underwater or aerials vehicles.

In alkali solution, the negative standard potential of aluminum (−2.35 V *versus* SHE) implies that hydrogen evolution occurs at the negative electrode before any electrodeposition of aluminum. Consequently Al–air systems are primary energy storage devices. However, it is possible to mechanically recharge the battery by replacing the aluminum after each electrochemical discharge.

This section aims to remind the electrochemical mechanisms that occur at the anode surface when the cell is operating. The parasitic self-discharge corrosion and the formation of hydroxide surface layer are the main drawbacks of this system. They are detailed and possible improvements on the design of the anode and on the electrolyte formulation are presented. Lastly, significant advances that might lead to rechargeable Al air are presented as open perspectives.

B. Electrolyte Solutions and Electrochemical Mechanisms

Extensive tables gathering various types of electrolytes and their performance have been presented by Mokhtar and coworkers.[202] In this part, a particular focus is pointed on alkali solutions as they are the most investigated electrolytes; however, other type of electrolytes such as saline or acidic ones are also described.

1. *Alkali Electrolytes*

As a metallic element, aluminum reacts in contact with air by forming a thin oxide layer on its surface. This protection layer is an advantage as the metal can be easily manipulated. However, this oxide film prevents from any electrochemical activity and thus needs to be removed. For this reason, concentrated alkali solutions are mainly used as electrolyte. During the unique discharge of an Al–air battery, the aluminum of the negative electrode is oxidized in alkaline media according to the following equation:

$$Al + 4OH^- \rightarrow Al(OH)_4^- + 3e^- \qquad (2)$$

Oxygen from the air reacts with the electrolyte and is electroreduced according to the reaction:

$$O_2 + 2H_2O + 4e^- \rightarrow 4HO^- \qquad (3)$$

Therefore, the ideal discharge reaction of an Al–air cell is

$$4Al + 3O_2 + 6H_2O \rightarrow 4\ Al(OH)_3. \qquad (4)$$

However, aluminum is not stable thermodynamically in alkaline electrolyte. Besides its electrochemical consumption, it also directly reacts with the electrolyte with hydrogen evolution:

$$H_2O + e^- \rightarrow 0.5H_2 + OH^- \qquad (5)$$

As the battery is discharged, the aluminate ion concentration in the electrolyte increases. Once the solubility limit is reached, aluminum trihydroxide precipitates and regenerates hydroxyl ions:

$$Al(OH)_4^- \rightarrow Al(OH)_3 + OH^- \qquad (6)$$

In order to optimize the electrochemical process alkaline solutions usually employed in Al–air cells with optimized hydroxide salt concentration, 4 and 7 mol/L for NaOH and KOH, respectively. In addition to a higher peak conductivity (0.7 compare to 0.39 S/cm), KOH-based solutions are also superior regarding the solubility limit of aluminum aluminate, thus preventing or delaying the formation of $Al(OH)_3$ precipitates that logically decrease the electrochemical performance by especially interacting with water and blocking pores. The oxygen reduction reaction (ORR) is also favorable with potassium hydroxide electrolytes in both thermodynamic and kinetic aspects (higher oxygen solubility and better oxygen diffusion coefficients).[203]

Theoretically al Al–air cell should exhibit high a working potential of 2.71 V regarding the potentials of both the aluminum dissolution and the oxygen electroreduction (Equations 2 and 3, at −2.31 V and 0.4 V *versus* NHE, respectively). However, the working potential is practically lowered to 1.2–1.6 V mainly by the corrosion reaction. The latter one can be seen as a Coulombic inefficiency, as it prevents from a complete use of aluminum, and therefore should be minimized. Activating the aluminum electrode through alloying, and optimizing the electrolyte formulation by corrosion inhibitors are mainly the two different but complementary strategies followed since the first studies on Al–air cells. They are further detailed separately later on this section even if they should logically be combined in order to get optimized performance.

2. *Neutral Saline and Other Types of Electrolytes*

Neutral saline electrolytes were also investigated. In this case, the anodic process of aluminum dissolution is described as follows:

$$Al \rightarrow Al_3^+ + 3e^- \tag{7}$$

NaCl solutions with concentrations ranging from 0.5 to 4 mol/L were tested in the literature. It was found that the potential value of the Al electrode significantly increases with the salt concentration as well as with the temperature.[202,204–207] Although such electrolytes are safer compare to caustic ones, both their conductivity and aluminum hydroxide solubility are lower, involving lower cell power and energy density. However, NaCl-based electrolytes can mimic seawater conditions, which might be interesting for marine applications such as underwater unmanned vehicle.

Acidic electrolytes such as concentrated hydrochloric or sulfuric acid solutions have been also investigated. Here, the carbonation reaction that occurs in alkali media is suppressed. However, the formation of thicker passivating layer is a problem. Addition of chloride or fluoride anions might break this thick film; however, they affect the oxygen reduction.

Non-aqueous electrolytes are already considered for Li–air cells. With the recent development of RTIL for Al-ion battery electrolytes, a logical transfer to Al–air cells is now operating. Revel and coworkers investigated EMImCl/$AlCl_3$ acidic solution for an Al–air device, while Gelman *et al.* directly used commercial EMIm-(HF)2.3 ionic liquid. In both cases, interesting performance were observed (0.6 and 1.5 mA/cm^2, respectively), confirming the potential interest of such electrolyte for future Al–air cells.[208,209]

C. Alloying for Active Aluminum

Doping aluminum with other element aims to decrease the corrosion reaction. It enables to modulate the thickness of the oxide film on the electrode surface, to reduce the water reduction rate and also to tune the dissolution morphology. The choice of the alloying element needs to fill several requirements. The second element should exhibit a lower melting point that aluminum and a good solubility into the aluminum matrix for ensuring an efficient diffusion during the high-temperature alloying, and thus get a monophasic solid-solution type alloy. Indeed, the presence of precipitates might enhance the electrode corrosion. The second element must also be nobler than aluminum in the galvanic series. A dissolution–deposition mechanism is often used to describe the anode dissolution: as the electrode is polarized, both aluminum and atoms of the second element are oxidized and thus dissolve into the electrolyte as cations. Because alloying elements are cathodic to aluminum, their cations plate out on the aluminum surface, which locally break the passivating layer and shifts down the anode potential.[210] Extensive studies have been carried out to activate aluminum by alloying it with lot of different elements including Mg, Zn, Pb, Sn, Ga, In, Mn, Hg, following or not the abovementioned criteria. Here only the impact of few of them is summarized as examples.

Tin is the one of the most investigated alloying element for activating aluminum. According to the Al–Sn phase diagram, the solubility of tin is limited to 0.12 wt%. Heat treatments at 600°C followed by quench enable to prepare monophasic Al/0.12Sn alloys. In comparison with highly pure aluminum, various groups evidenced an enhanced anodic behavior for Al/0.12Sn.[211–213] Following, the abovementioned mechanism oxidized tin dissolves as stannate ions that are subsequently reduced as long as the electrode potential is more negative than the Sn/SnO_3^{2-} reversible potential.

Gallium is also known to be an attractive candidate for aluminum activation. Tuck and coworkers investigated different Al/Ga alloys — prepared by heat treatment at 600°C — in both alkaline and saline electrolytes. It was found that activation occurs only when the alloy electrode is preliminary polarized below the Ga/GaO_3^{3-} reduction voltage. Localized sites related to Ga aggregates initiate the activation with extremely thin passivation film, which promotes the aluminum diffusion. Here also the temperature strongly influences the needed amount of gallium: from 0.055% at 25°C to at least 0.1% at 60°C[214] (Figure 26). Adding indium to pure aluminum is limited by the solubility, with a maximum of 0.16%. Al/In at this composition are

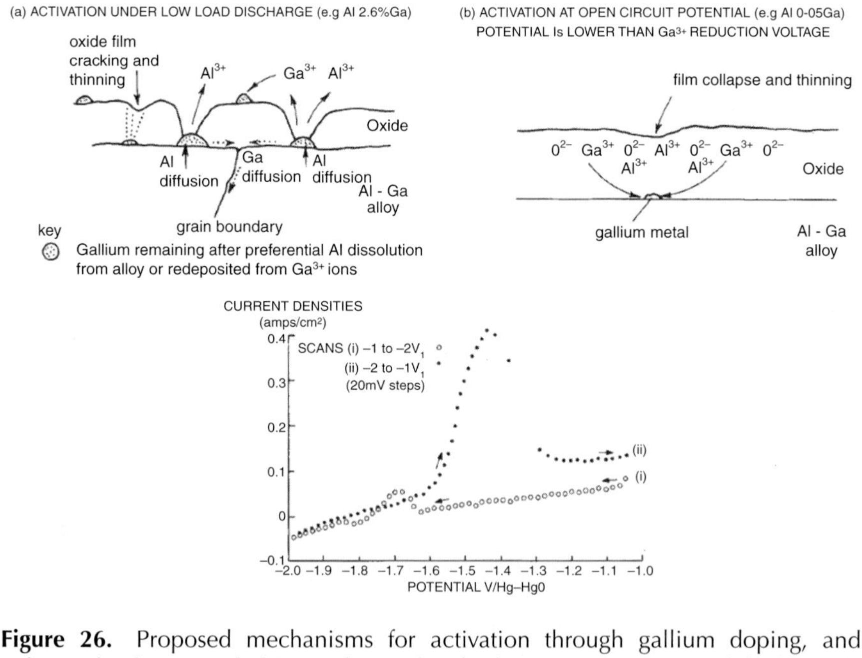

Figure 26. Proposed mechanisms for activation through gallium doping, and enhanced polarization behavior obtained with Al/Ga0.05 as a rotating disk electrode in 4 M NaOH solution. Adopted from Ref. 214.

obtained by heat treatment at 640°C and exhibit the highest anodic behavior in alkali electrolytes at RT. Activation with lower indium content required electrolytes heated at 60°C.

In all cases, overtaking the solubility limit does not provide positive effect. Indeed, the presence of Sn, Ga, or In precipitates as minor secondary phases aggravate the self-corrosion and significantly decrease the efficiency. The operating temperature also plays a key role in the electrode performance. As an example, it was found that some Al/Sn alloys needs an electrolyte at 60°C to be activated, especially because the passivating film is more dissolved than at RT. It is important to mention that other elements were also tested, such as Mn (0.15 wt%), Fe (0.04 wt%), or Mg (0.81 wt%). Ternary or quaternary alloys were also investigated with the purpose of enhancing some a specific property of binary alloys (such as better anodic current for Al/Sn alloys). Exhaustive tables of binary alloys and their corresponding electrochemical performance are presented by Egan and coworkers in a recent review.[199]

D. Improvements through Electrolyte Additives

Playing on the electrolyte composition might be a good alternative to activate aluminum electrode by alloying. Indeed, this last strategy is expensive especially when rare elements such as indium or gallium are considered. The purpose of adding chemical inhibitors in the electrolyte is — as for the electrode activation — to reduce the hydrogen evolution, that is, the parasitic corrosion reaction (Equation 5). Meanwhile, reporting the pioneering studies on Al–air cells, Zaromb and coworkers already identified that adding inhibitors or complexing agents can render the electrolytes less corrosive.[200,201] The following part briefly shows some example of corrosion inhibitors, how they are added to the electrolyte and how they influence the performance.

1. *Ions of Alloying Elements in Alkali Electrolytes*

Following the dissolution–deposition mechanism, adding corresponding ions of alloying elements in the electrolyte enables to promote the plating of these elements on the electrode surface during battery operation. Then the plated elements inhibit the water reduction reaction.

The addition of Na_2SnO_3 in alkaline electrolytes was found to significantly tune the corrosion and dissolution of aluminum.[215,216] Here also the additive concentration is a key parameter that needs to be controlled. Indeed, concentrations exceeding 10^{-2} mol/L might involve tin dendritic growth and consequently short circuit. In a manner to Al/In alloys, the addition of indium hydroxide in alkali electrolyte can enhance the electrochemical behavior of an aluminum electrode. MacDonald and English combined SnO_3^{2-} and $In(OH)_3$ with others inhibitors such as BiO_3^{3-}, $Ga(OH)_4^-$, or $Mn_2O_4^{2-}$. They evidenced the particular effect of potassium manganate at both high and low discharge rates.[215] Other inhibitors such as calcium oxide, calcium chloride, sodium chloride, sodium citrate, or zinc oxide have also been tested. The addition of ZnO was particularly interesting as it diminishes the temperature rise during the aluminum corrosion.[6,199,217–219]

2. *Organic Inhibitors and Cationic Surfactants*

Natural products have been identified as potential corrosion inhibitors for aluminum in alkaline solutions. For instance, lupine seeds extract controls both the anodic dissolution of aluminum and the hydrogen gas evolution Indeed, the seed extracts contains some chemicals constituents that are chemically adsorbed on the electrode surface.[220,221] A wide variety of other naturals products were also reported as green inhibitors for aluminum corrosion in alkaline solutions, as well as in saline or an acidic electrolytes. For most of them, it was noticed that the inhibition efficiency was improved for high inhibitor concentration, whereas it decreases with increasing temperatures.[202] Plant extracts are interesting as low-cost substitute to toxic chemicals; however, it was also reported that they might affect the anodic discharge of aluminum. Consequently and despite interesting reports, current directions on Al–air cells developments do not seem to include such green additives.

Cationic surfactants such as cetyltrimethylammonium bromide (CTAB) or chloride (CTAC) can be employed in the same manner of the abovementioned inhibitors. Indeed, they can preferentially block the cathodic sites of the aluminum electrode and thus inhibit the corrosion reaction[221,222] (Figure 27). However, the inhibiting efficiency reported so

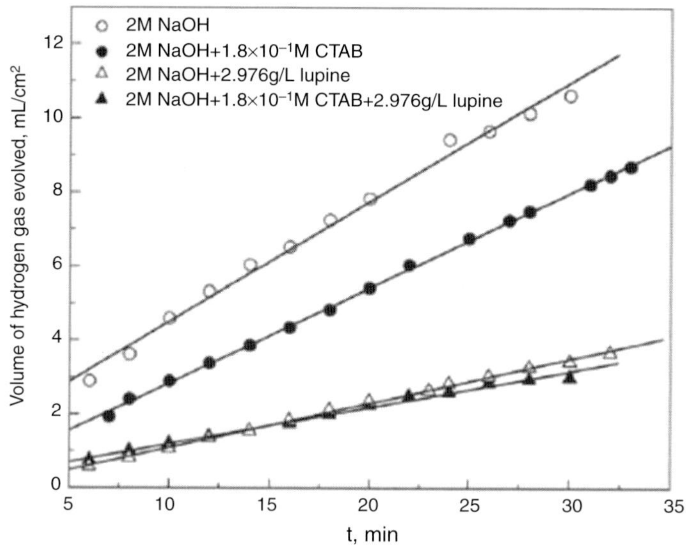

Figure 27. Influence of plants extracts as corrosion inhibitors: (top) volume variation of hydrogen gas evolved for Al in 2 M NaOH with various content of lupine seed extracts. Taken from Ref. 221 with permission of Elsevier.

far are poor. Another drawback is that the additives were also reported to restrict the active area and then significantly reducing the aluminum dissolution process. It is interesting to note that no synergistic action was evidenced with addition of both cationic surfactants and plant extracts.[221]

E. Cathodes Materials

Alike other metal–air electrochemical systems, the air cathode plays a key role in the overall performance of the Al–air cell. Indeed, the reaction at the cathode — the ORR — needs an efficient gas diffusion to be continued. However, this diffusion can be altered by pore clopping caused by insolubility of chemicals by-product of the metal dissolution. Consequently, the design of air cathode material is very important as it largely impacts the general performance. It is generally composed of a gas diffusion layer and an active layer. The gas diffusion layer usually made from porous PTFE brings the oxygen to the active layer and while limiting the liquid electrolyte from passing through the electrode. The active layer contains a catalyst that improves the ORR.

Actually, there is no much work especially devoted to air cathode material for Al–air cells. Indeed, most studies have been performed for applications in fuel cells or Li–air systems.

F. Conclusion

Despite their theoretical high performance and all the advantages of the use of aluminum metal, significant improvements are still necessary to rather develop Al–air batteries and to widen its applications. The major problem of the parasitic corrosion leading to hydrogen evolution has been identified since the pioneering works, as well as the two distinct strategies to circumvent it. Indeed, shifting to expensive super-pure aluminum to aluminum alloys as anode and adding inhibitor additives into the electrolyte enable to reduce the parasitic reaction. Nevertheless, there is to date no significant breakthrough in this field and the future of Al air seems still narrowed by very specific application, such as marine conditions, like in the case of underwater unmanned vehicles. However, the coming up of energy companies specially focused on the development of Al–air cell, such as Phinergy or Fuji Pigment, might open the door of a wider range of applications.

XIII. References

1. Peled, E. *J. Electrochem. Soc.* **1979**, *126*, 2047.
2. Verma, P.; Maire, P.; Novák, P. *Electrochim. Acta* **2010**, *55*, 6332.
3. Meitav, a. *J. Electrochem. Soc.* **1981**, *128*, 825.
4. Ely, Y. E.; Aurbach, D. *Langmuir* **1992**, *8*, 1845.
5. Hitzig, J.; Jüttner, K.; Lorenz, W.; Paatsch, W. *J. Electrochem. Soc.* **1986**, *133*, 887.
6. Li, Q.; Bjerrum, N. J. *J. Power Sources* **2002**, *110*, 1.
7. Sherif, E. -S. M.; Mohammed, J. A.; Abdo, H. S.; Almajid, A. A. *Int. J. Electrochem. Sci.* **2016**, *11*, 1355.
8. Rybalka, K. V.; Beketaeva, L. A.; Davydov, A. D. *Russ. J. Electrochem.* **2016**, *52*, 463.
9. Capuano, G. A.; Davenport, W. G. *J. Electrochem. Soc.* **1984**, *131*, 2595.
10. Peled, E.; Gileadi, E. *J. Electrochem. Soc.* **1976**, *123*, 15.
11. Walker Jr, C. W. *A Review of the Calcium Thionyl Chloride Electrochemical System*, DTIC Document, 1991.

12. Aurbach, D.; Skaletsky, R.; Gofer, Y. *J. Electrochem. Soc.* **1991**, *138*, 3536.
13. Ponrouch, A.; Frontera, C.; Barde, F.; Palacin, M. R. *Nat. Mater.* **2016**, *15*, 169.
14. Aurbach, D.; Lu, Z.; Schechter, A.; Gofer, Y.; Gizbar, H.; Turgeman, R.; Cohen, Y.; Moshkovich, M.; Levi, E. *Nature* **2000**, *407*, 724.
15. Muldoon, J.; Bucur, C. B.; Oliver, A. G.; Sugimoto, T.; Matsui, M.; Kim, H. S.; Allred, G. D.; Zajicek, J.; Kotani, Y. *Energy Environ. Sci.* **2012**, *5*, 5941.
16. Pour, N.; Gofer, Y.; Major, D. T.; Aurbach, D. *J. Am. Chem. Soc.* **2011**, *133*, 6270.
17. Mohtadi, R.; Matsui, M.; Arthur, T. S.; Hwang, S. -J. *Angew. Chem., Int. Ed.* **2012**, *51*, 9780.
18. Ha, S. -Y.; Lee, Y. -W.; Woo, S. W.; Koo, B.; Kim, J. -S.; Cho, J.; Lee, K. T.; Choi, N.-S. *ACS Appl. Mater. Interfaces* **2014**, *6*, 4063.
19. Shterenberg, I.; Salama, M.; Yoo, H. D.; Gofer, Y.; Park, J. -B.; Sun, Y. -K.; Aurbach, D. *J. Electrochem. Soc.* **2015**, *162*, A7118.
20. Tobias, C. W., Ed. *Advances in Electrochemistry and Electrochemical Engineering, Vol. 5: Electrochemical Engineering*; Interscience, 1967.
21. Brenner, A.; Sligh, J. *Trans. Inst. Metal Finishing* **1971**, *49*, 71.
22. Auborn, J. J.; French, K. W.; Lieberman, S. I.; Shah, V. K.; Heller, A. *J. Electrochem. Soc.* **1973**, *120*, 1613.
23. Behl, W. K.; Christopulos, J. A.; Ramirez, M.; Gilman, S. *J. Electrochem. Soc.* **1973**, *120*, 1619.
24. Peled, E.; Straze, H. *J. Electrochem. Soc.* **1977**, *124*, 1030.
25. Staniewicz, R. J. *J. Electrochem. Soc.* **1980**, *127*, 782.
26. Meitav, A.; Peled, E. *J. Electrochem. Soc.* **1982**, *129*, 451.
27. Meitav, A.; Peled, E. *Electrochim. Acta* **1988**, *33*, 1111.
28. Coagley, D. R.; Butler, J. N. *J. Phys. Chem.* **1968**, *72*, 4568.
29. Meitav, A.; Peled, E. *J. Electroanal. Chem. Interfacial Electrochem.* **1982**, *134*, 49.
30. Meitav, A.; Peled, E. *J. Electrochem. Soc.* **1981**, *128*, 825.
31. Avigal, Y.; Peled, E. *J. Electroanal. Chem. Interfacial Electrochem.* **1977**, *76*, 135.
32. Peled, E.; Elster, E.; Tulman, R.; Kimel, J. *J. Power Sources* **1985**, *14*, 93.
33. Peled, E.; Meitav, A.; Brand, M. *J. Electrochem. Soc.* **1981**, *128*, 1936.
34. Peled, E.; Cohen, R.; Melman, A.; Lavi, Y. *J. Electrochem. Soc.* **1992**, *139*, 1836.
35. Gaddum, L. W.; French, H. E. *J. Am. Chem. Soc.* **1927**, *49*, 1295.
36. Liebenow, C.; Yang, Z.; Wagner, M. W.; Lobitz, P. *Proc. — Electrochem. Soc.* **1997**, *97–18*, 597.
37. Gizbar, H.; Vestfrid, Y.; Chusid, O.; Gofer, Y.; Gottlieb, H. E.; Marks, V.; Aurbach, D. *Organometallics* **2004**, *23*, 3826.
38. Gofer, Y.; Chusid, O.; Gizbar, H.; Viestfrid, Y.; Gottlieb, H. E.; Marks, V.; Aurbach, D. *Electrochem. Solid-State Lett.* **2006**, *9*, A257.
39. Yoo, H. D.; Shterenberg, I.; Gofer, Y.; Gershinsky, G.; Pour, N.; Aurbach, D. *Energy Environ. Sci.* **2013**, *6*, 2265.
40. Shterenberg, I.; Salama, M.; Gofer, Y.; Levi, E.; Aurbach, D. *Mrs Bulletin* **2014**, *39*, 453.
41. Gregory, T. D.; Hoffman, R. J.; Winterton, R. C. *J. Electrochem. Soc.* **1990**, *137*, 775.

42. Novák, P.; Imhof, R.; Haas, O. *Electrochim. Acta* **1999**, *45*, 351.
43. Le, D. B.; Passerini, S.; Coustier, F.; Guo, J.; Soderstrom, T.; Owens, B. B.; Smyrl, W. H. *Chem. Mater.* **1998**, *10*, 682.
44. Liebenow, C.; Yang, Z.; Lobitz, P. *Electrochem. Commun.* **2000**, *2*, 641.
45. Kim, H. S.; Arthur, T. S.; Allred, G. D.; Zajicek, J.; Newman, J. G.; Rodnyansky, A. E.; Oliver, A. G.; Boggess, W. C.; Muldoon, J. *Nat. Commun.* **2011**, *2*, 1435/1.
46. Lancry, E.; Levi, E.; Gofer, Y.; Levi, M. D.; Aurbach, D. *J. Solid State Electrochem.* **2005**, *9*, 259.
47. Levi, M. D.; Lancri, E.; Levi, E.; Gizbar, H.; Gofer, Y.; Aurbach, D. *Solid State Ionics* **2005**, *176*, 1695.
48. Levi, E.; Gershinsky, G.; Aurbach, D.; Isnard, O. *Inorg. Chem.* **2009**, *48*, 8751.
49. Levi, M. D.; Lancry, E.; Gizbar, H.; Lu, Z.; Levi, E.; Gofer, Y.; Aurbach, D. *J. Electrochem. Soc.* **2004**, *151*, A1044.
50. Mitelman, A.; Levi, M. D.; Lancry, E.; Levi, E.; Aurbach, D. *Chem. Commun.* **2007**, 4212.
51. Levi, E.; Gershinsky, G.; Aurbach, D.; Isnard, O.; Ceder, G. *Chem. Mater.* **2009**, *21*, 1390.
52. Mitelman, A.; Levi, E.; Lancry, E.; Aurbach, D. *ECS Trans.* **2007**, *3*, 109.
53. Levi, E.; Lancry, E.; Mitelman, A.; Aurbach, D.; Ceder, G.; Morgan, D.; Isnard, O. *Chem. Mater.* **2006**, *18*, 5492.
54. Levi, E.; Mitelman, A.; Isnard, O.; Brunelli, M.; Aurbach, D. *Inorg. Chem.* **2008**, *47*, 1975.
55. Doe, R. E.; Han, R.; Hwang, J.; Gmitter, A. J.; Shterenberg, I.; Yoo, H. D.; Pour, N.; Aurbach, D. *Chem. Commun.* **2014**, *50*, 243.
56. Tutusaus, O.; Mohtadi, R.; Arthur, T. S.; Mizuno, F.; Nelson, E. G.; Sevryugina, Y. V. *Angew. Chem., Int. Ed.* **2015**, *54*, 7900.
57. Barile, C. J.; Barile, E. C.; Zavadil, K. R.; Nuzzo, R. G.; Gewirth, A. A. *J. Phys. Chem. C* **2014**, *118*, 27623.
58. Lu, Z.; Schechter, A.; Moshkovich, M.; Aurbach, D. *J. Electroanal. Chem.* **1999**, *466*, 203.
59. Erickson, E. M.; Markevich, E.; Salitra, G.; Sharon, D.; Hirshberg, D.; de la Llave, E.; Shterenberg, I.; Rozenman, A.; Frimer, A.; Aurbach, D. *J. Electrochem. Soc.* **2015**, *162*, A2424.
60. Aurbach, D.; Suresh, G. S.; Levi, E.; Mitelman, A.; Mizrahi, O.; Chusid, O.; Brunelli, M. *Adv. Mater.* **2007**, *19*, 4260.
61. Viestfrid, Y.; Levi, M. D.; Gofer, Y.; Aurbach, D. *J. Electroanal. Chem.* **2005**, *576*, 183.
62. Mohtadi, R.; Mizuno, F. *Beilstein J. Nanotechnol.* **2014**, *5*, 1291.
63. Mohtadi, R.; Matsui, M.; Arthur, T. S.; Hwang, S. -J. *Angew. Chemie* **2012**, *124*, 9918.
64. Muldoon, J.; Bucur, C. B.; Gregory, T. *Chem. Rev.* **2014**, *114*, 11683.
65. Benmayza, A.; Ramanathan, M.; Arthur, T. S.; Matsui, M.; Mizuno, F.; Guo, J.; Glans, P.-A.; Prakash, J. *J. Phys. Chem. C* **2013**, *117*, 26881.
66. Xu, K. *Chem. Rev.* **2004**, *104*, 4303.

67. Ponrouch, A.; Monti, D.; Boschin, A.; Steen, B.; Johansson, P.; Palacin, M. R. *J. Mater. Chem. A* **2015**, *3*, 22.
68. Aurbach, D.; Schechter, A.; Moshkovich, M.; Cohen , Y. *J. Electrochem. Soc.* **2001**, *148*, A1004.
69. Jäckle, M.; Groß, A. *J. Chem. Phys.* **2014**, *141*, 174710.
70. Canepa, P.; Gautam, G. S.; Malik, R.; Jayaraman, S.; Rong, Z.; Zavadil, K. R.; Persson, K.; Ceder, G. *Chem. Mater.* **2015**, *27*, 3317.
71. Okoshi, M.; Yamada, Y.; Yamada, A.; Nakai, H. *J. Electrochem. Soc.* **2013**, *160*, A2160.
72. Shakourian-Fard, M.; Kamath, G.; Sankaranarayanan, S. K. R. S. *ChemPhysChem* **2015**, *16*, 3607.
73. Xu, K.; von Cresce, A.; Lee, U. *Langmuir* **2010**, *26*, 11538.
74. Huggins, R. A. In *Handbook of Battery Materials*; Wiley-VCH Verlag GmbH: 2007, p 359.
75. Chianelli, R. R.; Scanlon, J. C.; Rao, B. M. L. *J. Solid State Chem.* **1979**, *29*, 323.
76. Li, Z.; Huang, J.; Yann Liaw, B.; Metzler, V.; Zhang, J. *J. Power Sources* **2014**, *254*, 168.
77. Kim, H.; Jeong, G.; Kim, Y. -U.; Kim, J. -H.; Park, C. -M.; Sohn, H. -J. *Chem. Soc. Rev.* **2013**, *42*, 9011.
78. Rosso, M.; Brissot, C.; Teyssot, A.; Dollé, M.; Sannier, L.; Tarascon, J. -M.; Bouchet, R.; Lascaud, S. *Electrochim. Acta* **2006**, *51*, 5334.
79. Gibbs, J. W. *Trans. Conn. Acad. Arts Sci.* **1873**, *2*, 382.
80. Hyde, M. E.; Compton, R. G. *J. Electroanal. Chem.* **2003**, *549*, 1.
81. Lian, G.; Gerko, O.; Aleksandar, R.; Peter, M. H.; Peter, C. S. *J. Phys. D: Appl. Phys* **2011**, *44*, 443001.
82. Mogi, R.; Inaba, M.; Jeong, S. -K.; Iriyama, Y.; Abe, T.; Ogumi, Z. *J. Electrochem. Soc.* **2002**, *149*, A1578.
83. Monroe, C.; Newman, J. *J. Electrochem. Soc.* **2005**, *152*, A396.
84. Matsui, M. *J. Power Sources* **2011**, *196*, 7048.
85. Aurbach, D.; Cohen, Y.; Moshkovich , M. *Electrochem. Solid-State Lett.* **2001**, *4*, A113.
86. Ling, C.; Banerjee, D.; Matsui, M. *Electrochim. Acta* **2012**, *76*, 270.
87. Pereira-Ramos, J. P.; Messina, R.; Perichon, J. *J. Electrochem. Soc.* **1988**, *135*, 3050.
88. Novak, P.; Desilvestro, J. *J. Electrochem. Soc.* **1993**, *140*, 140.
89. Amatucci, G. G.; Badway, F.; Singhal, A.; Beaudoin, B.; Skandan, G.; Bowmer, T.; Plitz, I.; Pereira, N.; Chapman, T.; Jaworski, R. *J. Electrochem. Soc.* **2001**, *148*, A940.
90. Cheng, Y.; Shao, Y.; Raju, V.; Ji, X.; Mehdi, B. L.; Han, K. S.; Engelhard, M. H.; Li, G.; Browning, N. D.; Mueller, K. T.; Liu, J. *Adv. Funct. Mater.* **2016**, *26*, 3446.
91. Zhang, R.; Yu, X.; Nam, K. -W.; Ling, C.; Arthur, T. S.; Song, W.; Knapp, A. M.; Ehrlich, S. N.; Yang, X. -Q.; Matsui, M. *Electrochem. Commun.* **2012**, *23*, 110.
92. Arthur, T. S.; Zhang, R.; Ling, C.; Glans, P. -A.; Fan, X.; Guo, J.; Mizuno, F. *ACS Appl. Mater. Interfaces* **2014**, *6*, 7004.

93. Kim, J. -S.; Chang, W. -S.; Kim, R. -H.; Kim, D. -Y.; Han, D. -W.; Lee, K. -H.; Lee, S. -S.; Doo, S. -G. *J. Power Sources* **2015**, *273*, 210.
94. Nam, K. W.; Kim, S.; Lee, S.; Salama, M.; Shterenberg, I.; Gofer, Y.; Kim, J. -S.; Yang, E.; Park, C. S.; Kim, J. -S.; Lee, S. -S.; Chang, W. -S.; Doo, S. -G.; Jo, Y. N.; Jung, Y.; Aurbach, D.; Choi, J. W. *Nano Lett.* **2015**, *15*, 4071.
95. Novak, P.; Imhof, R.; Haas, O. *Electrochim. Acta* **1999**, *45*, 351.
96. Sun, X.; Duffort, V.; Mehdi, B. L.; Browning, N. D.; Nazar, L. F. *Chem. Mater.* **2015**, *28*, 534.
97. Feng, Z.; Yang, J.; NuLi, Y.; Wang, J. *J. Power Sources* **2008**, *184*, 604.
98. NuLi, Y.; Yang, J.; Wang, J.; Li, Y. *J. Phys. Chem. C* **2009**, *113*, 12594.
99. NuLi, Y.; Yang, J.; Li, Y.; Wang, J. *Chem. Commun.* **2010**, *46*, 3794.
100. NuLi, Y.; Zheng, Y.; Wang, Y.; Yang, J.; Wang, J. *J. Mater. Chem.* **2011**, *21*, 12437.
101. Li, Y.; Nuli, Y.; Yang, J.; Yilinuer, T.; Wang, J. *Chin. Sci. Bull.* **2011**, *56*, 386.
102. Zheng, Y.; NuLi, Y.; Chen, Q.; Wang, Y.; Yang, J.; Wang, J. *Electrochim. Acta* **2012**, *66*, 75.
103. Orikasa, Y.; Masese, T.; Koyama, Y.; Mori, T.; Hattori, M.; Yamamoto, K.; Okado, T.; Huang, Z. -D.; Minato, T.; Tassel, C.; Kim, J.; Kobayashi, Y.; Abe, T.; Kageyama, H.; Uchimoto, Y. *Sci. Rep.* **2014**, *4*, 5622.
104. Wu, J.; Gao, G.; Wu, G.; Liu, B.; Yang, H.; Zhou, X.; Wang, J. *RSC Adv.* **2014**, *4*, 15014.
105. Wu, J.; Gao, G.; Wu, G.; Liu, B.; Yang, H.; Zhou, X.; Wang, J. *Phys. Chem. Chem. Phys.* **2014**, *16*, 22974.
106. Huang, Z. -D.; Masese, T.; Orikasa, Y.; Mori, T.; Minato, T.; Tassel, C.; Kobayashi, Y.; Kageyama, H.; Uchimoto, Y. *J. Mater. Chem. A* **2014**, *2*, 11578.
107. Van Noorden, R. *Nature* **2014**, *507*, 26.
108. Zhao-Karger, Z.; Zhao, X.; Wang, D.; Diemant, T.; Behm, R. J.; Fichtner, M. *Adv. Energy Mater.* **2015**, *5*, 1401155.
109. Vinayan, B. P.; Zhao-Karger, Z.; Diemant, T.; Chakravadhanula, V. S. K.; Schwarzburger, N. I.; Cambaz, M. A.; Behm, R. J.; Kubel, C.; Fichtner, M. *Nanoscale* **2016**, *8*, 3296.
110. Zhao-Karger, Z.; Lin, X. -M.; Bonatto Minella, C.; Wang, D.; Diemant, T.; Behm, R. J. r.; Fichtner, M. *J. Power Sources* **2016**, *323*, 213.
111. Lightfoot, P.; Krok, F.; Nowinski, J. L.; Bruce, P. G. *J. Mater. Chem.* **1992**, *2*, 139.
112. Bruce, P. G.; Krok, F.; Lightfoot, P.; Nowinski, J. L.; Gibson, V. C. *Solid State Ionics* **1992**, *53*, 351.
113. Tao, Z. -L.; Xu, L. -N.; Gou, X. -L.; Chen, J.; Yuan, H. -T. *Chem. Commun.* **2004**, 2080.
114. Sun, X.; Bonnick, P.; Nazar, L. F. *ACS Energy Lett.* **2016**, 297.
115. Sun, X.; Bonnick, P.; Duffort, V.; Liu, M.; Rong, Z.; Persson, K. A.; Ceder, G.; Nazar, L. F. *Energy Environ. Sci.* **2016**, *9*, 2273.
116. Liang, Y.; Tao, Z.; Chen, J. *Adv. Energy Mater.* **2012**, *2*, 742.

117. NuLi, Y.; Guo, Z.; Liu, H.; Yang, J. *Electrochem. Commun.* **2007**, *9*, 1913.
118. NuLi, Y.; Chen, Q.; Wang, W.; Wang, Y.; Yang, J.; Wang, J. *Sci. World J.* **2014**, *2014*, 7.
119. Song, Z.; Qian, Y.; Liu, X.; Zhang, T.; Zhu, Y.; Yu, H.; Otani, M.; Zhou, H. *Energy Environ. Sci.* **2014**, *7*, 4077.
120. Sano, H.; Senoh, H.; Yao, M.; Sakaebe, H.; Kiyobayashi, T. *Chem. Lett.* **2012**, *41*, 1594.
121. Bitenc, J.; Pirnat, K.; Bančič, T.; Gaberšček, M.; Genorio, B.; Randon-Vitanova, A.; Dominko, R. *ChemSusChem* **2015**, *8*, 4128.
122. Pan, B.; Huang, J.; Feng, Z.; Zeng, L.; He, M.; Zhang, L.; Vaughey, J. T.; Bedzyk, M. J.; Fenter, P.; Zhang, Z.; Burrell, A. K.; Liao, C. *Adv. Energy Mater.* **2016**, n/a.
123. Bitenc, J.; Pirnat, K.; Mali, G.; Novosel, B.; Randon Vitanova, A.; Dominko, R. *Electrochem. Commun.* **2016**, *69*, 1.
124. Cheng, Y.; Shao, Y.; Zhang, J. -G.; Sprenkle, V. L.; Liu, J.; Li, G. *Chem. Commun.* **2014**, *50*, 9644.
125. Cho, J. -H.; Aykol, M.; Kim, S.; Ha, J. -H.; Wolverton, C.; Chung, K. Y.; Kim, K. -B.; Cho, B. -W. *J. Am. Chem. Soc.* **2014**, *136*, 16116.
126. Gao, T.; Han, F.; Zhu, Y.; Suo, L.; Luo, C.; Xu, K.; Wang, C. *Adv. Energy Mater.* **2015**, *5*, n/a.
127. Yoo, H. D.; Liang, Y.; Li, Y.; Yao, Y. *ACS Appl. Mater. Interfaces* **2015**, *7*, 7001.
128. Yagi, S.; Ichitsubo, T.; Shirai, Y.; Yanai, S.; Doi, T.; Murase, K.; Matsubara, E. *J. Mater. Chem. A* **2014**, *2*, 1144.
129. Walter, M.; Kravchyk, K. V.; Ibáñez, M.; Kovalenko, M. V. *Chem. Mater.* **2015**, *27*, 7452.
130. Levi, E.; Gofer, Y.; Aurbach, D. *Chem. Mater.* **2010**, *22*, 860.
131. Gautam, G. S.; Canepa, P.; Malik, R.; Liu, M.; Persson, K.; Ceder, G. *Chem. Commun.* **2015**, *51*, 13619.
132. Liu, M.; Rong, Z.; Malik, R.; Canepa, P.; Jain, A.; Ceder, G.; Persson, K. A. *Energy Environ. Sci.* **2015**, *8*, 964.
133. Smeu, M.; Hossain, M. S.; Wang, Z.; Timoshevskii, V.; Bevan, K. H.; Zaghib, K. *J. Power Sources* **2016**, *306*, 431.
134. Arroyo-de Dompablo, M. E.; Krich, C.; Nava-Avendano, J.; Palacin, M. R.; Barde, F. *Phys. Chem. Chem. Phys.* **2016**, *18*, 19966.
135. Datta, D.; Li, J.; Shenoy, V. B. *ACS Appl. Mater. Interfaces* **2014**, *6*, 1788.
136. Er, D.; Li, J.; Naguib, M.; Gogotsi, Y.; Shenoy, V. B. *ACS Appl. Mater. Interfaces* **2014**, *6*, 11173.
137. Amatucci, G. G.; Badway, F.; Du Pasquier, A.; Zheng, T. *J. Electrochem. Soc.* **2001**, *148*, A930.
138. Bervas, M.; Klein, L. C.; Amatucci, G. G. *Solid State Ionics* **2005**, *176*, 2735.
139. Hayashi, M.; Arai, H.; Ohtsuka, H.; Sakurai, Y. *J. Power Sources* **2003**, *119*, 617.
140. Wang, R. Y.; Wessells, C. D.; Huggins, R. A.; Cui, Y. *Nano Letters* **2013**, *13*, 5748.
141. Lipson, A. L.; Pan, B.; Lapidus, S. H.; Liao, C.; Vaughey, J. T.; Ingram, B. J. *Chem. Mater.* **2015**, *27*, 8442.

142. Shiga, T.; Kondo, H.; Kato, Y.; Inoue, M. *J. Phys. Chem. C* **2015**, *119*, 27946.
143. Tojo, T.; Sugiura, Y.; Inada, R.; Sakurai, Y. *Electrochimica Acta* **2016**, *207*, 22.
144. Padigi, P.; Goncher, G.; Evans, D.; Solanki, R. *J. Power Sources* **2015**, *273*, 460.
145. He, Y.; Gu, M.; Xiao, H.; Luo, L.; Shao, Y.; Gao, F.; Du, Y.; Mao, S. X.; Wang, C. *Angew. Chemie* **2016**, *128*, 6352.
146. Ponrouch, A.; Tchitchekova, D.; Frontera, C.; Bardé, F.; Dompablo, M. E. A. -d.; Palacín, M. R. *Electrochem. Commun.* **2016**, *66*, 75.
147. See, K. A.; Gerbec, J. A.; Jun, Y. -S.; Wudl, F.; Stucky, G. D.; Seshadri, R. *Adv. Energy Mater.* **2013**, *3*, 1056.
148. Cabello, M.; Nacimiento, F.; González, J. R.; Ortiz, G.; Alcántara, R.; Lavela, P.; Pérez-Vicente, C.; Tirado, J. L. *Electrochem. Commun.* **2016**, *67*, 59.
149. Zhao, Y.; VanderNoot, T. J. *Electrochim. Acta* **1997**, *42*, 3.
150. Plotnikov, V. A. *J. Russ. Phys. Chem. Soc.* **1899**, *31*, 1020.
151. Brown, H. C.; Wallace, W. J. *J. Am. Chem. Soc.* **1953**, *75*, 6265.
152. Galova, M. *Surface Technology* **1980**, *11*, 357.
153. Eley, D. D.; King, P. J. *J. Chem. Soc.* **1952**, 2517.
154. Capuano, G. A.; Davenport, W. G. *J. Electrochem. Soc.* **1971**, *118*, 1688.
155. Capuano, G. A.; Davenport, W. G. *Plating* **1973**, *60*, 251.
156. Peled, E.; Gileadi, E. *Plating* **1975**, *62*, 342.
157. Ziegler, K.; Gellert, H. G.; Zosel, K.; Lehmkuhl, W.; Pfohl, W. *Angew. Chemie* **1955**, *67*, 424.
158. Ziegler, K.; Lehmkuhl, H. *Z. Anorg. Allg. Chem.* **1956**, *283*, 414.
159. Couch, D. E.; Brenner, A. *J. Electrochem. Soc.* **1952**, *99*, 234.
160. Ishibashi, N.; Yoshio, M. *Electrochim. Acta* **1972**, *17*, 1343.
161. Yoshio, M.; Ishibashi, N. *J. Appl. Electrochem.* **1973**, *3*, 321.
162. Nakayama, Y.; Senda, Y.; Kawasaki, H.; Koshitani, N.; Hosoi, S.; Kudo, Y.; Morioka, H.; Nagamine, M. *Phys. Chem. Chem. Phys.* **2015**, *17*, 5758.
163. Torsi, G.; Mamantov, G. *Inorg. Chem.* **1972**, *11*, 1439.
164. Robinson, J.; Bugle, R. C.; Chum, H. L.; Koran, D.; Osteryoung, R. A. *J. Am. Chem. Soc.* **1979**, *101*, 3776.
165. Robinson, J.; Osteryoung, R. A. *J. Am. Chem. Soc.* **1979**, *101*, 323.
166. Robinson, J.; Osteryoung, R. A. *J. Am. Chem. Soc.* **1980**, *102*, 4415.
167. Gale, R. J.; Osteryoung, R. A. *Inorg. Chem.* **1979**, *18*, 1603.
168. Gale, R. J.; Osteryoung, R. A. *J. Electrochem. Soc.* **1980**, *127*, 2167.
169. Lai, P. K.; Skyllas-Kazacos, M. *J. Electroanal. Chem. Interfacial Electrochem.* **1988**, *248*, 431.
170. Carlin, R. T.; Crawford, W.; Bersch, M. *J. Electrochem. Soc.* **1992**, *139*, 2720.
171. Liao, Q.; Pitner, W. R.; Stewart, G.; Hussey, C. L.; Stafford, G. R. *J. Electrochem. Soc.* **1997**, *144*, 936.
172. Jiang, T.; Chollier Brym, M. J.; Dubé, G.; Lasia, A.; Brisard, G. M. *Surf. Coat. Technol.* **2006**, *201*, 1.

173. Jones, S. D.; Blomgren, G. E. *J. Electrochem. Soc.* **1989**, *136*, 424.
174. Rogosic, J.; Sadoway, D. R. 2012.
175. Geng, L.; Lv, G.; Xing, X.; Guo, J. *Chem. Mater.* **2015**, *27*, 4926.
176. Jayaprakash, N.; Das, S. K.; Archer, L. A. *Chem. Commun.* **2011**, *47*, 12610.
177. Reed, L. D.; Menke, E. *J. Electrochem. Soc.* **2013**, *160*, A915.
178. Chiku, M.; Takeda, H.; Kunisawa, T.; Higuchi, E.; Inoue, H. *Meeting Abstr.* **2013**, *MA2013-02*, 521.
179. Wang, H.; Bai, Y.; Chen, S.; Luo, X.; Wu, C.; Wu, F.; Lu, J.; Amine, K. *ACS Appl. Mater. Interfaces* **2014**, *7*, 80.
180. Wang, W.; Jiang, B.; Xiong, W.; Sun, H.; Lin, Z.; Hu, L.; Tu, J.; Hou, J.; Zhu, H.; Jiao, S. *Sci. Rep.* **2013**, *3*.
181. Liu, S.; Hu, J. J.; Yan, N. F.; Pan, G. L.; Li, G. R.; Gao, X. P. *Energy Environ. Sci.* **2012**, *5*, 9743.
182. He, Y. J.; Peng, J. F.; Chu, W.; Li, Y. Z.; Tong, D. G. *J. Mater. Chem. A* **2014**, *2*, 1721.
183. Levitin, G.; Yarnitzky, C.; Licht, S. *Electrochem. Solid-State Lett.* **2002**, *5*, A160.
184. Vatsala Rani, J.; Kanakaiah, V.; Dadmal, T.; Rao, M. S.; Bhavanarushi, S. *J. Electrochem. Soc.* **2013**, *160*, A1781.
185. Lin, M. -C.; Gong, M.; Lu, B.; Wu, Y.; Wang, D. -Y.; Guan, M.; Angell, M.; Chen, C.; Yang, J.; Hwang, B. -J.; Dai, H. *Nature* **2015**, *520*, 324.
186. Sun, H.; Wang, W.; Yu, Z.; Yuan, Y.; Wang, S.; Jiao, S. *Chem. Commun.* **2015**, *51*, 11892.
187. Nie, P.; Shen, L.; Luo, H.; Ding, B.; Xu, G.; Wang, J.; Zhang, X. *J. Mater. Chem. A* **2014**.
188. Lu, Y.; Wang, L.; Cheng, J.; Goodenough, J. B. *Chem. Commun.* **2012**, *48*, 6544.
189. Liu, S.; Pan, G. L.; Li, G. R.; Gao, X. P. *J. Mater. Chem. A* **2015**, *3*, 959.
190. Pickup, P. G.; Osteryoung, R. A. *J. Electroanal Chem. Interfacial Electrochem* **1985**, *195*, 271.
191. Janiszewska, L.; Osteryoung, R. A. *J. Electrochem. Soc.* **1987**, *134*, 2787.
192. Tang, J.; Osteryoung, R. A. *Synth. Met.* **1991**, *45*, 1.
193. Takeishi, K.; Koura, N. *Denki Kagaku* **1995**, *63*, 947.
194. Ui, K.; Kuma, Y.; Koura, N. *Electrochemistry* **2006**, *74*, 536.
195. Koura, N.; Ejiri, H.; Takeishi, K. *Denki Kagaku (1961?€"1998)* **1991**, *59*, 74.
196. Koura, N.; Ejiri, H.; Takeishi, K. *J. Electrochem. Soc.* **1993**, *140*, 602.
197. Hudak, N. S. *J. Phys. Chem. C* **2014**, *118*, 5203.
198. Cheng, F.; Chen, J. *Chem. Soc. Rev.* **2012**, *41*, 2172.
199. Egan, D. R.; Ponce de Leon, C.; Wood, R. J. K.; Jones, R. L.; Stokes, K. R.; Walsh, F. C. *J. Power Sources* **2013**, *236*, 293.
200. Zaromb, S. *J. Electrochem. Soc.* **1962**, *109*, 1125.
201. Bockstie, L.; Trevethan, D.; Zaromb, S. *J. Electrochem. Soc.* **1963**, *110*, 267.
202. Mokhtar, M.; Talib, M. Z. M.; Majlan, E. H.; Tasirin, S. M.; Ramli, W. M. F. W.; Daud, W. R. W.; Sahari, J. *J. Ind. Eng. Chem.* **2015**, *32*, 1.
203. Jin, W.; Du, H.; Zheng, S.; Xu, H.; Zhang, Y. *J. Phys. Chem. B* **2010**, *114*, 6542.

204. El Shayeb, H. A. E.; Wahab, F. M. A. E.; Abedin, S. Z. E. *J. Appl. Electrochem.* **1999**, *29*, 473.
205. El Shayeb, H. A.; Abd El Wahab, F. M.; Zein El Abedin, S. *Cor. Sci.* **2001**, *43*, 655.
206. Nestoridi, M.; Pletcher, D.; Wood, R. J. K.; Wang, S.; Jones, R. L.; Stokes, K. R.; Wilcock, I. *J. Power Sources* **2008**, *178*, 445.
207. Gudic, S.; Smoljko, I.; Kliskic, M. *Mater. Chem. Phys.* **2010**, *121*, 561.
208. Revel, R.; Audichon, T.; Gonzalez, S. *J. Power Sources* **2014**, *272*, 415.
209. Gelman, D.; Shvartsev, B.; Ein-Eli, Y. *J. Mater. Chem. A* **2014**, *2*, 20237.
210. Reboul, M. C.; Gimenez, P.; Rameau, J. J. *Corrosion* **1984**, *40*, 366.
211. Doche, M. L.; Rameau, J. J.; Durand, R.; Novel-Cattin, F. *Cor. Sci.* **1999**, *41*, 805.
212. Hunter, J. A., University of Oxford, 1989.
213. Lee, K. -K.; Kim, K. -B. *Cor. Sci.* **2001**, *43*, 561.
214. Tuck, C. D. S.; Hunter, J. A.; Scamans, G. M. *J. Electrochem. Soc.* **1987**, *134*, 2970.
215. Macdonald, D. D.; English, C. *J. Appl. Electrochem.* **1990**, *20*, 405.
216. Doche, M. L.; Novel-Cattin, F.; Durand, R.; Rameau, J. J. *J. Power Sources* **1997**, *65*, 197.
217. Albert, I. J.; Kulandainathan, M. A.; Ganesan, M.; Kapali, V. *J. Appl. Electrochem.* **1989**, *19*, 547.
218. Kapali, V.; Venkatakrishna Iyer, S.; Balaramachandran, V.; Sarangapani, K. B.; Ganesan, M.; Anbu Kulandainathan, M.; Sheik Mideen, A. *J. Power Sources* **1992**, *39*, 263.
219. Patnaik, R. S. M.; Ganesh, S.; Ashok, G.; Ganesan, M.; Kapali, V. *J. Power Sources* **1994**, *50*, 331.
220. Abdel-Gaber, A. M.; Abd-El-Nabey, B. A.; Saadawy, M. *Cor. Sci.* **2009**, *51*, 1038.
221. Abdel-Gaber, A. M.; Khamis, E.; Abo-Eldahab, H.; Adeel, S. *Mater. Chem. Phys.* **2010**, *124*, 773.
222. Al-Rawashdeh, N. A. F.; Maayta, A. K. *Anti-Corros. Meth. Mater.* **2005**, *52*, 160.

6 Lithium Polymer Electrolytes and Batteries

Gebrekidan Gebresilassie Eshetu,*,†,‡ Michel Armand†,§ and Stefano Passerini*

*Helmholtz Institute Ulm (HIU), Electrochemical Energy Storage, Karlsruhe Institute of Technology, Ulm, Germany
†CICEnergiGUNE, Albert Einstein 48, 0150, Minano, Spain
‡Department of Chemistry,
College of Natural and Computational Sciences,
Mekelle University, Mekelle, P.O. Box-231, Ethiopia
§marmand@cicenergigune.com

Table of Contents

List of Abbreviations

PEO	poly(ethylene oxide)
PPO	poly(propylene oxide)
PMMA	poly(methyl methacrylate)
PAN	polyacrylonitrile
PVDF	poly(vinylidene fluoride)
HFP	hexafluorpropylen
PVC	poly(vinyl chloride)
PCL	polycaprolactone
MEEP	poly[bis(2-(2-methoxyethoxy) ethoxy)phosphazene]
PEG	polyethylene glycol
TrFE	trifluoroethylene
PFPE	perfluorpolyether
PEI	polyethylenimine
PUA	poly(urethane acrylate)
PE	polyethylene
MEEGE	2-(2-methoxyethoxy) ethyl glycidyl ether]
PTMC	poly(trimethylene carbonate)
SPE	solid polymer electrolyte
CPE	composite polymer electrolyte
$LiPF_6$	lithium hexafluorophosphate
$LiBF_4$	lithium tetrafluoroborate
$LiClO_4$	lithium perchlorate
LiTFSI	lithium bis(trifluoromethanesulfonyl)imide
EC	ethylene carbonate
DMC	dimethyl carbonate
DEC	diethyl carbonate
EIS	electrochemical impedance spectroscopy
EMC	ethyl methyl carbonate

Mw	Molecular weight
PC	propylene carbonate
$[BMI][PF_6]$	1-butyl-3-methylimidazolium hexafluorophosphate
PAMPSLi	poly(lithium 2-acrylamido-2-methyl propanesulfonate)
TCM	tricyanomethanide
SEI	solid electrolyte interphase
T_{Li+}	transference number
WoS	web of science
XPS	X-ray photoelectron spectroscopy

I. Introduction to Lithium-Polymer Electrolytes and Batteries

There is a huge global demand toward the development of safer and advanced energy storage technologies with higher energy and power densities to meet the emerging innovative large-scale applications, such as in the electro mobility (xEVs: EVs/HEVs/PHEVs), renewable energies (solar and wind), and in utility/grid management (e.g., load leveling).[1–3] The impetus behind such transition aims to improve fuel economy, reduce dependence on exhaustible fossil fuels, trim down emission of greenhouse gases to mitigate global warming, and ultimately protect the deteriorating environment and its occupants.[4,5] In this regard, lithium-based rechargeable batteries are very promising candidates for the abovementioned applications because of their high gravimetric and volumetric energies with respect to other cell chemistries.[5–7] Conventional Li-ion rechargeable batteries utilize organic liquid electrolytes; hence they pose potential thermal and chemical threats associated to the electrolytes' flammability/combustibility. Thus, despite the remarkable progress in the research and development over the last two decades, the stringent safety requirements remain a major hurdle, slowing down their expansion for the foreseen staggering energy storage demands in general, and that of the electro mobility sector in particular.[8] Henceforth, substantial improvements are vital with regard to the prevailing conventional lithium battery technology so as to fully electrify societies' transportation. Among all battery components, the electrolyte is viewed as the prime contributor to the safety and thereby hindering the mass deployment of the Li-based rechargeable battery systems.[9] Accordingly, much research has been focused on investigating new and/or optimizing electrolyte systems that could bypass the abovementioned intrinsic limitations of organic liquid electrolytes.[2] One of the viable strategies consists in replacing the existing state-of-the-art, alkyl carbonate-based liquid electrolytes, with solid electrolytes, enabling the building up of intrinsically safer, possibly eco-friendly lithium battery systems. Ideal technological candidates for achieving such tough requirements are polymer electrolytes, leading to the fabrication of all solid-state Li-based batteries. The use of polymer electrolytes, in fact, avoids cell leakage and strongly reduces the risk of flame/combustion (main shortcomings of organic solvent-based liquid electrolytes), offers flexible battery design, satisfactory mechanical stability, high temperature use, easier manipulation (including fabrication of ultrathin films), and continuous cell manufacturing process, for instance, by using extrusion/lamination technology with large gains in productivity and cost.

Herein, a detailed survey in the evolution and status of the research on polymer electrolytes and their industrial-scale applications, as electrolytes in practical large-scale batteries is reported. Moreover, thoughts on the existing challenges and future outlooks are presented.

II. Lithium-Polymer Electrolytes

The launch of polymer electrolytes, as new solid-state materials for use in electrochemical energy storage devices, has been one important aspect in solid-state electrochemistry, as witnessed by the massive research work. Although the complexing ability of these polymers (for instance, oligoethers) had been known for some time, the first measurement of the ionic conductivity in polymer–Na salt complexes was carried out by Wright *et al.*[10] in 1973. A few years later, Armand *et al.* realized that the polymer/LiX complexes could be deployed as solid electrolytes well matching the intercalation electrodes for lithium batteries that had been introduced in the early 1970s.[11] These earliest polymer electrolytes consist of a salt dissolved in a high molecular weight (MW) polymer matrix, leading to a true solid solution.[12] A bibliometric analysis of the publication trends on polymer electrolytes using the web of science (WoS) was carried using title, topic, abstract, or key words, starting since 1973. The number of published materials in the past 40 years clearly reveals the steady growth of the research work in the field of polymer electrolytes (Figure 1). Besides,

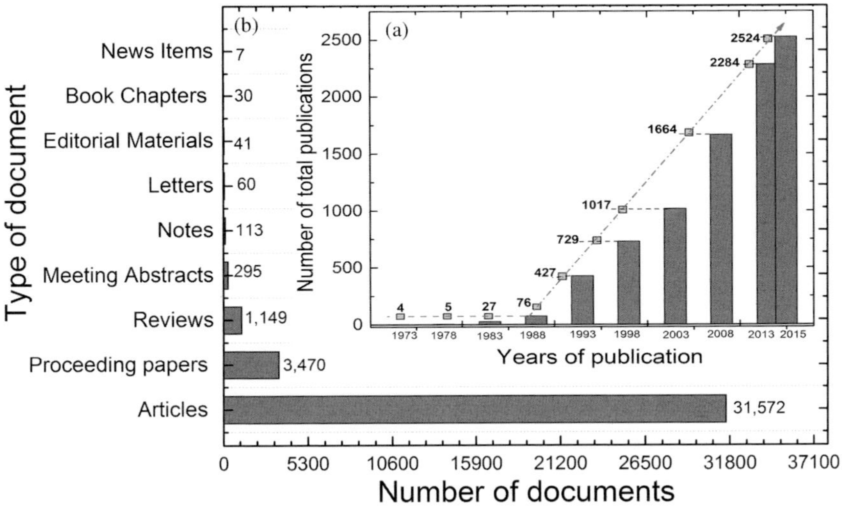

Figure 1. (a) Number of publications on polymer electrolytes for Li-rechargeable batteries and (b) type (diversity) of published materials on polymer electrolytes. (Data accessed from WoS on November 23, 2016.)

the analysis conveys the existence of wide-range diversity in the published materials, including articles, reviews, proceeding papers, and so on.

To act as a successful host matrix by effectively solvating the lithium salt, the polymer, or active part of a copolymer should generally possess some characteristics:

(i) Electron pair donor site(s): host polymers should possess high concentration of polar atoms or group of atoms on their backbone, such as ether (–O–), sulfide (–S–), amine (–N–), carbonyl (C=O), phosphine (–P–), and cyano (C≡N), capable of acting as ligands for coordinating the cations of the dissolved salt, by providing the necessary solvation energy for the polymer–lithium complex.

(ii) High dielectric constant (ε): host polymers should possess high dielectric constant to assist a greater ionization of lithium salt and thus provide a high concentration of charge carriers.

(iii) Appropriate distance between the coordinating centers because the formation of multiple intrapolymer ion bonds appears to be crucial.

(iv) A flexible backbone and lower barrier to bond rotation so that the segmental motion of the polymer chain can takes place readily.

The preparation of polymer electrolytes requires at least one host polymer acting as a base matrix prior to the incorporation of any other ingredients. Extensive research efforts have been devoted toward exploring several polymer hosts for practical applications in Li-based rechargeable batteries, including poly(ethylene oxide) (PEO),[12–21] poly(propylene oxide) (PPO),[22–24] poly(methyl methacrylate) (PMMA),[2,25,26] polyacrylonitrile (PAN),[27,28] poly(vinylidene fluoride) (PVDF)[29,30] and its copolymers with hexafluoropropylene (HFP), poly(vinyl chloride) (PVC),[2] polyesters-like polycaprolactone-based polycarbonates,[31–33] and so on. As mentioned above, the simplest polymer electrolyte (a.k.a. binary polymer electrolyte) is composed of two components: (1) the solid solvent (polymer matrix) and (2) the active ion source (e.g., lithium salt). Both critically affect the ionic conductivity and other add-on characteristics of polymer electrolytes. Thus, not all polymers can serve appropriately. Strict requirements exist also for the lithium conducting salt. The first and most important requirement is its solubility in the polymer matrix; a direct reflection of its capability to dissociate and thereby supply free ions. Obviously, it is the Li^+ cation mobility that really matters, however, in general, the bulkier is the salt anion; the higher is the resulting ionic conductivity, inferring that a salt anion with a more delocalized negative charge and low basicity is preferred for enhancing the conductivity. In this regard, the most widely used lithium salts are those with the ClO_4^-, AsF_6^-, $CF_3SO_3^-$, BF_4^-, and $N(CF_3SO_2)_2^-$) anions. Their associated ions mobility and dissociation constants vary in the following orders[34]:

$$\text{Li}^+ \text{ ion mobility: } LiBF_4 > LiClO_4 > LiPF_6 > LiAsF_6 > LiCF_3SO_3 > LiN(CF_3SO_2)_2;$$

$$\text{Dissociation constant: } LiN(CF_3SO_2)_2 > LiAsF_6 > LiPF_6 > LiClO_4 > LiBF_4 > LiCF_3SO_3.$$

Nevertheless, it should be mentioned that being the Li^+ cation coordinated by the polymer is usually the less mobile ionic species.

An overview of the most commonly employed polymer matrices for use in polymer electrolytes, along with their physicochemical properties, is presented in Table 1.

Table 1. Commonly investigated polymers and their physicochemical characteristics[2,35,36]

Polymer Matrix	Polymer Unit	T_g/°C	T_m/°C	Dielectric Constant (ε)[a]	Li^+-Solvating Site
PEO	$-(CH_2CH_2O)_n-$	−64	65	10–15	$-(CH_2CH_2O)_n-$
PPO	$-(CH(CH_3)CH_2O)_n-$	−60	[b]		$-(CH(-CH_3)CH_2O)_n-$
PAN	$-(CH_2CH(CN))_n-$	125	317	5.5	C≡N
PMMA	$-(CH_2C(CH_3)(CO_2CH_3))_n-$	120	[b]	3.0	$C{=}O_2$
PVDF	$-(CF_2CH_2)_n-$	−30	171	8.2–10.5	C–F[c]
P(VDF–HFP)	$-(CF_2CH_2)_n-(CF_2CF(CF_3))_m-$	−90	135	7.9–10.0	C–F[c]
PVC	$-(CH_2CHCl)_n-$	80	220	3.3–3.9	—
PCL	$-(CH_2)_5C(=O)O-$	−60	65	3.3	C=O

[a] At 1kHz, [b] amorphous, [c] contributes to anodic stability, but does not solvate Li^+-cations. HFP = hexafluorpropylen, PAN = polyacrylonitrile, PCL = polycaprolactone, PEO = poly (ethylene oxide), PMMA = poly (methyl methacrylate), PPO = poly (propylene oxide), PVC = poly (vinyl chloride), PVDF = poly (vinylidene fluoride), VDF = vinylidene fluoride.

It should also be mentioned that some of these polymers are unable to fully dissociate, that is, dissolve, the lithium salt. For such a reason, molecular or ionic (molten salts) solvents have been frequently added to form ternary or quaternary systems (see later).

A. Basic Requirements of Polymer Electrolytes

From the practical application point of view, a polymer electrolyte for rechargeable Li-based batteries should inherently possess high ionic conductivity[27] and, possibly, high lithium transference number ($T_{Li}{}^+$) in order to minimize concentration polarizations during the charge/discharge process and thus yield higher power density and energy efficiencies. Ideally, the Li^+ (cationic) transference number should be close to unity. Anchoring the anions to the polymer backbone or introducing receptors selectively blocking or reducing the mobility of anions can greatly increase the Li^+ transference number of the electrolyte system.[36–40]

Polymer electrolytes should offer good mechanical performance to comply not only with the electrodes' volume changes upon cell operation, but also with all processes involved in the cell manufacturing, transport, and storage. Other important requirements are the thermal, chemical, and electrochemical stabilities enabling operation of the cell under safe conditions.

B. Categories of Polymer Electrolytes and Accompanying Historical Evolution

Since their inception in the early 1970s, nearly 20 years prior to the commercialization of lithium-ion battery (LIB) by Sony, the research progress in polymer electrolytes has passed *via* several developmental stages.[41,42] Moreover, since the term polymer can be applied to a large family of structural systems, there is lack of a universally accepted classification standard of the polymer electrolytes. However, based on the physical state (morphology) and composition, three main families can be identified: (1) solid polymer electrolytes (SPEs), (2) plasticized polymer electrolytes, and (3) composite polymer electrolytes (CPEs).

These polymer electrolyte families are characterized by rather different characteristics with respect to their use in Li-rechargeable batteries, for example, ease of ion transport, quality of electrode/polymer interfaces, electrochemical, mechanical, and thermal stabilities, low-temperature performance, and global safety requirements (i.e., thermal and chemical threats).

PEs may contain some additives to boost their property, for example, ion mobility, that is, ionic conductivity. These additives can consist of molecular (organic) liquids, ionic liquids (ILs), inorganic fillers, or their combinations. The effect of such additives on the polymer electrolyte properties is summarized in Table 2.

Aside from the discussion presented below, there are many classical as well as historical review papers[2,27,43–86] and books[7,12,13,17–21,87–103] reporting the evolution, governing properties, and characteristics of polymer electrolytes to which the readers are recommended to refer for more details.

1. *Solid Polymer Electrolytes*

Ever since Armand *et al.* verified the viability of the technological application of polyether(PEO)/lithium salt complexes in electrochemical devices, SPEs have been extensively studied for applications in electrochromic and

Table 2. Overview of polymer electrolytes, their optimized versions, and their physicochemical properties

	Polymer-Based Electrolyte Systems				
Electrolyte Properties	SPEs	GPEs	SPE–IL	SPE–OL–IL	CPEs
Ionic conductivity	o^{-}	o	o*	o*	o^{+}
Interfacial properties	o^{-}	o	o*	o*	o^{+}
Electrochemical stability	o^{+}	o^{-}	o*	o*	o^{+}
Low temperature performance	o^{-}	o	o*	o*	o^{-}
Thermal stability	o^{+}	o^{-}	o*	o*	o^{+}
Mechanical strength	o^{+}	o	o	o	o^{++}
Safety	o^{+}	o	o*	o*	o^{+}

o^{++} = Very high, o^{+} = high/good; o = medium; o^{-} = poor, o* = composition and condition dependent, SPE = solid polymer electrolytes, GPE = gel-polymer electrolyte, IL = ionic liquid, OL = organic liquid, CPE = composite polymer electrolyte.

electrochemical devices such as solid-state batteries, fuel cells, solar cells, electrochemical capacitors, dye-sensitized solar cells (DSSCs), actuators, electrochemical sensors, analog memory devices, and thermoelectric generators.[84] The advent of such all solid-state electrolytes exemplify a breakthrough in the field of lithium battery electrolytes, offering improved safety and engineering benefits such as mechanical and design flexibility, possibility to be fabricated into desired thicknesses and shapes, not allowed with liquid electrolytes.

SPEs, above defined as a lithium salt dissolved solely in a polymer, are also called "dry systems." The lithium salt dissolves in the polymer matrix, because Li^{+} cations are strongly coordinated by electron donor atoms on the polymer chain. Additionally, they form ion pairs with the anions. The lithium-ion conduction results from hopping of lithium ions along the polymer chains (intrachain mobility) and, taking advantage from the polymer's segmental motion, between the polymer chains (interchain mobility).[35,67,104] The Mw of the polymer and the nature of the lithium salt anion play a key role in governing the global lithium-ion conductivity.[35,104]

Polyethers appear to be the best candidates for these electrolytes polymer matrices based on their ability to dissolve lithium salts and their

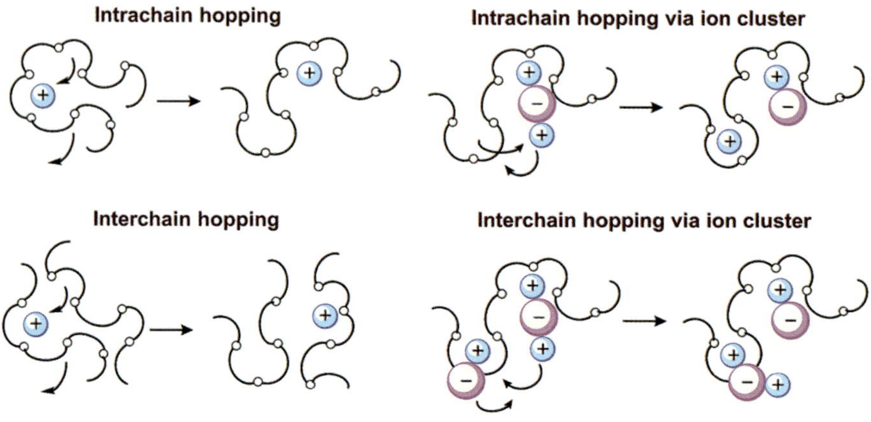

Figure 2. Ion transport mechanism in PEO-based solid polymer electrolyte. Taken from Ref. 48.

electrochemical stability window.[51] The ion transport in PEO-based electrolytes is the most investigated one among all polymer electrolyte systems.[83] The ether oxygen in the polymer unit (–(CH_2CH_2O)–) has a high donor number, that is, optimal for Li^+ dissociation, and high chain flexibility resulting in rapid ion transport. Figure 2 shows the most common ion transport mechanism in PEO matrix, in which lithium ions are coordinated by several ether oxygens (–O–) of the PEO chain in a similar fashion to their complexation in alkyl carbonate solvents. Thus, ion transport is facilitated by intra- or interchain hopping *via* the Li–O bonds breaking/forming.[105]

At room temperature (RT), PEO/LiX electrolytes are usually composed of crystalline (e.g., PEO and $P(EO)_6LiX$) and amorphous phases, resulting in two distinct ion transport mechanisms. These are ion hopping (rather low conductivity, i.e., $\leq 10^{-6}$ S cm^{-1}) from (defective) sites in crystalline regions and migration of ionic species (much faster process, $>10^{-4}$ S cm^{-1}) in the amorphous regions, associated to micro-Brownian segmental motion of the PEO chains.

The PEO/LiTFSI system is the most studied among several others, for example, PEO/LiI or PEO/$LiClO_4$, as it provides an amorphous phase at the lowest temperatures.[106]

Despite the inherent safety of such solvent-free, polymer electrolytes, their ionic conductivity remains usually in the range of 10^{-8}–10^{-3} S cm^{-1},

Table 3. Ionic conductivities of some typical dry solid polymer electrolytes

Polymer Matrix	Li-Salt	Temp (°C)	Ionic Conductivity ($S\ cm^{-1}$)	Reference
$P(EO)_{20}$	$LiBF_4$	27	6.372×10^{-7}	107
PEO	5% $LiPF_6$	25	1.20×10^{-6}	108
$P(EO)_{20}$	$LiClO_4$	27	2.28×10^{-7}	107
$P(EO)_{24}$	$LiN(CF_3SO_2)_2$	50	3.84×10^{-4}	109
MEEP	25% $LiCF_3SO_3$	30	2.70×10^{-5}	110
PEO/PVDF	LiTFSI	30	4.9×10^{-3}	111
PEO/PVDF-TrFE	$LiClO_4$	25	7.0×10^{-4}	112
PEG/PFPE	LiTFSI	30	$\sim 10^{-4}$	113
PEO/PEI	$LiPF_6$	25	$\sim 10^{-4}$	114

the latter being achieved only at higher temperature, around 80°C–100°C. Thus, conductivities fitting to the requirement for practical applications can be only achieved in the range of 60°C–80°C.[48]

A short list of typical solvent-free, polymer electrolytes is given in Table 3, together with their ionic conductivity near RT, showing their main limitation.

Such a low RT ionic conductivity of dry polymers has put pressure on the scientific community, resulting in a variety of research efforts. The approaches include modification of the host polymer (*via* developing branched and cross-linked polymers, blended polymer matrices, etc.) and the salt's anion (e.g., large anion with delocalized charge such as LiTFSI, LiBETI, and LiBOB) mainly to hinder the polymer chain crystallization and thereby extending the domain of the amorphous phases. Addition of micro- and nanosized fillers, for example, ceramic and other inorganic materials, and, especially, the incorporation of plasticizers (e.g., alkyl carbonates and ILs) into the host polymer matrix have also been pursued.

2. *Gel-polymer Electrolytes*

Amid the various approaches to address the RT ionic conductivity limitation of SPEs, the incorporation of liquid plasticizers is certainly the most common. It involves the addition of low MW (MW = 400–500) polar polymers, organic alkyl carbonates like ethylene carbonate (EC), dimethyl carbonate (DMC), diethylene carbonate (DEC), propylene carbonate (PC), ethyl methyl carbonate (EMC), leading to ionic conductors commonly known as gel-polymer electrolytes (GPEs).

In these materials, a liquid solution (composed of lithium salt, e.g., $LiPF_6$, $LiClO_4$, and aprotic polar solvents, e.g., EC, DMC, DEC, EMC, and PC) is entrapped in a polymeric matrix, which may or may not interact with the lithium salt itself. Thus, the dielectric constant of the plasticizer dictates the degree of salt dissociation. Accordingly, the conductivity increases in the order of DEC < DMC < PC < EC. That is, the maximum conductivity value corresponds to EC because of its high dielectric constant as compared to the other plasticizers.

GPEs keep some of the beneficial features of polymer matrix, that is, good mechanical and structural stabilities, high flexibility, reduced leakage, and improved operation safety, but are subjected to solvent evaporation and chemical reactivity at the electrode interface. However, they inherit the better conductivity of organic liquids.[2] In this regard, GPEs share both diffusive and cohesive properties of liquids and solids, respectively.[35] Moreover, the encapsulation of liquid electrolytes in solid polymer host matrix enhances the electrode/electrolyte interfaces, especially with graphite negative electrodes. Such kinds of polymer electrolytes, as schematized in Figure 3, are characterized by polymer host (e.g., PEO)–solvent (e.g., EC)–lithium salt (e.g., LiTFSI) interactions, dictating the overall properties of the ternary systems. Depending on the nature of the host polymer matrix, the polymer–solvent–Li^+ interactions can be classified as strong or weak. From infrared spectroscopy, the solvating ability of polymer matrix toward Li^+ was found to decrease in the following order of PEO > PMMA > PAN >> P(VDF-HFP).[115] Irrespective of the dielectric constant, the observed tendency is found strongly correlated to the polymer donor number. Thus, polymer matrices such as PEO and PMMA result in strong interactions, attributed to the interaction of Li^+ with electron donating atoms/group of atoms.

A number of researchers have extensively investigated gel polymers of high MW PEO, obtaining very high ionic conductivities approaching 10^{-3} S cm^{-1} even at lower temperatures (−10°C).[116–121] Also, Appetecchi *et al.*, reported fast transport properties resulting in ionic conductivities in the

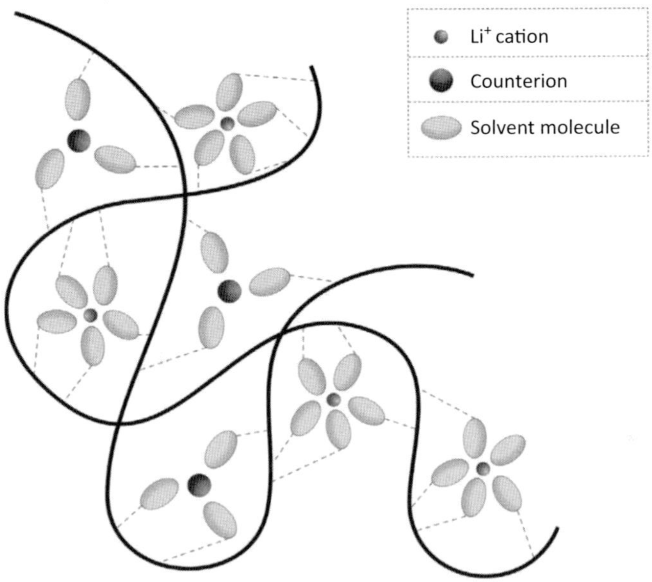

Figure 3. Schematic representation of a gel-polymer electrolyte. Taken from Ref. 12.

order of 4 × 10^{-3} S cm^{-1} at RT in combination with very high electrochemical stability window (>5 V) for PAN-based GPEs.[122] Although exhibiting lower ion transport properties, PMMA showed higher chemical stability attributed to the stronger polymer–solvent interactions with respect to PAN, in line with the abovementioned DN-based trend.[123–126] The thermally cross-linked poly(fluoroalkyl-siloxane)-ethylene oxide matrix demonstrated ion conductivities higher than 10^{-3} S cm^{-1} at ambient temperature, while at same time exhibiting higher electrochemical stability (>5 V) and superior mechanical properties.[127,128] For further conductivity improvement of such electrolyte systems keeping the amount of the plasticizer at lower amount, a number of strategies have been employed, such as replacing the "inactive" polymer matrix with a (co)polymer, which itself contributes to the ion transport. This includes (co)polymers containing polar electron donor species such as oxygen, –O– (e.g., PEO, PPO, PMMA or nitrogen, –N– (e.g., PAN), and so on.[35,36,67,85,86] For instance, the highly polar C≡N side chain can participate in the solvation of Li^+ through dipole interactions, rendering the polymer a very suitable gel-forming polymer matrix with ionic conductivities in the range of 10^{-4} to 10^{-2} S cm^{-1}.[35,36,61]

However, although there are almost unlimited possibilities of designing polymers and/or copolymers, the final products do not completely fulfill the abovementioned requirements as SPEs do.[2] In particular, PEO and PMMA-based GPEs suffer of poor mechanical properties, although these latter can be improved by the addition of nanosize SiO_2.[117]

In polymer electrolytes exhibiting weak polymer–solvent and polymer–salt interactions, the polymer does not contribute to the lithium salt dissolution/dissociation.[35,129] In this regard, PVDF and its analogs are the worthiest polymers to mention. Since the first report of GPEs based on polyvinyl formal of PVDF copolymer swollen with liquid electrolyte in 1975 by Feuillade and Perche,[130] they have attracted a great deal of attention of both the scientific and industrial communities. Tarascon *et al.* developed a GPE based on poly (vinylidene difluoride-co-hexafluropropylene) (PVDF-HFP) swollen with a solution of $LiPF_6$ in EC/DMC, out of which the first lab-scale practical Li-ion plastic battery was fabricated.[129,131] Owing to the high crystallinity of linear PVDF, PVDF-HFP, or PVDF-TrFE, copolymers are usually employed to improve the ionic conductivity. Moreover, the incorporation of nanometric fillers and the design of blended polymer hosts, among many other approaches, improve the mechanical stability and ionic conductivity (well above 10^{-3} S cm^{-1}) of PVDF-based gel electrolytes. However, such electrolytes based on weak interactions are chemically instable due to easy phase separation between the gel-forming polymer host and the liquid electrolyte (plasticizer), thus resulting in liquid leakage that depletes their intended benefits.

In general, compared to SPEs, GPEs are more realistic for RT applications. However, their performance not only heavily depends on the characteristics of the plasticizer, such as the dielectric constant, donor number, but also their volatility and, thus, flammability. Table 4a presents some selected GPEs exhibiting high conductivity, that is, in the range for practical applications.

3. *IL-containing SPEs*

Although the incorporation of aprotic organic solvents into solid polymer host matrices increases ionic conductivity, the approach does not guarantee superior safety characteristics. In fact, the combustibility of GPEs is

Table 4. Ionic conductivity of some plasticized solid polymer electrolytes

Electrolyte Composition	Ionic Conductivity (S cm^{-1})	Electrochemical Window (V)	Temp (°C)	Reference
(a)				
PAN/1M $LiPF_6$/EC + DMC + DEC (1/1/1, wt%)	>1 × 10^{-3}	4.5	RT	162
PAN/1M $LiBF_4$/EC + DMC (1/1, Vol%)	2.8 × 10^{-3}	5.0	RT	163
PVDF/1M $LiPF_6$/EC + DMC (1/1, wt%)	1 × 10^{-3}	4.5	RT	164
PVDF/1M $LiClO_4$/EC + PC (1/1, Vol%)	>1.4 × 10^{-3}	>4.5	20	165
P(VDF-HFP)/1M $LiPF_6$/ EC+DMC+DEC(1/1/1,wt%)	1.43 × 10^{-3}	–	RT	166
P(VDF-HFP)/1M $LiPF_6$/EC + DEC(1/1, Vol%)	1 × 10^{-3}	4.5	RT	167
PVDF/1M $LiPF_6$/EC+DMC + DEC (1/1/1, wt%)	1 × 10^{-3}	5.0	RT	168
PEO/P(VDF-HFP)/1M LiClO4/ EC + PC (1/1, wt%)	1.25 × 10^{-3}		30	169
PEO/LiTFSI/SCN	2.9 × 10^{-3}		25	170
$(PEO)_8$/$LiCF_3SO_3$/PEG	~10^{-3}		25	171
$(PEO)_8$/$LiClO_4$/EC + PC, 20 mol%)	1 × 10^{-3}		20	77
PMMA/LiClO4/EC + PC	~10^{-3}		25	172
PAN/$LiClO_4$/EC + PC	4 × 10^{-3}		25	173
PAN/$LiClO_4$/EC + DEC	4 × 10^{-3}		RT	174
PVDF/$LiBF_4$/EC + PC	6 × 10^{-3}		RT	174
PMMA/$LiN(CF_3SO_2)_2$/EC + PC + BL	1 × 10^{-3}		RT	175
21PAN/8% $LiClO_4$/38%EC + 33% PC	1.1 × 10^{-3}		−10	176 and 177
	1.2 × 10^{-3}		0	
	1.7 × 10^{-3}		20	

(*Continued*)

Table 4. (*Continued*)

Electrolyte Composition	Ionic Conductivity (S cm^{-1})	Electrochemical Window (V)	Temp (°C)	Reference
3PEDGA/14LiClO$_4$/68%EC + 15%PC	1.2×10^{-3}		−10	176 and 177
	2.1×10^{-3}		0	
	4.1×10^{-3}		20	
(b)				
PEO/LiPF$_6$/MMPIPF$_6$	1.13×10^{-3}		25	178
PEO/LiBF$_4$/MMPIBF$_4$	2.06×10^{-3}		25	178
P(VDF-HFP)/LiN(C$_2$F$_5$SO$_2$)$_2$/EMI-N(C$_2$F$_5$SO$_2$)$_2$	$\sim 10^{-3}$	5.5	25	179
P(VDF-HFP)/LiPF$_6$/EMI-PF$_6$	1.6×10^{-3}	5.0	20	61
P(VDF-HFP)LiN(CF$_3$SO$_2$)$_2$/MPP-N(CF$_3$SO$_2$)$_2$	0.3×10^{-3}	5.75	20	180
PEOEMA/LiPF$_6$/BMI-PF$_6$	0.9×10^{-3}	4.30	25	181
PEO$_{20}$LiTFSI$_2$/[Pyr$_{12O1}$TFSI]$_2$	2.5×10^{-4}		20	182
PEO$_{20}$LiTFSI$_2$/[Pyr$_{12O1}$TFSI]$_2$	1.8×10^{-3}		60	182
PEO$_{20\%/[}$Li(G4)]TFSI$_{80\%}$	4.0×10^{-4}		20	183
PEO$_{20\%/[}$Li(G4)]TFSI$_{80\%}$	2.5×10^{-3}		60	183
P([AMPS-Li]$_1$−VF$_9$)$_{10\%[}$IM$_{12}$TCM]$_{90\%}$	5.0×10^{-3}		20	184
P([AMPS-Li]$_1$−VF$_9$)$_{10\%[}$IM$_{12}$TCM]$_{90\%}$	2.0×10^{-2}		60	184

largely determined by the high vapor pressure of the most volatile component (i.e., the organic solvent). To overcome this safety issue while maintaining their beneficial features, a wealth of research efforts has been devoted toward replacing the organic liquid plasticizer with nonvolatile, nonflammable, and electrochemically and thermally stable compounds. In this regard, ILs appear to be auspicious candidates to optimize SPEs. Ternary systems formed by the incorporation of ILs into polymer/salt systems have been reported by Watanabe *et al.* since 1995.[132–134] Depending on their chemistry and nature, ILs are endowed with negligible vapor pressure, very low (if not negligible) flammability, wide liquidus phase range (−100 to +400, avoiding possible evaporation or crystallization), wide electrochemical stability window (>5 V for certain cation–anion combinations), and so on, eventually making them suitable alternatives as plasticizers for solvent-free polymer composite electrolytes.[135–139] The detailed mechanism is not yet fully comprehended, but in strongly interacting polymer–salt complexes, it could be related to the plasticizing effect of the IL anion on the polymer matrix rather than the dissolution of the salt into the IL. As a matter of fact, the addition of ILs to SPEs leads to a significant improvement in their ionic conductivity although with transference numbers (TLi^+) still comparable to the parent polymer/salt complex (≈0.2) involved.[140,141] Such improvement can only be attributed to the plasticizing effect of the ILs, ensuring fast relaxation of the polymer chains and thus lowering the glass transition temperature (T_g).[142,143] Thus, the aptitude of ILs to modify the ion transport mechanism appears to differ from that of conventional, nonionic plasticizers.[144] Moreover, IL-incorporated polymer electrolytes exhibit wider electrochemical window and excellent thermal stability, a quite decisive requirement for practical application in batteries. Importantly, the nature of the IL and the structure of the polymer matrix play a significant role, leading to improve further the performance of IL-containing SPEs. Higher conductivities (in the range of 10^{-4}–10^{-3}) exceeding the well-known limits of "dry" SPEs, but yet lower than that of conventional liquid electrolytes were reported after incorporation of various ILs into Li-salt doped PEO and/or P(VDF-HFP).[81,141,145–149] The free energy of the IL-SPE is impacted by the nature of the IL ions, especially the anion. Fisher *et al.*[150] and Yeon *et al.*[151] showed the effect of different anions for a fixed cation and disparity of the conductivity of polymer electrolytes incorporating ILs from the intrinsic properties of the Ils. The above observations clearly indicate that the Li^+ mobility in such systems is linked to a number of factors associated to solvation dynamics, strength of ionic interactions, polymer relaxation motions, and others. One of the stern demand with the IL-SPE using PEO and P(VDF-HFP) is to synergistically improve both the electrochemical and mechanical properties of the resulting polymer electrolyte, that is, to increase the IL content enhancing the ionic conductivity without losing the mechanical integrity/stability. As schematized in

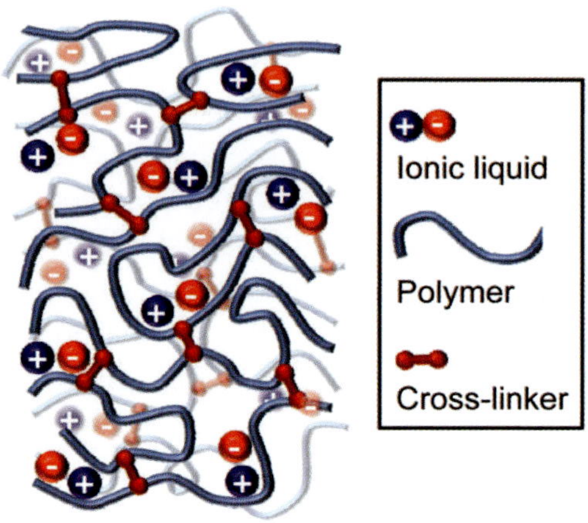

Figure 4. Schematic illustration of cross-linked IL-incorporated polymer electrolytes. Taken from Ref. 46.

Figure 4, the use of cross-linked IL-SPE results in the improved electrochemical performance of the Li/cross-linked IL-SPEs/MWCNT-coated $LiFePO_4$. The cross-linked material demonstrated better capacity, rate capability, and cyclability in the tested batteries, surpassing that of the neat IL, while preserving the mechanical strength.[152–155] As the ultimate motive to use SPE is to develop Li-metal-based batteries, tests enlisting the compatibility of IL-SPEs with cathode active materials are lagging behind. Accordingly, a number of studies employing various cathode active materials, including $LiMn_2O_4$,[156] V_2O_5,[157] $LiFePO_4$,[158,159] and LiV_3O_8[160] in IL-SPEs demonstrated an improved electrochemical performance, with reduced cathode active materials dissolution, one of the severe problem with conventional liquid electrolytes.

In general, high ionic conductivity, high thermal stability, appropriate mechanical stability, and good mechanical adhesion characteristics of IL-incorporated polymer electrolytes could make them promising electrolyte materials. As indicated in Table 4b, the ionic conductivity is quite promising and they could be treated as strong contenders to other electrolyte systems.

From the abovementioned plasticizers (i.e., organic liquid carbonates and ILs), one can take advantages of each and find an optimized electrolyte system, consisting of organic carbonate, IL, and polymer and lithium salt.[2] A quaternary electrolyte system formulated by combining poly(2-ethoxyethyl methacrylate) as the polymer host matrix, 1-butyl-3-methylimidazolium hexafluorophosphate ([BMI][PF_6]) as IL, the 1:1 mixture of PC and EC as organic solvents, and $LiClO_4$ as salt (in the ratio of 49.2:19.6:9.8:1.8 mol%), revealed a very high RT ionic conductivity $\approx 0.94 \times 10^{-3}$ S cm^{-1}, promising electrochemical stability (>4.4 V *versus* Li/Li^+), as well as good thermal and chemical characterstics.[161] Similarly, Scrosati *et al.* reported an enhanced ionic conductivity of about 8.1×10^{-3} S cm^{-1} for a quaternary electrolyte system based on (PVDF-HFP)-EC/PC-Pyr_{24}TFSI-LiTFSI system, a much higher value compared to EC/PC-free ternary system of conductivity about 3.7×10^{-3}S cm^{-1}. Nevertheless, the presence of volatile compounds is, once more, raising safety issues, only slightly reduced by the flame retardant effect of the IL.

4. *Composite Polymer Electrolytes*

In an attempt to overcome the limitations of SPEs (poor ionic conductivity at RT) and GPEs (poor mechanical and thermal stabilities), inorganic material-containing polymer electrolytes, herein called CPEs, were also developed in the last two decades. These solid electrolytes are now encountering a renewed interest due to the availability of inorganic solid materials (fillers) offering high Li^+ ionic conductivity. CPEs may, in fact, display advantages from both components, such as ease of processing, flexibility, improved electrode/electrolyte compatibility, enhanced ionic conductivity, reduction of crystallinity and glass transition temperature of the polymer matrix, wider electrochemical window, and transference number.[185] Additionally, the inorganic material may compensate the deterioration of mechanical properties resulting from the decrease in crystallinity.

In the following, only CPEs entailing SPEs together with inorganic solid additives, called fillers, are presented. These materials involve the dispersion of ceramic or glass powders into the polymer–salt matrix, resulting in the formation of CPEs.[49] In general terms, an increase in conductivity is observed with such systems, which could be explained by the reduction in crystallinity of the polymer domains and the creation of new conduction pathways at the polymer/inorganic interface. High conductivities are generally reported for fillers' content of 10%–20% wt%. Higher filler contents result in phase discontinuity with consequent ionic conductivity decrease.[19] It is noteworthy to mention that addition of ceramic fillers can also impart negative attributes, such as an increase in T_g and inhibition of the transfer of ions from polymer chain segments resulting in reduced ionic conductivities. Hence, appropriate amount of filler needs to be added to obtain enhanced features compared to the constituents of the CPEs. Aside from the amount, the filler's particle size is another critical factor that affects the conductivity of CPEs. Depending on their nature, fillers are broadly divided into active and passive (inert) additives. Table 5 presents the conductivities of some selected CPEs with either inert or active inorganic fillers.

a. *Passive Fillers*

These are electrochemically inactive materials, which influence the ion transport phenomenon in polymers *via* mechanisms, such as plasticizing the polymer, that is, inhibiting polymer crystallization, increasing the free volume and thereby facilitating segmental dynamics, among others,

Table 5. Ionic conductivity of some selected composite electrolyte polymers

Composite Polymer Electrolyte	Temp (°C)	σ (S cm^{-1})	Battery Configuration	Reference
PAN/$LiClO_4$/SiO_2	25	6×10^{-4}	$LiFePO_4$/Li	237
PAN/$LiPF_6$/PC/Al_2O_3	30	8.0×10^{-3}	$LiNi_{0.8}Co_{0.16}Al_{0.04}O_2$/Li	238
PAN/PC/EC/α-Al_2O_3	—	6.0×10^{-4}	$LiCoO_2$/Li	153
PEO/$LiCF_3SO_3$/Al_2O_3	25	1.0×10^{-5}	–	239
PEO/$LiClO_4$/TiO_2	30	2.3×10^{-5}	–	240
PEO/$LiClO_4$/8%$ZnAl_2O_4$	25	1.0×10^{-5}	$LiFePO_4$/Li	241
PUA/LiTFSI/SiO_2	60	1.1×10^{-4}	$Li_{0.33}MnO_2$/Li	242
PEO/$LiCF_3SO_3$/92 wt% Li_3N	30	1.2×10^{-4}	–	243
PE/$LiCF_3SO_3$/84.3 vol% $1.2Li_2S$–$1.6LiI$–B_2S_3	25	4.8×10^{-5}	–	244

without getting involved in the conduction process. Since the pioneering work of Weston and Steele,[186] reporting on the enhancement of both the mechanical properties and the ionic conductivity of $(PEO)_8$-$LiClO_4$ by doping with an inert filler (α-alumina, α-Al_2O_3) in 1982, several research teams have proposed the incorporation of micro- and nanosize fillers into solid as well as GPEs. Inert ceramic fillers including Al_2O_3,[72,186–204] SiO_2,[54,72,147,205–209] ZrO_2,[210–214] TiO_2,[187,193,215–220] a-Fe_2O_3,[221] $ZnAl_2O_4$,[222] and B_2O_3,[223] particularly when controlled at nanoscale, have been found to augment the ionic conductivity of SPEs. Additionally, graphene,[224] carbon nanotubes (CNTs),[225] microporous molecular sieves,[226] metal-organic frameworks,[227,228] and zeolites,[229] enlisting specific structures and constituents, can also act as potential inert additives to improve the global conductivity of SPEs.

Scrosati *et al.*[193] observed an increase in conductivity by 1-2 orders of magnitude in PEO-based electrolytes *via* the addition of ceramic fillers such as Al_2O_3 or TiO_2. The enhancement in conductivity was attributed to the influence of the surface groups in the nanoparticles, facilitating the lithium salt dissociation by the formation of hydrogen bonds between the surface moieties on the ceramic filler (e.g., –OH groups on the SiO_2 surface) and polymer chain (e.g., $–CH_2–CH_2–O–$ unit of PEO). Besides to the increase in ionic conductivity, the specific Lewis acid–base interactions of the lithium salt anions with the surface groups of ceramic nanofillers were proposed to result in increased lithium transference number (Figure 5). For example, the addition of Al_2O_3 nanoparticles is reported to increase T_{Li+} by 30%.[230] Moreover, an increase in the Young's modulus and ultimate tensile strength and yield stress was reported for the PEO-$LiClO_4$ electrolyte reinforced by Al_2O_3 or TiO_2. Zhou *et al.* developed a composite electrolyte comprised of low-MW PEO, lithium salt, and fumed silica as filler, resulting in high conductivity ($>10^{-3}$ S cm^{-1} at RT), mechanical strength (elastic modulus $>10^5$ Pa), and interfacial stability toward lithium metal (up to 5.5 V *versus* Li/Li^+ on the anodic side).[231]

b. *Active Fillers*

Unlike passive fillers, these are electrochemically active in which they participate in the ion transport by providing additional pathways for Li^+

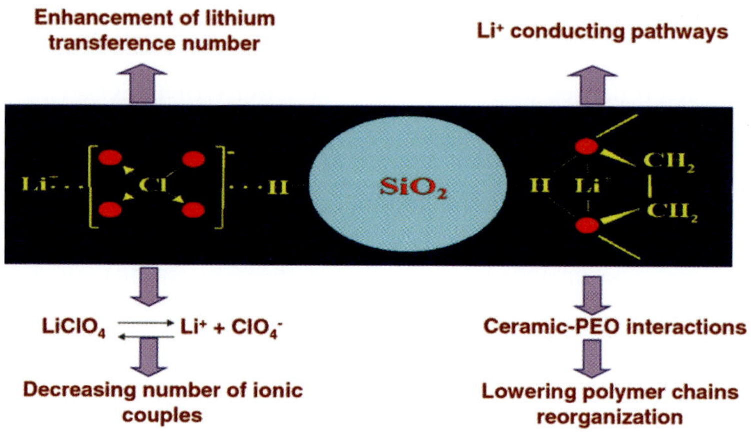

Figure 5. Schematic illustration of the Lewis acid–base interactions of the lithium salt anions with the surface groups of ceramic nanofillers. Taken from Ref. 49.

ion, or acting as Li^+ source.[232] The first study on active fillers was reported by Skaarup *et al.*[233] on a PEO-$LiCF_3SO_3$ electrolyte system containing Li_3N as active filler. The incorporation of a small amount of Li_3N was found to increase the conductivity by a factor of ~1000 larger than the pure SPE, PEO-$LiCF_3SO_3$. Unfortunately, Li_3N has a poor electrochemical stability, ~2.5 V *versus* Li/Li^+. Due to its larger defect-rich grain boundaries, γ-$LiAlO_2$ exhibits high conductivity when it is incorporated into $(PEO)_8$-$LiClO_4$.[234] Moreover, Li/$(PEO)_8$-$LiClO_4$-γ-$LiAlO_2$ showed boosted interfacial stability and mechanical property, compared with $(PEO)_8$-$LiClO_4$ and Li/PAN-EC-PC-$LiClO_4$. Besides, it offers improved mechanical properties and interfacial stability.[235] A composite electrolyte consisting of PAN-$LiClO_4$ containing 15% $Li_{0.33}La_{0.557}TiO_3$ nanowires displayed an ionic conductivity of 2.4×10^{-4} S cm^{-1} at 25°C.[236]

In summary, in virtue of the above-enumerated properties, CPEs could be regarded as valid materials for the development of advanced Li-based rechargeable batteries. Electrochemical tests have performed on various battery configurations employing various SPEs incorporating ceramic fillers and cathode materials including $LiFePO_4$,[245] $Li_{1.2}V_3O_8$,[246] $LiCr_{0.01}Mn_{1.99}O_4$,[247] $LiFePO_4$,[248] $Li_{0.33}MnO_2$,[249] and so on.

III. Interfaces in Lithium-Polymer Batteries

The development of intrinsically safer electrolyte systems, that is, SPEs, for applications in large-format LIBs has been focused on the ionic conductivity enhancement, so far. However, this improvement may not necessarily result in well performing battery cells. In fact, in addition to the electrolyte bulk properties (noticeably the ion transport, which is linked to the polymer–salt interactions), the electrodes/polymer electrolyte interfaces play key roles in dictating the overall performance of the lithium metal polymer electrolyte batteries (LMPBs), particularly when considering large-scale industrial applications. The electrode/polymer electrolyte interfaces have been well recognized to be decisive in governing the kinetics of the overall cell reactions (noticeably rate capability), reversible capacity, safety, and life span within the battery system.[82] The formation of the solid electrolyte interface (SEI) at the Li-metal anode/electrolyte interface was proposed by Peled already in 1979. Since then, however, the largest efforts have been directed at mastering the electrified interface at the vicinity of lithiated graphite (Li_xC_6) in contact with liquid organic electrolytes.[250] Nonetheless, the SEI layer remained still to be the least understood element of the battery system. SEI formation at electrode/electrolyte interfaces, at both the cathode and the anode, originates from the mismatch between the electrochemical stability window (i.e., reduction/oxidation limits) of the electrolytes and the electrochemical potential of the electrodes.[251] Although SPEs are expected to be less reactive due to their intrinsic chemical/thermal inertness and poor flowability to reach the electrochemically active electrode surfaces, there are many evidences for the formation of SEIs at the interfaces between the anode/cathode electrodes and polymer-based electrolytes.[252] In PEs, it is postulated that at least two separate, but competitive reactions occur simultaneously to from a passivation layer; the reaction of lithium with contaminants (e.g., H_2O) and with the conducting salt and polymer matrix.[21] So far, the research work in the SEI on the anode and cathode (this latter sometimes called solid permeable interphase[SPI]) has been predominantly devoted to liquid electrolytes and, to a lesser extent, gel electrolytes.[82,253–255] Thus, compared to LIBs, the interfacial chemistry in Li-polymer batteries (LPBs) using SPEs has been much less explored. This may be due to the common belief that solid polymers are generally more chemically and electrochemically stable than liquid solvents. However, the technical difficulties to expose the electrode surface to probe the SEI composition at electrode/SPEs interfaces because of the strong adhesion of the two layers played a role. The SEI formed by polymer electrolytes in contact with lithium metal anodes are reported to be affected by various parameters such as the nature of the polymer matrix (e.g., hygroscopic *versus* hydrophobic), lithium conducting salt (i.e., stability and hygroscopic nature), viscosity, electrolyte preparation procedure and environment, that is, presence of various contaminants, additives (e.g., ceramic fillers), cell assembly procedure/protocol, and so on.[256,256–259] Armand *et al.* reported the stability domain of polyether (PEO and PPO)-based SPEs using cycling voltammetry (CV) in 1981.[260] The polymers were found to be inert up to 4.0 V and the stability range was chiefly limited by the anion of the dissolved salts, that is, I^- and $CF_3SO_3^-$ for PEO and PPO, respectively. Following this early work, Christie *et al.*[261] and Olsen *et al.*[262] investigated the lithium plating/stripping, deposition/electro dissolution of lithium from SPEs; Bonino *et al.*[263] studied the lithium/hybrid polymer interfacial properties. Watanabe *et al.* studied the interfacial property of SPEs based on high Mw, comb-shaped polyethers, poly[ethylene oxide–co-2-(2-methoxyethoxy)ethyl glycidyl either] [P(EO/MEEGE)] and $LiClO_4$ conducting lithium salt.[264] The study conveyed a decrease in the internal resistance (R_i) with increasing ionic conductivity of the polymer, which, in turn, get affected by the MEEGE composition and lithium salt species complexed with the copolymers. The solid-state lithium metal polymer (LMP) cell Li/P(EO/MEEGE) = 91/9)-$LiClO_4$/$LiCoO_2$ achieved a discharge life for over 7 h at 0.1 mA/cm^2 and low cell impedance of about 96 Ω cm^2 after full charge at 60°C. Peled *et al.* reported the formation of SEI layer on the lithium surface in CPE batteries, but it was not possible to analyze its composition using the employed technique.[265,266] Appetecchi *et al.* conducted an in-depth investigation on the impact of electrode preparation, addition of inorganic filler, and cell assembly on the stability of interfaces formed by Li/PEO_{20}-$LiCF_3SO_3$ symmetric cells. It was found that the solvent-free, high purity of PEO-based electrolyte lead to excellent interfacial compatibility with the Li anode. While addition of fillers slightly improve the interfacial compatibility, the cell assembling protocol, including the hot lamination step was found to be highly critical toward building a very stable lithium/polymer electrolyte interface. In situ electrochemical impedance spectroscopy (EIS) analysis showed an increase in the evolution of the interfacial resistance at the Li/SPE interface, indicating the formation and advancement of interphase layers in LPBs with electrochemical cycling.[258,259,267] In literature, it is well documented that moisture (e.g., due to the hygroscopic nature of some polymers and/or salts) plays an important role in contributing to the passivation layer, usually indicated by the presence of large quantity of LiOH, which may be resulted from the reduction of traces of water as indicated below.

$$H_2O\left(\text{Polymer or salt}\right) + Li^{+} + e^{-} \rightarrow LiOH + 1/2H_2 \qquad (1)$$

Le Granvalent-Mancini *et al.*[268] investigated the passivation film formed on Li/PEO-$LiCF_3SO_3$ SPE *via* atomic force microscopy (AFM), concluding that the SEI layer exhibits low conductivity. Ismali *et al.*[269] utilized X-ray photoelectron spectroscopy (XPS) to investigate the surface layer on native lithium and after contact with SPEs, observing LiF, resulting from the salt decomposition, and some native film compounds such as Li_2CO_3, LiOH, and Li_2O. Xu *et al.* conducted a comparative XPS analysis on Li||graphite cells with PEO-LiTFSI (cycled at 50°C) and a conventional liquid electrolyte (i.e., EC/DEC at 2/1, 1M LiTFSI).[270] While severe salt degradation was observed at the graphite/PEO interface after the first discharge, the Li/PEO interface was found more stable even when using LiTFSI-based liquid electrolyte. LiOH was detected on the SEI layer derived from PEO-based Li/graphite cells, while at the Li/PEO, newly produced compounds, namely LiF, and $ROCO_2Li$ were identified. Moreover, significant differences were observed between the SEI layers formed in PEO_{25}-LiTFSI and 1M LiTFSI/EC-DEC (1:2)-based cells, where the main decomposition products such as carbonate species, PEO-type polymers, and so on were detected. The same authors investigated the role of the host polymer using a hydrophobic polymer, poly(trimethylene carbonate) (PTMC), doped with LiTFSI salt.[271] The results were compared to that of the SEI layer formed with conventional PEO-based electrolytes. At the graphite/PTMC-LiTFSI interface, the SEI layer was found to contain LiF, Li_2S, Li_2SO_3, and Li_3N. Both PEO and PTMC-based electrolytes displayed polymer degradation products on the anode side, and PTMC-derived $ROCO_2Li$ and ROLi were proposed as the principal products at the graphite/PTMC-LiTFSI interface, due to series of electrochemical/chemical degradation processes similar to those occurring with linear carbonate solvents. Very importantly, a large quantity of LiOH, resulting from moisture contamination, was found in PEO-based cells, but not for the hydrophobic PTMC. Thus, the nature of the host polymer impacts the characteristics of the interface, but might also offer a protective layer on the active anode material. On the cathode side, the interface of $LiFePO_4$/SPE appeared to be stable, with only small amounts of LiF and salt decomposition products identified.

At the end, the interface stability could be improved by the addition of appropriate ceramic fillers to the base SPEs.[256,258,259] For instance, Kumar and Scanlo[272] investigated the effect of a number of ceramic fillers on the lithium/polymer interfaces and found that ceramic fillers such as nitrides (aluminum nitride [A1N], boron nitride [BN]) exhibit better interfacial properties than SiO_2 or Al_2O_3, attributed to the formation of Li_3N that facilitates for high ionic conductivity.

IV. Practical Lithium-Polymer Batteries

A key advantage of the LPB concept consist in the absence of (flammable) liquids, which facilitates the production of rugged (i.e., tolerant to vibration, shock, and mechanical deformation), and leak proof cells. Additional advantages are simplicity and flexibility of fabrication (i.e., in shape and geometry) and the possibility of packaging in lightweight plastic containers unlike conventional lithium batteries, which require metallic casing.[103] The commercial development of the first generation LPBs took place at the turn of the century.[130] These lithium-ion polymer battery cells were fabricated with GPEs for good RT conductivity. The first attempt to commercialize lithium-polymer cells was based on the technology developed by Bellcore laboratory, referred to as the Bellcore process.[273] Since the first patent involving the use of GPEs for the manufacture of lithium-ion polymer batteries in 1994, the progress in the research of these type of electrolytes has been very rapid, marking the beginning of active research on lithium-ion polymer batteries. The

Bellcore technology focused on the extraction–activation process to facilitate the industrial production of lithium-polymer cells.[131] The basis of the Bellcore LPB technology made use of a continuous production process in which a stack comprising the two electrode layers composed of carbon and $LiMn_2O_4$-based as the negative and positive electrodes, respectively, separated by a polymeric layer, all fused together by hot-lamination to form a unified tri-layer strip. The polymeric layer as well as the composite electrodes contained a polymer plasticized with diocyl phthalate (DOP), which was removed by extraction with an organic solvent, thus leaving the necessary porosity for the filling of liquid electrolyte. Prior to this, the cell laminate was housed in the coffee-bag envelop, usually constituted by polypropylene polyesthers. Further, the appropriate amount of the electrolyte solution, usually $LiPF_6$ in EC/DMC (2:1), was injected into the cell followed by the final thermal sealing of the package. From the mass production perspective, however, Bellcore's technology was relatively complicated in that the elimination procedure of the plasticizer with volatile diethyl ether was mandatorily complete because its residues lead to poor cyclability of the cells and DOP is toxic. Additionally, the first polymer battery was withdrawn from the market owing to the liquid electrolyte leakage reported to occur.

By the end of 1980s, Sony started to develop LPBs with GPEs, considering several polymer matrices such as PEO, PAN, and P(VDF-HFP) were examined for this purpose. Out of the tested GPEs, P(VDF-HFP) was found to outperform others, yielding high ionic conductivity (up to 9×10^{-3} S cm^{-1} at RT up on optimization of the salt concentration, PC/EC ratio, VDF/HFP ratio, and MW of polymer matrix), excellent adhesive property to the active materials, and sufficient mechanical strength. Sony's concept for the development of LPBs has proved to lead to simple manufacturing process, and by far favorable production (Figure 6).

A cyclability up to 1500 cycles to a cut-off of 80% of the initial capacity at 1C rate at RT and energy density in the order of about 130 Wh/kg and 300 Wh/dm^3 have been reported by then.

As described above, the Bellcore and Sony technologies have enjoyed some success in the lithium-ion polymer electrolyte configurations, however, they cannot be utilized with the lithium metal anode because of the instability of fluorinated polymers against Li.[274] Besides, as they still use

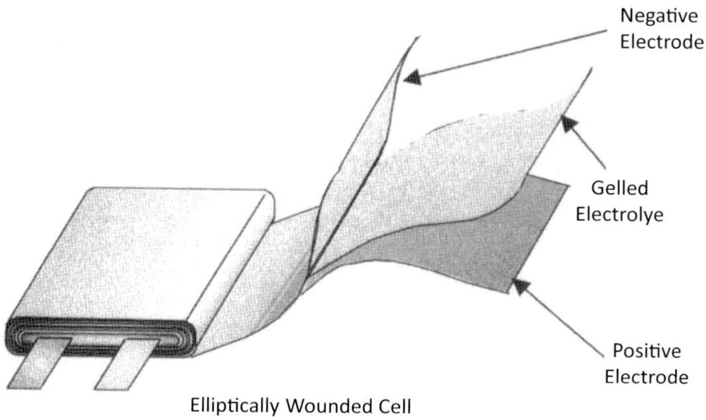

Figure 6. A schematic drawing of the Sony's LPB. Taken from Ref. 20.

flammable/combustible liquid electrolytes as plasticizers, the stern safety requirement could not be accomplished. In the quest for a battery chemistry that can make electrified means of transportation more competitive with internal combustion, coupling lithium metal anodes with solid-state electrolytes has huge potentiality to offer, including the elimination of flammable liquid electrolytes, inhibition of lithium dendrites, three-dimensional cell designs, and simpler packaging requirements. Li-metal batteries where metallic lithium foil serves as anode and current collector offer higher specific energies since the theoretical capacity of lithium is 3860 Ah Kg^{-1}.

Though, the prospect of using lithium metal looked challenging, in 1985, Moli Energy developed Li||MoS_2||batteries with liquid electrolyte with excellent electrochemical performance in the early cycles, but the batteries were recalled in 1989 due to runaway reaction. This failure came as a one more demonstration that a metallic lithium anode cannot be used with conventional liquid electrolytes or GPEs, due to dendritic growth occurring on repetitive charging/discharging, leading to ultimately, electrode short circuit and thereby explosion of the battery. All attempts to develop Li-metal rechargeable batteries were short lived owing to the safety hazards that occurred in a series of Li-metal batteries starting from the middle of 1970s to early 1990s.[275]

To overcome the aforesaid drawbacks, "dry" polymer electrolytes were used instead of liquid electrolytes, following the pioneering work of Armand *et al.* in the R&D efforts at Hydro-Québec since 1978.[276,277]

Besides building solventless solid-state polymer electrolytes, the thin film cell configuration was demonstrated to prevent short circuit linked to dendrite growth on cycling in optimized conditions.[276–278] The reasons are not only the mechanical resilience of the polymer to push back dendrite, but also as a result of thin film construction, the surface capacities (<4 mAh cm^{-2}), and hence the current densities are lower, a favorable feature for smooth metal plating.

In 1993, Bollore technologies, EDF, and CEA/CEREM started a program aimed at developing a 40-Ah all solid-state LPB, with an optimized process offering low cost, high reproducibility and without any liquid plasticizer.[276] After 3 years of R&D, 40 Ah single cells, built using metallic lithium, VO_x, and PEO/fluorinated Li-salt as anode, cathode, and electrolytes respectively were produced. Performance (energy density, specific power, and cyclability), thermal modeling, and series reproducibility tests all confirmed that the lithium-polymer technology could be a serious candidate for an advanced all solid-state battery technology for xEVs application.

In the late 1990s, all solid-state large laminated battery modules based on cells manufactured from lithium (foil) anode, modified PEO-based electrolyte layer (i.e., containing either $LiClO_4$ or $LiN(CF_3SO_2)_2$) as slat, and vanadium-based oxides (either VO_x or LiV_3O_8) cathode (see Figure 7) were developed jointly by Hydro Québec (Canada) and 3M (United States) (Figure 7). The fabricated battery module had excellent performance, with about 155 Wh/Kg specific energy and a cycle life of about 600 cycles at 80% depth of charge and was proposed as a power source for electric vehicles.[102] However, despite the successful demonstration tests, the project went into financial difficulties.

$$\mathrm{Li(s)|M\text{-}PEO\text{-}LiClO_4} \quad \text{or} \quad \mathrm{LiN(CF_3SO_2)_2(s)|LiV_3O_8(composite)(s)}$$

The solid-state Li-polymer technology initiated by Hydro-Quebec, which is based on VOx cathode, was further exploited by Bolloré 30 years later, which lead to the commercialization of the current LMPBs making use of $LiFePO_4$ (LFP) as cathode electrode material.[7] This expertise, added to the over 20 years of in-house R&D, enabled Bolloré to successfully introduce a range of e-mobility applications (cars, buses, tramways, and boats) as indicated in Figure 8.

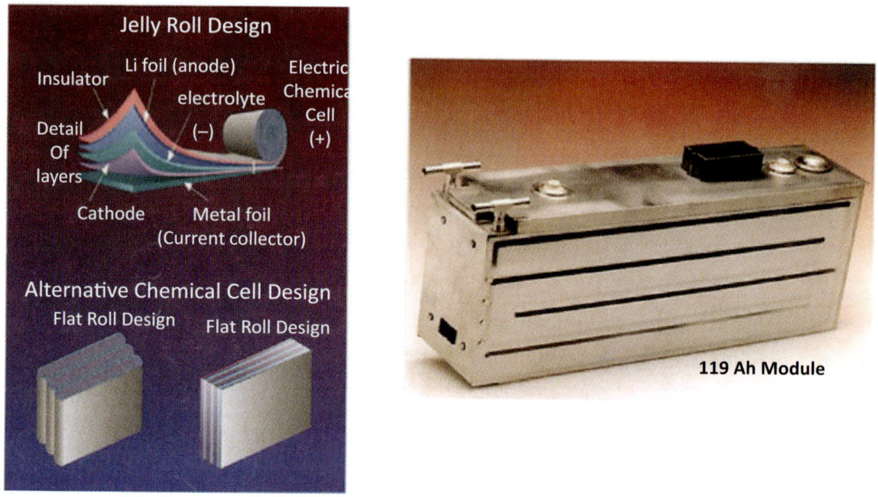

Figure 7. Concept and a 119-Ah module of all-solid advanced electric vehicle battery using PEO dry polymer and a metallic metal anode (developed by 3M and Hydro Québec). Taken from Ref. 102.

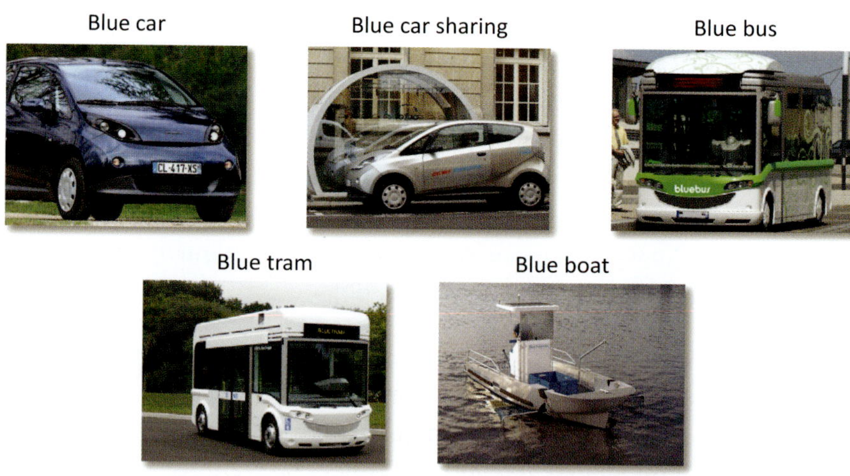

Figure 8. E-mobility application domains of Bollore LMP batteries. (Courtesy of Bolloré.)

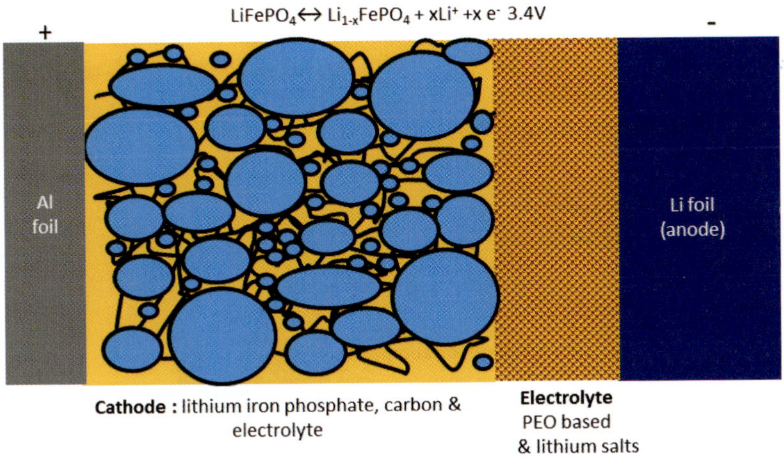

Figure 9. All solid-state battery based on lithium metal polymer battery (LMPB) of Bolloré technology. (Courtesy of Bolloré.)

The commercial production of LMP batteries consists of a laminate of four ultrathin materials: (1) metallic Li foil, acting as both the Li source and the anode current collector; (2) SPE composed of the lithium salt (LiTFSI) dissolved in the solvating polymer, PEO; (3) cathode composed of lithium iron phosphate ($LiFePO_4$), carbon, and the polymer electrolyte as above to form a plastic composite coated on both side of the aluminum foil current collector (Figures 9 and 11a).

The fabrication process for the LMP batteries' ultrathin films is based on extrusion (Figure 10), which confers Blue Solutions' batteries several advantages for the industrial production of batteries compared to the commonly used coating process, including:

- no use of polluting chemical or solvent during the battery manufacturing, enabling lower costs and ecological benefits;
- production of consistent and high quality components;
- high production yields, making rapid industrialization.

The extruded films are all solid design, environmentally friendly (no solvent, no rare earth element, highly recyclable or reusable), altogether allowing the safe energy storage technology operating within the external

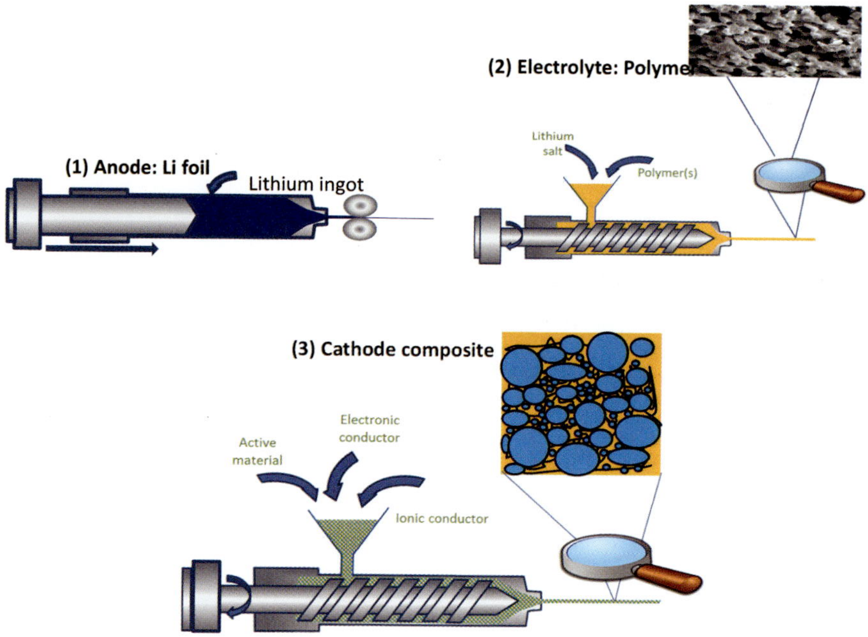

Figure 10. Blue Solution's extrusion process for the production of LMP® components. (Courtesy of Bolloré.)

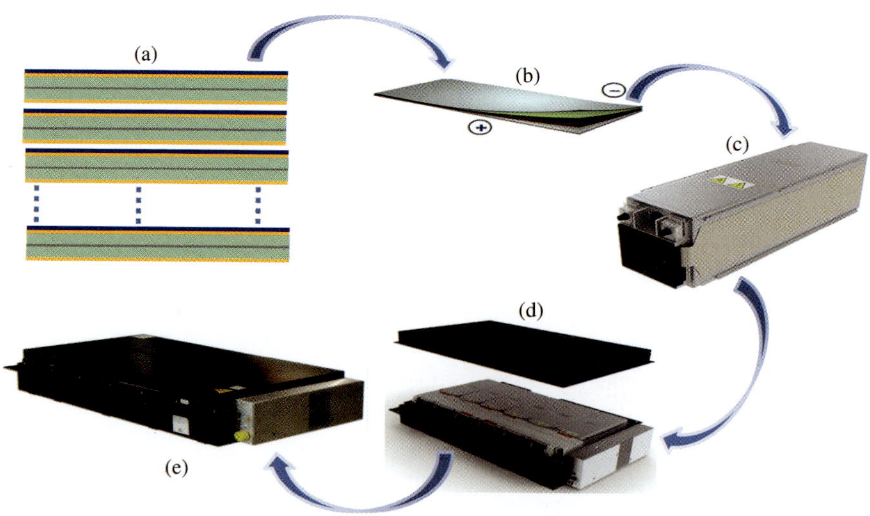

Figure 11. Blue Solutions lithium-metal-polymer battery assembly strategy: (a) ultrathin layered films, (b) elementary cell, (c) module, (d) pack, and (e) 30 kWh Bollore LMP battery. (Courtesy of Bolloré.)

temperature range of −20°C to +60°C and internal temperature of 60°C–80°C.

As schematically presented in Figure 11, the thin films are coiled, compressed into a prismatic shape, and assembled into elementary cells (Figure 11b), which are then assembled in series into modules (Figure 11c), then to packs (Figure 11d), and finally into solid-state LMP battery configurations (Figure 11e). The battery offers a capacity of 75 Ah at C/4 and energy storage is 30 kWh with nominal voltage of 410 V, maximum power delivery of 45 kW, and energy density of 100 Wh/kg and 100 Wh/l at pack level.

Using such a battery, for which it holds more than 400 patents, Blue Solutions (Bolloré) developed the BlueCar used in Autolib car sharing as part of the urban project in Paris since 2011.[279] Since then, the LMP battery internally operating at 60°C–80°C has been implemented in several practical applications (Figure 12).

In particular, the BlueCar is equipped with the 30 kWh $LiFePO_4$/SPE/Li-metal battery pack offering 250 km (160 miles) driving range (under normal urban driving conditions) and 150 km (93 miles) in highways on one charge with a C-rate of C/4, a much higher autonomy compared to

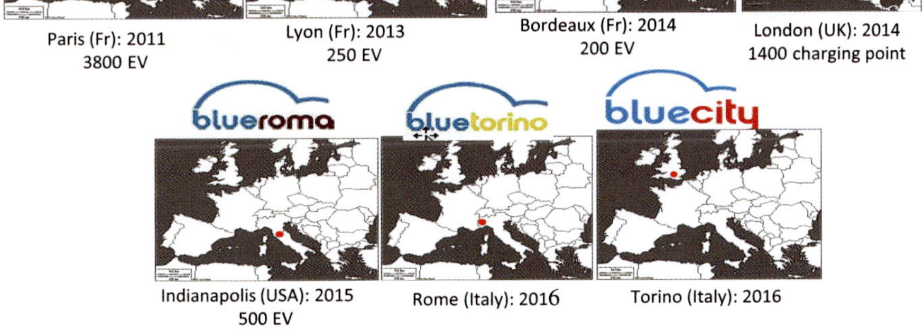

Figure 12. Ballore's blue solution car sharing program. (Courtesy of Bolloré.)

their lithium-ion counterparts and a maximum speed of 130 km h^{-1} (81 mph). The total weight of the BlueCar is about 1.1 ton, including the 300 kg (660 lb) LMP battery.

Additionally, automotive developments include the BlueBus, which is equipped with a 90-kWh LMP battery and guarantees more than 120 km of driving autonomy. Bluetrams, on the other hand, have infinite driving autonomy and are equipped with a disruptive technology based on super capacitors, which are capable of recharging the vehicle in 20s at each station. In this application, an LMP pack serves as safety emergency power source.

Besides to e-mobility, since 2014, Blue Solutions has introduced LMP batteries for stationary energy storage solutions (in combination with PV) in two African cities: Abidjan (Republic of Cote d'Ivoire) and Yaounde (Cameroon), and one Asian city, in Angkor (Cambodia).

As evidenced by the various applications above, among which BlueCar automobiles are the most eminent, LMPBs are safe and reliable. So far, more than 3000 cars have been driven over 10 million miles without any major difficulty. However, there are still challenges and opportunities that need to be exploited so as to appraise the full-scale deployment of LMPBs for electro mobility and related innovative applications. These include as follows:

- The large Li-metal excess (threefold) needed to ensure the aforementioned performance, which certainly calls for improvement in terms of the energy density.
- The poor electrolyte ionic conductivity at RT and below, which forces to keep the battery plugged to maintain the operating internal temperature of 60°C–80°C when not in use.
- Limited electrochemical stability window of the polymer electrolyte, limiting the positive electrode material choice to $LiFePO_4$.

V. Conclusions and Future Outlooks

Technological advancements are accelerating the transition toward an era dominated by sustainable e-mobility as means of electrified means of transportation. According to the forecast made by the International

Transport Forum in 2011, the number of cars worldwide would triple from 850 million to 2.5 billion by 2050.[280] In this regard, solid-state polymer electrolytes are expected to meet the stringent requirements of xEVs. SPEs have received large attention due their unique characteristics, such as no leakage, negligible flammability, excellent process ability, good flexibility, high safety, and superior thermal stability, are certainly the most important. The Li-metal solid polymer electrolyte (LMP) battery technology is among the most promising technologies, as demonstrated by the outstanding performance of the few thousands Bluecars (Bollore) used for car sharing services worldwide. These LMP batteries demonstrated their robustness, offering high energy, superior safety, insensitivity to climatic conditions, compactness, lightweight, and reasonable autonomy. In order to improve further this technology, increasing the polymer conductivity for the 50°C operation would reduce the loss of energy for heating. Increasing the Li^+ transport number to unity will have a double advantage of building a better metal/electrolyte interface, and if conductivity is kept or improved, allow increasing the areal capacity with a benefit for the specific energy. Reducing the lithium excess would represent another important gain.

VI. Acknowledgment

G. G. Eshetu and S. Passerini gratefully acknowledge the financial support from the Karlsruhe Institute of Technology, Helmholtz Institute Ulm (HIU), Germany and G. G. Eshetu and M. Armand for the financial support of CIC energy cooperative center, Spain. The authors would further thank Bolloré for providing some of the figures and publishers mentioned inside the text for their generous permission to reproduce their figures.

VII. References

1. Eshetu, G. G.; Grugeon, S.; Laruelle, S.; Boyanov, S.; Lecocq, A.; Bertrand, J. -P.; Marlair, G. *Phys. Chem. Chem. Phys.* **2013**, *15* (23), 9145–9155.
2. Kalhoff, J.; Eshetu, G. G.; Bresser, D.; Passerini, S. *ChemSusChem* **2015**, *8* (13), 2154–2175.
3. Armand, M.; Tarascon, J. -M. *Nature* **2008**, *451* (7179), 652–657.

4. Eshetu, G. G.; Bertrand, J. -P.; Lecocq, A.; Grugeon, S.; Laruelle, S.; Armand, M.; Marlair, G. *J. Power Sources* **2014**, *269*, 804–811.
5. Tarascon, J.-M.; Armand, M. *Nature* **2001**, *414* (6861), 359–367.
6. Deng, D. *Energy Sci. Eng.* **2015**, *3* (5), 385–418.
7. Tarascon, J. M.; Simon, P. *Eletrochemical Energy Storage*; John Wiley & Sons, Inc: Hoboken, NJ, 2015; Vol. 1, pp 1–76.
8. Eshetu, G. G.; Grugeon, S.; Laruelle, S.; Boyanov, S.; Lecocq, A.; Bertrand, J.-P.; Marlair, G. *Phys. Chem. Chem. Phys.* **2013**, *15* (23), 9145–9155.
9. Lecocq, A.; Eshetu, G. G.; Grugeon, S.; Martin, N.; Laruelle, S.; Marlair, G. *J. Power Sources* **2016**, *316*, 197–206.
10. Wright, P. V. *Br. Polym. J.* **1975**, *7* (5), 319–327.
11. Armand, M.; Vincent, C. A. Second International Meeting on Solid Electrolytes, Scotland, St Andrews, September 20–22,1978.
12. Franco, A. Alejandro, *Rechargeable Lithium Batteries*; Elsevier Ltd: Oxford, UK, 2015, pp 73–116.
13. Bruce; W, Duncan O'Hare, D, Walton, Richard, I. *Energy Materials*; John Wiley & Sons, Ltd, 2011; pp 1–27.
14. Khurana, R.; Schaefer, J. L.; Archer, L. A.; Coates, G. W. *J. Am. Chem. Soc.* **2014**, *136* (20), 7395–7402.
15. Yuan, R.; Teran, A. A.; Gurevitch, I.; Mullin, S. A.; Wanakule, N. S.; Balsara, N. P. *Macromolecules* **2013**, *46* (3), 914–921.
16. Feng, S.; Shi, D.; Liu, F.; Zheng, L.; Nie, J.; Feng, W.; Huang, X.; Armand, M.; Zhou, Z. *Electrochim. Acta* **2013**, *93*, 254–263.
17. Ozawa, K. *Lithium ion Rechargeable Batteries*;WILEY-VCH Verlag GmbH & Co. KGaA: Weinheim, 2009; pp 2013–250.
18. Gray, F.; Armand, M. *Handb. Batter. Mater.* **2011**, 989.
19. Wu;Y. *Lithium-Ion Batteries Fundamentals and Applications*; Taylor & Francis Group: Boca Raton, FL, 2015; pp 273–399.
20. Schalkwijk, W. Van; Scrosati, B. *Advances in Lithium-Ion Batteries*; Kluwer Academic Publishers: New York, NY, 2002; pp 155–267.
21. O. Besenhard; J. *Handbook of Batterz Materials*; WILEY-VCH: Weinheim, Germany, 1999; pp 449–550.
22. Li, H.; Zhang, H.; Liang, Z. Y.; Chen, Y. M.; Zhu, B. K.; Zhu, L. P. *Electrochim. Acta* **2014**, *116*, 413–420.
23. Miranda, D. F.; Versek, C.; Tuominen, M. T.; Russell, T. P.; Watkins, J. J. *Macromolecules* **2013**, *46* (23), 9313–9323.
24. Ai, Y.; Zhang, J.; Zhao, F.; Zeng, B. *J. Chromatogr. A* **2015**, *1407*, 52–57.
25. Liao, Y.; Sun, C.; Hu, S.; Li, W. *Electrochim. Acta* **2013**, *89*, 461–468.
26. Cui, W. W.; Tang, D. Y.; Gong, Z. L. *J. Power Sources* **2013**, *223*, 206–213.
27. Hu, P.; Chai, J.; Duan, Y.; Liu, Z.; Cui, G.; Chen, L. *J. Mater. Chem. A* **2016**, 10070–10083.
28. Kurc, B. *Electrochim. Acta* **2014**, *125*, 415–420.

29. Zhu, Y.; Xiao, S.; Shi, Y.; Yang, Y.; Hou, Y.; Wu, Y. *Adv. Energy Mater.* **2014**, *4* (1), 7790–7797.
30. Zhu, Y.; Wang, F.; Liu, L.; Xiao, S.; Chang, Z.; Wu, Y. *Energy Environ. Sci.* **2013**, *6*, 618.
31. Tominaga, Y.; Yamazaki, K. *Chem. Commun. (Camb).* **2014**, *50*, 4448–4450.
32. Sun, B.; Mindemark, J.; Edström, K.; Brandell, D. *Solid State Ionics* **2014**, *262*, 738–742.
33. Mindemark, J.; Sun, B.; Törmä, E.; Brandell, D. *J. Power Sources* **2015**, *298*, 166–170.
34. Ue, M. *J. Electrochem. Soc.* **1994**, *141* (12), 3336–3342.
35. Stephan, A. M. *Eur. Polym. J.* **2006**, *42* (1), 21–42.
36. Song, J. Y.; Wang, Y. Y.; Wan, C. C. *J. Power Sources* **1999**, *77* (2), 183–197.
37. Lee, H. S.; Ma, Z. F.; Yang, X. Q.; Sun, X.; McBreen, J. *J. Electrochem. Soc.* **2004**, *151* (9), A1429.
38. Lee, H. S. *J. Electrochem. Soc.* **1998**, *145* (8), 2813.
39. Meziane, R.; Bonnet, J. P.; Courty, M.; Djellab, K.; Armand, M. *Electrochim. Acta* **2011**, *57* (1), 14–19.
40. Cowie, J. M. G.; Spence, G. H. *Solid State Ionics* **1999**, *123* (1), 233–242.
41. Nishi, Y. *Chem. Rec.* **2001**, *1* (5), 406–413.
42. Nishi, Y. *J. Power Sources* **2001**, *100* (1–2), 101–106.
43. Quartarone, E. *Solid State Ionics* **1998**, *110* (1–2), 1–14.
44. Ahmad, S. *Ionics (Kiel).* **2009**, *15* (3), 309–321.
45. Hallinan, D. T.; Balsara, N. P. *Annu. Rev. Mater Res* **2013**, *43*, 503–525.
46. Park, M. J.; Choi, I.; Hong, J.; Kim, O. *J. Appl. Polym. Sci.* **2013**, *129* (5), 2363–2376.
47. Xu, W.; Wang, J.; Ding, F.; Chen, X.; Nasybulin, E.; Zhang, Y.; Zhang, J. -G. *Energy Environ. Sci.* **2014**, *7* (2), 513–537.
48. Xue, Z.; He, D.; Xie, X. *J. Mater. Chem. A* **2015**, *3* (38), 19218–19253.
49. Hassoun, J.; Scrosati, B. *J. Electrochem. Soc.* **2015**, *162* (14), A2582–A2588.
50. Muldoon, J.; Bucur, C. B.; Boaretto, N.; Gregory, T.; Noto, V. Di. *Polym. Rev.* **2015**, *55* (2), 208–246.
51. Long, L.; Wang, S.; Xiao, M.; Meng, Y. *J. Mater. Chem. A Mater. Energy Sustain.* **2016**, *4*, 10038–10069.
52. Takeda, Y.; Yamamoto, O.; Imanishi, N. *Electrochemistry* **2016**, *84* (4), 210–218.
53. Hu, P.; Chai, J.; Duan, Y.; Liu, Z.; Cui, G.; Chen, L. *J. Mater. Chem. A* **2016**, 9–12.
54. Manuel Stephan, A.; Nahm, K. S. *Polymer (Guildf).* **2006**, *47* (16), 5952–5964.
55. Zhang, R. Advanced gel electrolyte for lithium-ion polymer batteries. Master Thesis, Iowa State University, 2013.
56. Zheng, Y.; Lui, J.; Ungar, G.; Wright, P. V. *Chem. Rec.* **2004**, *4* (3), 176–191.
57. Giridhar, P.; Prasad, K. A.; Kalaiselvi, N.; Gopalakrishnan, K.; Ganesan, M.; Veluchamy, A. *Bull. Electrochem.* **1999**, *15* (9–10), 414–418.
58. Quartarone, E.; Mustarelli, P. *Chem. Soc. Rev. Chem. Soc. Rev* **2011**, *40* (40), 2525–2540.
59. Scrosati, B. *Polym. Int.* **1998**, *47*, 50–55.

60. Zhao, Y.; Zhang, Y.; Gosselink, D.; Doan, T. N. L.; Sadhu, M.; Cheang, H. J.; Chen, P. *Membranes (Basel).* **2012**, *2* (3), 553–564.
61. Baskakova, Y. V; Yarmolenko, O. V; Efimov, O. N. *Russ. Chem. Rev.* **2012**, *81* (4), 367–380.
62. Wang, Y.; Zhong, W. H. *ChemElectroChem.* **2015**, *2* (1), 22–36.
63. Zhang, H.; Zhou, M. -Y.; Lin, C. -E.; Zhu, B. -K. *RSC Adv.* **2015**, *5* (109), 89848–89860.
64. Gadjourova, Z.; Andreev, Y. G.; Tunstall, D. P.; Bruce, P. G. *Nature* **2001**, *412* (6846), 520–523.
65. Wang, H.; Matsui, M.; Takeda, Y.; Yamamoto, O.; Im, D.; Lee, D.; Imanishi, N. *Membranes (Basel).* **2013**, *3* (4), 298–310.
66. Owens, B. B. Rechargeable Lithium Polymer Electrolyte Batteries, Power Sources Symposium, IEEE 35th International, June 22–25, 1992
67. Scrosati, B.; Vincent, C. A. *MRS Bull.* **2000**, *25* (March), 28.
68. Marcinek, M.; Syzdek, J.; Marczewski, M.; Piszcz, M.; Niedzicki, L.; Kalita, M.; Plewa-Marczewska, A.; Bitner, A.; Wieczorek, P.; Trzeciak, T.; Kasprzyk, M.; Łęzak, P.; Zukowska, Z.; Zalewska, A.; Wieczorek, W. *Solid State Ionics* **2015**, *276*, 107–126.
69. Young, W. S.; Kuan, W. F.; Epps, T. H. *J. Polym. Sci. Part B Polym. Phys.* **2014**, *52* (1), 1–16.
70. Dias, F. *J. Power Sources* **2000**, *88* (2), 169–191.
71. Murata, K.; Izuchi, S.; Yoshihisa, Y. *Electrochim. Acta* **2000**, *45* (8–9), 1501–1508.
72. Scrosati, B.; Croce, F.; Panero, S. *J. Power Sources* **2001**, *100* (1–2), 93–100.
73. Fauteux, D.; Massucco, A.; McLin, M.; Van Buren, M.; Shi, J. *Electrochim. Acta* **1995**, *40* (13–14), 2185–2190.
74. de Vries, H.; Jeong, S.; Passerini, S. *RSC Adv.* **2015**, *5* (18), 13598–13606.
75. Hallinan, D. T.; Balsara, N. P. *Annu. Rev. Mater. Res.* **2013**, *43* (17), 503–525.
76. Ye, Y. -S.; Rick, J.; Hwang, B. -J. *J. Mater. Chem. A* **2013**, *1* (8), 2719–2743.
77. Agrawal, R. C.; Pandey, G. P. *J. Phys. D. Appl. Phys.* **2008**, *41* (22), 223001.
78. Golodnitsky, D.; Strauss, E.; Peled, E.; Greenbaum, S. *J. Electrochem. Soc.* **2015**, *162* (14), A2551–A2566.
79. K., K.; R., A.; S., A.; S., B.; X., S. S. *Mater. Sci. Forum* **2015**, *807* (March 2016), 41–63.
80. Armand, M. *Solid State Ionics* **1994**, *69* (3–4), 309–319.
81. Meyer, W. H. *Adv. Mater.* **1998**, *10* (6), 439–448.
82. Xu, K. *Chem. Rev.* **2014**, *114* (23), 11503–11618.
83. Osada, I.; De Vries, H.; Scrosati, B.; Passerini, S. *Angew. Chemie. Int. Ed.* **2016**, *55* (2), 500–513.
84. Ngai, K. S.; Ramesh, S.; Ramesh, K.; Juan, J. C. *Ionics (Kiel).* **2016**, *22* (8), 1259–1279.
85. Fergus, J. W. *J. Power Sources* **2010**, *195* (15), 4554–4569.
86. Di Noto, V.; Lavina, S.; Giffin, G. A.; Negro, E.; Scrosati, B. *Electrochim. Acta* **2011**, *57* (1), 4–13.

87. Liu, W.; Janik, M. J.; Colby, R. H. *Polymers for Energy Storage and Delivery: Polyelectrolytes for Batteries and Fuel Cells.* ACS Symposium Series 1096, Washington, 2012.
88. Mittal, V. *Polymer for Energy Storage and Convers Conversion*; John Wiley & Sons, Inc.: Hoboken, New Jersey and Scrivener Publishing LLC: Salem, Massachusetts, 2013; pp 27–99.
89. Yoshio, M.; R.J. Brodd, A. Kozawa. Lithium-Ion Batteries Science and Technologies; Springer: New York, 2009; pp 413–426.
90. Eiamlamai, P. Polymer Electrolytes Based on Ionic Liquids for Lithium Batteries. Ph.D.Thesis, Universite de Grenoble, August 7, 2006.
91. Gibbs, A. **2013**, 1–67.
92. Li, L.; Sharp, J.; Wong-ng, W. Polymer Composite for Energy Harvesting, Conversion, and Storage, ACS Symposium 1161, Washington DC, 2014.
93. Scrosati, B. *Applications of Electroactive Polymers*; Springer-Science+Business Media, B.V, 2012.
94. Abu-Lebdeh, Y.; Davidson, I. *Nanotechnology for Lithium-Ion Batteries*; Springer: New York, 2013; pp 222–244.
95. Jow, T. R.; Xu, K.; Borodin, O.; Ue, M. *Electrolytes for Lithium Ion Batteries;* Springer: New York, 2014; pp 209–308.
96. Himadri B. Bohidar; Paul Dubin, Yoshihito Osada, Polymer Gels: Fundamentals and applications, ACS Symposium Series 833,Washington, DC, 2002.
97. Sequeira, C.; Santos, D. *Polymer Electrolytes: Fundamentals and Applications*; Woodhead Publishing Limited: Cambrige, U.K., 2010; pp 550–578.
98. Abdullah, M.; Lenggoro, W.; Okuyama, K. *Encylopedia of Nanoscience and Nanotechnology,* 2004; Vol. 8, pp.731–762.
99. Armand, M.; Sanchez, J. Y.; Gauthier, M.; Choquette, Y. Polymeric Materials for Lithium Batteries. In *Electrochemistry of Novel Materials*; Lipkowski, J., Ross, P. N., Eds.; VCH, 2000; pp 1–45.
100. Ter-Gazarian, A. G. *Energy Storage for Power Systems*; Peter Peregrinus Ltd: London, U.K., 2011.
101. Vincent. C.A; Scrosati. B. Modern Batteries; Butterworth-Heinemann, Oxford, 1984.
102. Scrosati, B.; Abraham, K. M.; van Schalkwijk, W.; Hassoun, J. *Lithium Batteries: Advanced Technologies and Applications*; John Wiley & Sons. Inc, Hoboken, NJ, 2013.
103. Minami. T; Tatsumisago. M.; Twakura. C.; Wakihera. M.; Tanaka. I. Solid State Ionics for Batteries; Springer, Tokyo, Japan, 2005.
104. Bruce, P. G. *Electrochim. Acta* **1995**, *40* (13), 2077–2085.
105. Xu, K. *Chem. Rev.* **2004**, *104* (10), 4303–4418.
106. Bouridah, A.; Dalard, F.; Deroo, D.; Armand, M. B. *J. Appl. Electrochem.* **1987**, *17*, 625–634.
107. Choudhary, S.; Sengwa, R. J. *Ionics (Kiel).* **2012**, *18* (4), 379–384.
108. Ibrahim, S.; Johan, M. R.; Materials, A. **2011**, *6*, 5565–5587.

109. Lascaud, S.; Perrier, M.; Vallee, A.; Besner, S.; Prud'homme, J.; Armand, M. *Macromolecules* **1994**, *27* (25), 7469–7477.
110. Blonsky, P. M.; Shriver, D. F.; Austin, P.; Allcock, H. R. *Solid State Ionics* **1986**, *18*, 258–264.
111. Xue, Z.; He, D.; Xie, X. *J. Mater. Chem. A* **2015**, *3* (38), 19218–19253.
112. Correia, D. M.; Costa, C. M.; Nunes-Pereira, J.; Silva, M. M.; Botelho, G.; Ribelles, J. L. G.; Lanceros-Méndez, S. *Solid State Ionics* **2014**, *268, Part*, 54–67.
113. Wong, D. H. C.; Vitale, A.; Devaux, D.; Taylor, A.; Pandya, A. A.; Hallinan, D. T.; Thelen, J. L.; Mecham, S. J.; Lux, S. F.; Lapides, A. M.; Resnick, P. R.; Meyer, T. J.; Kostecki, R. M.; Balsara, N. P.; DeSimone, J. M. *Chem. Mater.* **2015**, *27* (2), 597–603.
114. Tanaka, R.; Sakurai, M.; Sekiguchi, H.; Mori, H.; Murayama, T.; Ooyama, T. *Electrochim. Acta* **2001**, *46* (10–11), 1709–1715.
115. Kim, C. S.; Oh, S. M. *Electrochim. Acta* **2000**, *45* (13), 2101–2109.
116. Choi, B.; Shin, K. H.; Kim, Y. W. *Solid State Ionics* **1998**, *113–115* (1–2), 123–127.
117. Appetecchi, G. B.; Aihara, Y.; Scrosati, B. *Solid State Ionics* **2004**, *170* (1–2), 63–72.
118. Appetecchi, G. B.; Aihara, Y.; Scrosati, B. *J. Electrochem. Soc.* **2003**, *150* (3), A301.
119. Aihara, Y.; Appetecchi, G. B.; Scrosati, B.; Hayamizu, K. *Phys. Chem. Chem. Phys.* **2002**, *4* (14), 3443–3447.
120. Aihara, Y.; Appetecchi, G. B.; Scrosati, B. *J. Electrochem. Soc.* **2002**, *149* (7), A849–A854.
121. Kang, Y.; Kim, H. J.; Kim, E.; Oh, B.; Cho, J. H. *J. Power Sources* **2001**, *92* (1–2), 255–259.
122. Groce, F.; Gerace, F.; Dautzemberg, G.; Passerini, S.; Appetecchi, G. B.; Scrosati, B. *Electrochim. Acta* **1994**, *39* (14), 2187–2194.
123. Interactionpoly, S. O. F. I.; Gel, B.; By, E.; Spectroscopy, R. **1995**, *40* (13), 2379–2382.
124. Appetecchi, G. B.; Croce, F.; Scrosati, B. *Electrochim. Acta* **1995**, *40* (8), 991–997.
125. Appetecchi, G. B.; Croce, F.; Scrosati, B. *J. Power Sources* **1997**, *66* (1–2), 77–82.
126. Cazzanelli, E.; Mariotto, G.; Appetecchi, G. B.; Croce, F. *Ionics (Kiel).* **1996**, *2* (2), 81–87.
127. Appetecchi, G. B.; Alessandrini, F.; Passerini, S.; Caporiccio, G.; Boutevin, B.; Guida-Pietrasanta, F. *Electrochim. Acta* **2005**, *50* (22), 4396–4404.
128. Appetecchi, G. B.; Alessandrini, F.; Passerini, S.; Caporiccio, G.; Boutevin, B.; Guida-Pietrasanta, F. *Electrochim. Acta* **2004**, *50* (1), 149–158.
129. Du Pasquier, A.; Warren, P. C.; Culver, D.; Gozdz, A. S.; Amatucci, G. G.; Tarascon, J. M. *Solid State Ionics* **2000**, *135* (1–4), 249–257.
130. Feuillade, G.; Perche, P. *J. Appl. Electrochem.* **1975**, *5* (1), 63–69.
131. Tarascon, J. M.; Gozdz, A. S.; Schmutz, C.; Shokoohi, F.; Warren, P. C. *Solid State Ionics* **1996**, *86–88* (PART 1), 49–54.
132. Watanabe, M.; Yamada, S.; ichiro; Ogata, N. *Electrochim. Acta* **1995**, *40* (13–14), 2285–2288.
133. Watanabe, M.; Mizumura, T. *Solid State Ionics* **1996**, *86–88* (PART 1), 353–356.

134. Fuller, J.; Breda, A. C.; Carlin, R. T. *J. Electrochem. Soc.* **1997**, *144* (4), L67–L70.
135. MacFarlane, D. R.; Tachikawa, N.; Forsyth, M.; Pringle, J. M.; Howlett, P. C.; Elliott, G. D.; Davis, J. H.; Watanabe, M.; Simon, P.; Angell, C. A. *Energy Environ. Sci.* **2014**, *7* (1), 232–250.
136. Passerini, S. S. *Energy Environ. Sci.* **2015**, *9*, 49–61.
137. Eshetu, G. G.; Armand, M.; Scrosati, B.; Passerini, S. *Angew. Chemie., Int. Ed.* **2014**, *53* (49), 13342–13359.
138. Ishikawa, M.; Sugimoto, T.; Kikuta, M.; Ishiko, E.; Kono, M. *J. Power Sources* **2006**, *162* (1), 658–662.
139. Armand, M.; Endres, F.; MacFarlane, D. R.; Ohno, H.; Scrosati, B. *Nat. Mater.* **2009**, *8* (8), 621–629.
140. Shin, J. -H.; Henderson, W. A.; Passerini, S. *Electrochem. Solid-State Lett.* **2005**, *8* (2), A125.
141. Shin, J. H.; Henderson, W. A.; Passerini, S. *Electrochem. Commun.* **2003**, *5* (12), 1016–1020.
142. Choi, I.; Ahn, H.; Park, M. J. *Macromolecules* **2011**, *44* (18), 7327–7334.
143. Nakajima, H.; Ohno, H. *Polymer (Guildf).* **2005**, *46* (25), 11499–11504.
144. Monteiro, M. J.; Bazito, F. F. C.; Siqueira, L. J. A.; Ribeiro, M. C. C.; Torresi, R. M. *J. Phys. Chem. B* **2008**, *112* (7), 2102–2109.
145. Fisher, A. S.; Khalid, M. B.; Widstrom, M.; Kofinas, P. *J. Power Sources* **2011**, *196* (22), 9767–9773.
146. Singh, B.; Sekhon, S. S. *J. Phys.Chem.B.* **2005**, 109,16539–16543.
147. Appetecchi, G. B.; Hassoun, J.; Scrosati, B.; Croce, F.; Cassel, F.; Salomon, M. *J. Power Sources* **2003**, *124* (1), 246–253.
148. Nishimoto, A.; Watanabe, M.; Ikeda, Y.; Kohjiya, S. *Electrochim. Acta* **1998**, *43* (10–11), 1177–1184.
149. Shin, J. -H.; Henderson, W. A.; Passerini, S. *J. Electrochem. Soc.* **2005**, *152* (5), A978.
150. Fisher, A. S.; Khalid, M. B.; Widstrom, M.; Kofinas, P. *J. Electrochem. Soc.* **2012**, *159* (5), A592–A597.
151. Yeon, S. -H.; Kim, K. -S.; Choi, S.; Cha, J. -H.; Lee, H. *J. Phys. Chem. B* **2005**, *109* (38), 17928–17935.
152. Tigelaar, D. M.; Meador, M. A. B.; Bennett, W. R. **2007**, 4159–4164.
153. Guo, C.; Zhou, L.; Lv, J. *Polym. Polym. Compos.* **2013**, *21* (7), 449–456.
154. Patel, M.; Gnanavel, M.; Bhattacharyya, A. J. *J. Mater. Chem.* **2011**, *21* (43), 17419.
155. Kim, G. T.; Appetecchi, G. B.; Carewska, M.; Joost, M.; Balducci, A.; Winter, M.; Passerini, S. *J. Power Sources* **2010**, *195* (18), 6130–6137.
156. Sato, T.; Morinaga, T.; Marukane, S.; Narutomi, T.; Igarashi, T.; Kawano, Y.; Ohno, K.; Fukuda, T.; Tsujii, Y. *Adv. Mater.* **2011**, *23* (42), 4868–4872.
157. Shin, J. -H.; Henderson, W. A.; Passerini, S. *J. Electrochem. Soc.* **2005**, *152* (5), A978–A983.
158. Shin, J. H.; Henderson, W. A.; Scaccia, S.; Prosini, P. P.; Passerini, S. *J. Power Sources* **2006**, *156* (2), 560–566.

159. Navarra, M. A.; Manzi, J.; Lombardo, L.; Panero, S.; Scrosati, B. *ChemSusChem* **2011**, *4* (1), 125–130.
160. Chew, S. Y.; Sun, J.; Wang, J.; Liu, H.; Forsyth, M.; MacFarlane, D. R. *Electrochim. Acta* **2008**, *53* (22), 6460–6463.
161. Reiter, J.; Vondrák, J.; Michálek, J.; Mička, Z. *Electrochim. Acta* **2006**, *52* (3), 1398–1408.
162. Choi, S. W.; Kim, J. R.; Jo, S. M.; Lee, W. S.; Kim, Y. -R. *J. Electrochem. Soc.* **2005**, *152* (5), A989–A995.
163. Min, H. -S.; Ko, J. -M.; Kim, D. -W. *J. Power Sources* **2003**, *119–121*, 469–472.
164. Choi, S. W.; Jo, S. M.; Lee, W. S.; Kim, Y. R. *Adv. Mater.* **2003**, *15* (23), 2027–2032.
165. Wang, X.; Gong, C.; He, D.; Xue, Z.; Chen, C.; Liao, Y.; Xie, X. *J. Memb. Sci.* **2014**, *454*, 298–304.
166. Li, Z. H.; Cheng, C.; Zhan, X. Y.; Wu, Y. P.; Zhou, X. D. *Electrochim. Acta* **2009**, *54* (18), 4403–4407.
167. Yan, P.; Huang, Z.; Lin, Y.; Wu, X.; Yang, Y.; Wang, D.; Chen, F.; Zhang, C.; He, D. *Ionics (Kiel).* **2015**, *21* (2), 593–599.
168. Kim, J. R.; Choi, S. W.; Jo, S. M.; Lee, W. S.; Kim, B. C. *Electrochim. Acta* **2004**, *50* (1), 69–75.
169. Fan, L.; Dang, Z.; Nan, C. -W.; Li, M. *Electrochim. Acta* **2002**, *48* (2), 205–209.
170. Echeverri, M.; Kim, N.; Kyu, T. *Macromolecules* **2012**, *45* (15), 6068–6077.
171. Nagasubramanian, G.; Di Stefano, S. *J. Electrochem. Soc.* **1990**, *137* (12), 3830–3835.
172. Appetecchi, G. B.; Croce, F.; Scrosati, B. *Electrochim. Acta* **1995**, *40* (8), 991–997.
173. Alamgir, M.; Abraham, K. M. *J. Power Sources* **1995**, *54* (1), 40–45.
174. Periasamy, P.; Tatsumi, K.; Shikano, M.; Fujieda, T.; Saito, Y.; Sakai, T.; Mizuhata, M.; Kajinami, A.; Deki, S. *J. Power Sources* **2000**, *88* (2), 269–273.
175. Sekhon, S. S.; Deepa; Agnihotry, S. A. *Solid State Ionics* **2000**, *136–137*, 1189–1192.
176. Alamgir, M.; Abraham, K. M. *J. Electrochem. Soc.* **1993**, *140* (6), L96–L97.
177. Manuel Stephan, A. *Eur. Polym. J.* **2006**, *42* (1), 21–42.
178. Sutto, T. E. *J. Electrochem. Soc.* **2007**, *154* (10), P101–P107.
179. Bansal, D.; Cassel, F.; Croce, F.; Hendrickson, M.; Plichta, E.; Salomon, M. *J. Phys. Chem. B* **2005**, *109* (10), 4492–4496.
180. Ye, H.; Huang, J.; Xu, J. J.; Khalfan, A.; Greenbaum, S. G. *J. Electrochem. Soc.* **2007**, *154* (11), A1048–A1057.
181. Reiter, J.; Vondrák, J.; Michálek, J.; Mička, Z. *Electrochim. Acta* **2006**, *52* (3), 1398–1408.
182. de Vries, H.; Jeong, S.; Passerini, S. *RSC Adv.* **2015**, *5* (18), 13598–13606.
183. Kido, R.; Ueno, K.; Iwata, K.; Kitazawa, Y.; Imaizumi, S.; Mandai, T.; Dokko, K.; Watanabe, M. *Electrochim. Acta* **2015**, *175*, 5–12.
184. Cha, E. H.; Lim, S. A.; Park, J. H.; Kim, D. W.; Macfarlane, D. R. *J. Power Sources* **2008**, *178* (2), 779–782.
185. Chen, R.; Qu, W.; Guo, X.; Li, L.; Wu, F. *Mater. Horiz.* **2016**, *3* (6), 487–516.
186. Weston, J. E.; Steele, B. C. H. *Solid State Ionics* **1982**, *7* (1), 75–79.
187. Best, A. S.; Adebahr, J.; Jacobsson, P.; MacFarlane, D. R.; Forsyth, M. *Macromolecules* **2001**, *34* (13), 4549–4555.

188. Pitawala, H. M. J. C.; Dissanayake, M. A. K. L.; Seneviratne, V. A.; Mellander, B. -E.; Albinson, I. *J. Solid State Electrochem.* **2008**, *12* (7–8), 783–789.
189 Appetecchi, G. B.; Croce, F.; Hassoun, J.; Scrosati, B.; Salomon, M.; Cassel, F. *J. Power Sources* **2003**, *114* (1), 105–112.
190. Bloise, A. C.; Tambelli, C. C.; Franco, R. W. A.; Donoso, J. P.; Magon, C. J.; Souza, M. F.; Rosario, A. V.; Pereira, E. C. *Electrochim. Acta* **2001**, *46* (10–11), 1571–1579.
191. Golodnitsky, D.; Ardel, G.; Peled, E. *Solid State Ionics* **2002**, *147* (1–2), 141–155.
192. Krawiec, W.; Scanlon, L. G.; Fellner, J. P.; Vaia, R. A.; Vasudevan, S.; Giannelis, E. P. *J. Power Sources* **1995**, *54* (2), 310–315.
193. Croce, F.; Appetecchi, G. B.; Persi, L.; Scrosat, B. *Nature* **1998**, *394* (July), 456.
194. Borkowska, R.; Reda, A.; Zalewska, A.; Wieczorek, W. *Electrochim. Acta* **2001**, *46* (10–11), 1737–1746.
195. Wieczorek, W.; Lipka, P.; Wycis, H. *J. Phys.Chem.B* **1998**, 102, 6968–6974.
196. Tokuda, H.; Watanabe, M. *Electrochim. Acta* **2003**, *48* (14–16 SPEC.), 2085–2091.
197. Syzdek, J.; Armand, M.; Marcinek, M.; Zalewska, A.; Zukowska, G.; Wieczorek, W. *Electrochim. Acta* **2010**, *55* (4), 1314–1322.
198. Croce, F.; Scrosati, B.; Mariotto, G. *Chem. Mater.* **1992**, *4* (6), 1134–1136.
199. Appetecchi, G.; Croce, F.; Persi, L.; Ronci, F.; Scrosati, B. *Electrochim. Acta* **2000**, *45* (8–9), 1481–1490.
200. Croce, F.; Persi, L.; Ronci, F.; Scrosati, B. *Solid State Ionics* **2000**, *135* (1–4), 47–52.
201. Croce, F.; Curini, R.; Martinelli, a.; Persi, L.; Ronci, F.; Scrosati, B.; Caminiti, R. *J. Phys. Chem. B* **1999**, *103* (48), 10632–10638.
202. Chung, S. H.; Wang, Y.; Greenbaum, S. G.; Marcinek, M.; Persi, L.; Croce, F.; Wieczorek, W.; Scrosati, B. *J. Phys. Condens. Matter* **2001**, *13* (13), 11763–11768.
203. Pitawala, H. M. J. C.; Dissanayake, M. A. K. L.; Seneviratne, V. A. *Solid State Ionics* **2007**, *178* (13–14), 885–888.
204. Dissanayake, M. A. K. L.; Jayathilaka, P. A. R. D.; Bokalawala, R. S. P.; Albinsson, I.; Mellander, B. E. *J. Power Sources* **2003**, *119–121*, 409–414.
205. Wetjen, M.; Navarra, M. A.; Panero, S.; Passerini, S.; Scrosati, B.; Hassoun, J. *ChemSusChem* **2013**, *6* (6), 1037–1043.
206. Mei, A.; Wang, X. L.; Feng, Y. C.; Zhao, S. J.; Li, G. J.; Geng, H. X.; Lin, Y. H.; Nan, C. W. *Solid State Ionics* **2008**, *179* (39), 2255–2259.
207. Forsyth, M.; MacFarlane, D. R.; Best, A.; Adebahr, J.; Jacobsson, P.; Hill, A. J. *Solid State Ionics* **2002**, *147* (3–4), 203–211.
208. Swierczynski, D.; Zalewska, A.; Wieczorek, W. *Chem. Mater.* **2001**, *13* (10), 1560–1564.
209. Livshits, V.; Blum, A.; Strauss, E.; Ardel, G.; Golodnitsky, D.; Peled, E. *J. Power Sources* **2001**, *97–98*, 782–785.
210. Croce, F.; Settimi, L.; Scrosati, B. *Electrochem. Commun.* **2006**, *8* (2), 364–368.
211. Croce, F.; Sacchetti, S.; Scrosati, B. *J. Power Sources* **2006**, *162* (1), 685–689.
212. Fiory, F. S.; Croce, F.; D'Epifanio, A.; Licoccia, S.; Scrosati, B.; Traversa, E. *J. Eur. Ceram. Soc.* **2004**, *24* (6), 1385–1387.
213. Epifanio, A D.; Fiory, F. S.; Licoccia, S.; Traversa, E.; Scrosati, B. **2004**, 403–408.

214. Panero, S.; Scrosati, B.; Sumathipala, H. H.; Wieczorek, W. *J. Power Sources* **2007**, *167* (2), 510–514.
215. Nagao, M.; Imade, Y.; Narisawa, H.; Kobayashi, T.; Watanabe, R.; Yokoi, T.; Tatsumi, T.; Kanno, R. *J. Power Sources* **2013**, *222*, 237–242.
216. Hassoun, J.; Scrosati, B. *Adv. Mater.* **2010**, *22* (45), 5198–5201.
217. Appetecchi, G. B.; Scrosati, B. *Electrochim. Acta* **1998**, *43* (9), 1105–1107.
218. Vignarooban, K.; Dissanayake, M. A. K. L.; Albinsson, I.; Mellander, B. -E. *Solid State Ionics* **2014**, *266*, 25–28.
219. Agrawal, R. C.; Mahipal, Y. K. *Int. J. Electrochem. Sci.* **2011**, *6* (4), 867–881.
220. Gurevitch, I.; Buonsanti, R.; Teran, A. A.; Gludovatz, B.; Ritchie, R. O.; Cabana, J.; Balsara, N. P. *J. Electrochem. Soc.* **2013**, *160160* (99), 1611–1617.
221. Kim, J.-K.; Aguilera, L.; Croce, F.; Ahn, J. -H. *J. Mater. Chem. A* **2014**, *2* (10), 3551–3556.
222. Wang, L.; Yang, W.; Wang, J.; Evans, D. G. *Solid State Ionics* **2009**, *180* (4–5), 392–397.
223. Jadhav, H. S.; Kalubarme, R. S.; Jang, S. -Y.; Jung, K. -N.; Shin, K. -H.; Park, C.-J. *Dalt. Trans.* **2014**, *43* (30), 11723–11727.
224. Gao, S.; Zhong, J.; Xue, G.; Wang, B. *J. Memb. Sci.* **2014**, *470*, 316–322.
225. Tang, C.; Hackenberg, K.; Fu, Q.; Ajayan, P. M.; Ardebili, H. *Nano Lett.* **2012**, *12* (3), 1152–1156.
226. Xi, J.; Bai, Y.; Qiu, X.; Zhu, W.; Chen, L.; Tang, X. *New J. Chem.* **2005**, *29* (11), 1454–1460.
227. Gerbaldi, C.; Nair, J. R.; Kulandainathan, M. A.; Kumar, R. S.; Ferrara, C.; Mustarelli, P.; Stephan, A. M. *J. Mater. Chem. A* **2014**, *6*, 9948–9954.
228. Angulakshmi, N.; Kumar, R. S.; Kulandainathan, M. A.; Stephan, A. M. *J. Phys. Chem. C* **2014**, *118* (42), 24240–24247.
229. Di Noto, V.; Lavina, S.; Giffin, G. A.; Negro, E.; Scrosati, B. *Electrochim. Acta* **2011**, *57*, 4–13.
230. Croce, F.; Appetecchi, G. B.; Persi, L.; Scrosati, B. *Nature* **1998**, *394* (6692), 456–458.
231. Long, L.; Wang, S.; Xiao, M.; Meng, Y. *J. Mater. Chem. A* **2016**, *4* (26), 10038–10069.
232. Srivastava, S.; Schaefer, J. L.; Yang, Z.; Tu, Z.; Archer, L. A. *Adv. Mater.* **2014**, *26*(2), 201–234.
233. Skaarup, S.; West, K.; Zachau-Christiansen, B. *Solid State Ionics* **1988**, *28*, 975–978.
234. Capuano, F.; Croce, F.; Scrosati, B. *J. Electrochem. Soc.* **1991**, *138* (7), 1918–1922.
235. Jung, S.; Kim, D. W.; Lee, S. D.; Cheong, M.; Nguyen, D. Q.; Cho, B. W.; Kim, H. S. *Bull. Korean Chem. Soc.* **2009**, *30* (10), 2355–2361.
236. Liu, W.; Liu, N.; Sun, J.; Hsu, P. -C.; Li, Y.; Lee, H. -W.; Cui, Y. *Nano Lett.* **2015**, *15*(4), 2740–2745.
237. Yue, L.; Ma, J.; Zhang, J.; Zhao, J.; Dong, S.; Liu, Z.; Cui, G.; Chen, L. *Energy Storage Mater.* **2016**, *5*, 139–164.
238. Panero, S.; Satolli, D.; D'Epifano, A.; Scrosati, B. *J. Electrochem. Soc.* **2002**, *149* (4), A414–A417.

239. Klongkan, S.; Pumchusak, J. *Electrochim. Acta* **2015**, *161*, 171–176.
240. Yue, L.; Ma, J.; Zhang, J.; Zhao, J.; Dong, S.; Liu, Z.; Cui, G.; Chen, L. *Energy Storage Mater.* **2016**, *5*, 139–164.
241. Wang, L.; Li, X.; Yang, W. *Electrochim. Acta* **2010**, *55* (6), 1895–1899.
242. Jiang, G.; Maeda, S.; Yang, H.; Saito, Y.; Tanase, S.; Sakai, T. *J. Power Sources* **2005**, *141* (1), 143–148.
243. Skaarup, S.; West, K.; Zachau-Christiansen, B. *Solid State Ionics* **1988**, *28–30* (PART 2), 975–978.
244. Skaarup, S.; West, K.; Julian, P. M.; Thomas, D. M. *Solid State Ionics* **1990**, *40*, 1021–1024.
245. Zewde, B. W.; Admassie, S.; Zimmermann, J.; Isfort, C. S.; Scrosati, B.; Hassoun, J. *ChemSusChem* **2013**, *6* (8), 1400–1405.
246. Hovington, P.; Lagacé, M.; Guerfi, A.; Bouchard, P.; Mauger, A.; Julien, C. M.; Armand, M.; Zaghib, K. *Nano Lett.* **2015**, *15* (4), 2671–2678.
247. Manuel Stephan, A.; Nahm, K. S. *Polymer.* **2006**, *47* (16), 5952–5964.
248. Wang, L.; Li, X.; Yang, W. *Electrochim. Acta* **2010**, *55* (6), 1895–1899.
249. Jiang, G.; Maeda, S.; Yang, H.; Saito, Y.; Tanase, S.; Sakai, T. *J. Power Sources* **2005**, *141* (1), 143–148.
250. Peled, E. *J. Electrochem. Soc.* **1979**, *126* (12), 2047–2051.
251. Goodenough, J. B.; Kim, Y. *Chem. Mater.* **2010**, *22* (3), 587–603.
252. Xu, C.; Sun, B.; Gustafsson, T.; Edström, K.; Brandell, D.; Hahlin, M. *J. Mater. Chem. A* **2014**, *2* (20), 7256.
253. Sloop, S. E.; Kerr, J. B.; Kinoshita, K. *J. Power Sources* **2003**, *119–121*, 330–337.
254. Eshetu, G. G.; Grugeon, S.; Kim, H.; Jeong, S.; Wu, L.; Gachot, G.; Laruelle, S.; Armand, M.; Passerini, S. *ChemSusChem* **2016**, *9* (5), 462–471.
255. Verma, P.; Maire, P.; Novák, P. *Electrochim. Acta* **2010**, *55* (22), 6332–6341.
256. Sun, H. -Y. *J. Electrochem. Soc.* **1999**, *146* (5), 1672.
257. Sun, B.; Xu, C.; Mindemark, J.; Gustafsson, T.; Edström, K.; Brandell, D. *J. Mater. Chem. A* **2015**, *3* (26), 13994–14000.
25. Appetecchi, G. B.; Scaccia, S.; Passerini, S. *J. Electrochem. Soc.* **2000**, *147* (12), 4448–4452.
259. Li, Q.; Sun, H. Y.; Takeda, Y.; Imanishi, N.; Yang, J.; Yamamoto, O. *Spec. Issue Interfacial Phenom. Batter.* **2001**, *94* (2), 201–205.
260. Armand, M. B.; Duclot, M. J.; Rigaud, P. *Solid State Ionics* **1981**, *3–4* (C), 429–430.
261. Christie, A. M.; Oleksiak, A. L.; Vincent, C. A. *Electrochim. Acta* **1995**, *40* (13), 2405–2411.
262. Olsen, I. I.; Barker, J.; Koksbang, R. *Solid State Ionics* **1996**, *83*, 125–133.
263. Bonino,F.; Scrosati, B.; Selvaggi, A. Solid State Ionics **1986**, *19*, 1050–1053.
264. Watanabe, M.; Endo, T.; Nishimoto, A.; Miura, K.; Yanagida, M. *J. Power Sources* **1999**, *81–82*, 786–789.
265. Peled, E.; Golodnitsky, D.; Ardel, G.; Eshkenazy, V. *Electrochim. Acta* **1995**, *40*(13–14), 2197–2204.

266. Peled, E. *J. Electrochem. Soc.* **1997**, *144* (8), L208.
267. Bouchet, R.; Lascaud, S.; Rosso, M. *J. Electrochem. Soc.* **2003**, *150* (10), A1385–A1389.
268. Granvalet-mancini, M. Le; Honeycutt, L.; Teeters, D. *Electrochem. Acta* **2000**, *45*, 1491–1500.
269. Ismail, I.; Noda, A.; Nishimoto, A.; Watanabe, M. *Electrochim. Acta* **2001**, *46*(10–11), 1595–1603.
270. Xu, C.; Sun, B.; Gustafsson, T.; Edstrom, K.; Brandell, D.; Hahlin, M. *J. Mater. Chem. A* **2014**, *2* (20), 7256–7264.
271. Sun, B.; Xu, C.; Mindemark, J.; Gustafsson, T.; Edstrom, K.; Brandell, D. *J. Mater. Chem. A* **2015**, *3* (26), 13994–14000.
272. Kumar, B.; Scanlon, L. G. *J. Power Sources* **1994**, *52* (2), 261–268.
273. Gozdz, A. S.; Schmutz, C. N.; Tarascon, J. M.; Warren, P. C. Polymeric electrolytic cell separator membrane. US Patent, 5,418,091, May 23, 1995.
274. Choe, H. S.; Giaccai, J.; Alamgir, M.; Abraham, K. M. *Electrochim. Acta* **1995**, *40* (13–14), 2289–2293.
275. Zhang, J. -G.; Xu, W.; Henderson, W. A. *Lithium Metal Anodes and Rechargeable Lithium Metal Batteries*; Springer Nature: Switherland, 2017; Vol. 249, pp 1–193.
276. Fauteux, D. *Electrochim. Acta* **1993**, *38* (9), 1199–1210.
277. Gauthier, M.; Bélanger, A.; Bouchard, P.; Kapfer, B.; Ricard, S.; Vassort, G.; Armand, M.; Sanchez, J. Y.; Krause, L. *J. Power Sources* **1995**, *54* (1), 163–169.
278. Baudry, P.; Lascaud, S.; Majastre, H.; Bloch, D. *J. Power Sources* **1997**, *68* (2), 432–435.
279. Motavalli, J. *Nature* **2015**, *526*, S96–S97.
280. Terrien, C.; Maniak, R.; Chen, B.; Shaheen, S. *Res. Transp. Bus. Manag.* **2016**, *20*, 20–32.

Index